PUNISHMENT AND POLITICS

For my mother and father

PUNISHMENT AND POLITICS

The Maximum Security Prison
in New Zealand

GREG NEWBOLD

AUCKLAND
OXFORD UNIVERSITY PRESS
MELBOURNE OXFORD NEW YORK

Oxford University Press
Oxford University Press, Walton Street, Oxford OX2 6DP

OXFORD   NEW YORK   TORONTO
DELHI   BOMBAY   CALCUTTA   MADRAS   KARACHI
PETALING JAYA   SINGAPORE   HONG KONG   TOKYO
NAIROBI   DAR ES SALAAM   CAPE TOWN
MELBOURNE   AUCKLAND
and associated companies in
BERLIN and IBADAN

Oxford is a trade mark of Oxford University Press

First published 1989
© Greg Newbold 1989

This book is copyright.
Apart from any fair dealing for the purpose of private study,
research, criticism, or review, as permitted under the Copyright
Act, no part may be reproduced by any process without the prior
permission of the Oxford University Press.

The assistance of the New Zealand Literary Fund
is gratefully acknowledged

ISBN 0 19 558179 2

Cover design by Hilary Ravenscroft
Photoset in Times by Saba Graphics
and printed in Hong Kong
Published by Oxford University Press
1A Matai Road, Greenlane, Auckland 3, New Zealand

CONTENTS

Preface   vi
Acknowledgements   viii
Chronology of Tenure   ix
Diagram of Legal and Administrative Responsibility
   in New Zealand Prisons   x

1:	Origins and Background of the Maximum Security Prison 1880-1949	1
2:	The Elements of Change 1949	17
3:	The Reformist's Zeal 1949-1954	28
4:	Progress and Controversy 1954-1957	42
5:	The Troubled Years 1957-1960	59
6:	Power and Politics at Mt. Eden 1950-1960	71
7:	The Prison Band 1952-1960	84
8:	The Decline of Haywood 1958-1964	99
9:	The Hanan-Robson Era 1960-1969	117
10:	The Buckley Administration 1963-1965	131
11:	Mt. Eden Prison in Crisis 1965	144
12:	Interim Security 1965-1969	165
13:	The Building of Paremoremo 1960-1969	175
14:	Habitation and Adjustment 1969	187
15:	Paremoremo in Crisis 1970-1971	202
16:	The Bubble Bursts 1970-1972	217
17:	A New Regime 1972	234
18:	Conflict and Change 1972	253
19:	The End of an Era 1973-1975	267
20:	1975 and After	286

Abbreviations   300
Bibliography   300
Index   307

PREFACE

One fine afternoon in the winter of 1975, detectives kicked the door down. The next day, sitting in the cold stone cell in the old gaol at Mt. Eden, I began to wonder about the history of the maximum security prison.

Until now nobody has ever documented the varied and intricate sequences which have informed the development of the security prison in New Zealand. Instruction about what has gone on in our prisons has been largely restricted to official reports on specific matters and the selective folklore of old timers. The years pass and the memories of the old lags and the screws who worked with them fade, as surviving numbers steadily dwindle. Piece by piece, the history of the prison is lost.

In the spring of 1975, not long after my entry to the top custody institution at Paremoremo, Auckland, I began the research which ended twelve years later with completion of this book. Five years in gaol talking to inmates and studying their culture for an MA degree, preceded six years of doctoral research at the University of Auckland. It was during the latter period that the bulk of material contained here was collected. Perusal of the century of parliamentary debates and departmental annual reports gave foundation to the reconstruction of the past. This was supplemented by coverage of significant events in the back-files of the nation's major press offices and data from official and archival files in Wellington. Some sixty former inmates were contacted, as well as prison staff, head office personnel, Ministers, and professional and voluntary workers. In all, over a hundred individuals were spoken to, sixty in formal recorded interviews. Transcripts from the interviews total more than two thousand pages; my personal notes another four thousand. It is from these that this book has been written.

When I set out to compile this history I did so with three objectives in mind. First of all, I wanted, for the first time, to chronicle the matrix of events which have taken place in the security prison of New Zealand, especially over the last forty years. Secondly, I wanted to situate those events in relation to complex interacting forces. The principal factors which motivate change in a penal system are those operating at inmate, local administrative, head office, and political levels. However, the correspondence of these, in turn, is sensitive to factors such as population pressure, economics, public opinion, international trends, and even a good measure of chance. An ambition of this project was to place these variables in contexts which could illustrate how each of them contributed to the development of the lock-up.

The third major goal of this research was, having painted a picture of the prison in evolution, to comment on New Zealand developments in relation to those overseas. While New Zealand has frequently looked abroad for directions in policy, it has also shown a willingness to experiment on its own. Likewise, although the social system of inmates shares much in common with those in Australia, England, and the United States, it has a strong local flavour as well. A purpose of this work was to indicate areas of similarity and difference and to consider why the situation is so.

It is probably commitments of time and expense that have prevented such an exercise from taking place before, and which make another unlikely in the future. However, the information I have collected is precious not only in that it encompasses an interesting aspect of New Zealand's mortal and degradable past, but because of its relevance to circumstances today. Without knowledge of a system's background, its course cannot be steered properly. Just as genetic engineering is impossible without a grasp of the mechanics of inheritance, so is the future of the prison unpredictable without awareness of its executive structure. In fact almost every sociological enquiry into prison unrest since 1950 has emphasized the importance of analysing the administrative foundations. The value of such attention has been recognized by the New Zealand Justice Department for almost two decades. In 1960 an official publication commented on the need for a study which would:

> ... involve prison administration from the most senior personnel to the prisoners themselves. Recent studies on the sociology of institutions have revealed the importance of personalities, interactions, tensions and role conceptions of the people involved. Studies of this nature would help iron out many difficulties and mutual misconceptions which traditionally exist between theorists, researchers and senior officials, and prison staff. (Dept. J. (1970b) 2.)

Currently the penal system of New Zealand is in a state of crisis, and perhaps at no other point has research of this type had more occasion. Prisons are so overcrowded that police cells have been gazetted as temporary gaols and inmates have been released before their time. In the system generally and in maximum security in particular, relationships have deteriorated to the point where almost one in four inmates needs some form of segregation and 9 per cent are in protective custody. The conditions I described in the book about my prison experiences, *The Big Huey*, have vanished. The rate of self-mutilation and suicide has rocketed. The Department of Justice can only speculate about the causes of this situation and is labouring hard to discover remedies. However, the prison is no stranger to adversity, and today's difficulties result partly from accelerating change. No change, it is said, can be made without inconvenience, and prisons are for ever evolving. The value of this critique, then, is that an unravelling of the past can help identify the hazards of the future. It might be too much to hope that durance vile will ever service inmate reform, but attention to historical detail may at least inspire policies which are practical to run.

Greg Newbold
October, 1988

ACKNOWLEDGEMENTS

To examine the history and politics of the New Zealand maximum security prison was an idea I first had while awaiting trial at Mt. Eden gaol in 1975. The desire was not fulfilled until nearly six years after my release, sixty-three months hence.

Failing the insight of a long period of confinement, this project could not have come about. However, inimitable though the instruction of experience might be, few scholars can organize lengthy research programmes without the significant help of others. Such has been very much the case here.

Contacts and friendships I made in gaol were crucial to the creation of this history, and although too numerous to nominate, I must acknowledge my deep indebtedness to the former inmates, Justice employees, politicians, and independent workers who made themselves available during the data-collection period. It is their personal contributions which have drawn the facts of the security prison's progress into lively, often dramatic, relief.

In addition to firsthand material, primary documentation has proven invaluable in reconstructing the eras with which I have been concerned. Accordingly I would also offer my special thanks to the library staff at *The New Zealand Herald*, at News Media Ltd., and at Wellington Newspapers Ltd. I could not have done without the generous access they gave me to back-file resources.

Conducting work of this nature inevitably requires assistance from people who are not involved with the investigation itself. Although, once more, these have been many, particular gratitude is due to Megan Clark and Dr Peter Donelan of Victoria University's Mathematics Department, for arranging office space in Wellington, and to Belinda Clark, Diana Corrigan, Andrew Murdoch, and Margaret Franken, for their kind assistance with accommodation. I am also thankful to Robyn Lawson, Kevin Steel, and Simon Oliver, who efficiently managed most of the word processing.

Finally, as the doctoral dissertation, from which this book derives, evolved from a series of scrappy drafts, I must record the helpful advice of those who commented on parts of the script. Professor Ian Carter, Associate Professor Bill Hodge, Dr Charles Crothers, Mr Nick Perry, and Mr John Jackson of the Justice Department have all been valuable in this regard, but more important than any have been my supervisors, Associate Professor Bernard Brown and Dr David Bedggood. These two patiently oversaw what must at times have seemed an endless procession of chapters and rewrites, and offered many suggestions for improvement. While their useful input has been abundant, their advice has not always been taken; my own judgement has prevailed and if any shortcomings in the book remain, responsibility is entirely mine.

Chronology of Tenure: Government and Senior Justice Personnel 1938–1988

Year	Maximum Security Superintendents	Permanent Heads (Secretary for Justice)	Ministers of Justice	Prime Ministers	Governments	Year
1938	W.T. Leggett 1935–1946	B.L.S. Dallard 1925–1949	H.G.R. Mason 1935–1949	M.J. Savage 1935–1940	First Labour 1935–1949	1938
1940				P. Fraser 1940–1949		1940
1942						1942
1944						1944
1946	J.J.H. Lauder 1946–1951					1946
1948						1948
1950		S.T. Barnett 1949–1960	T.C. Webb 1949–1954	S.G. Holland 1949–1957	First National 1949–1957	1950
1952	H.V. Haywood 1951–1963					1952
1954						1954
1956			J.R. Marshall 1954–1957	K.J. Holyoake 1957		1956
1958			H.G.R. Mason 1957–1960	W. Nash 1957–1960	Second Labour 1957–1960	1958
1960		J.L. Robson 1960–1970	J.R. Hanan 1960–1969	K.J. Holyoake 1960–1972	Second National 1960–1972	1960
1962						1962
1964	E.G. Buckley 1963–1972					1964
1966						1966
1968						1968
1970		E.A. Missen 1970–1974	D.J. Riddiford 1969–1972			1970
1972	J. Hobson 1972–1984		R.E. Jack 1972 A.M. Finlay 1972–1975	J.R. Marshall 1972 N.E. Kirk 1972–1974	Third Labour 1972–1975	1972
1974		G.S. Orr 1974–1978	D. Thomson 1975–1978	W.E. Rowling 1974–1975	Third National 1975–1984	1974
1976						1976
1978		J.F. Robertson 1978–1982	J.K. McLay 1978–1984	R.D. Muldoon 1975–1984		1978
1980						1980
1982		J. Callahan 1982–1986				1982
1984	S.G. Ward 1984–1985		G.W.R. Palmer 1984–	D.R. Lange 1984–	Fourth Labour 1984–	1984
1986	L. Hine 1985–1987	D. Oughton 1986–				1986
1988	M. Hindmarsh 1987–					1988

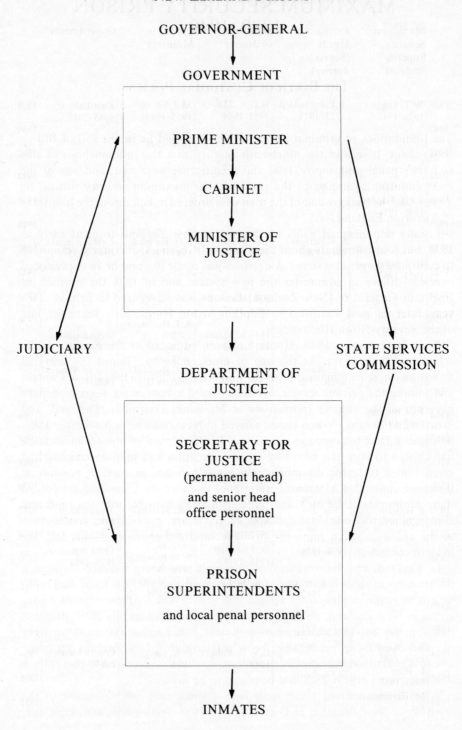

Diagram of Legal and Administrative Responsibility in New Zealand Prisons

1: ORIGINS AND BACKGROUND OF THE MAXIMUM SECURITY PRISON 1880–1949

THE BIRTH OF CUSTODIAL POLICY

The foundations of criminal justice in New Zealand lie in the soil of British colonialism. It was in the nineteenth century that the fundamentals of this country's penal philosophy, law, and architecture were cast, and one of the most enduring examples of the period is the maximum security prison. Its design, like the background of the man who ordered it, belongs to the traditions of Victorian England.

Locally administered gaols are known in New Zealand from as early as 1838, but sour comments about them led, in 1876, to a Government resolution to centralize its penal system.[1] A decision was made to appoint an experienced overseas officer to administer the new system and in 1878 the position of Inspector-General of [New Zealand] Prisons was advertized in Britain. Two years later the post was filled by Captain Arthur Hume, forty years old, late of the Seventy-Ninth Highlanders.

Born in Dublin in 1840, Hume had been educated at the English public school of Cheltenham. At the age of nineteen he had joined the army as an ensign where he remained until 1874 before leaving with the rank of Captain and joining the prisons service. He then served a further six years as deputy governor of the convict institutions at Millbank, Dartmoor, Portland, and Wormwood Scrubs.[2] When Hume arrived in New Zealand in November 1880, he came with a background which was hardly typical of the colonialists he had chosen to join. His boarding-school education and military training had given him a taste for discipline, and his experience in English prisons, at that time controlled by an autocratic puritan called du Cane, had reinforced these sentiments. Like du Cane, Hume's approach to his job was rigid and closely allied to traditional concepts of crime and punishment. Lawlessness in the colony was an imposing problem, and the solution, Hume felt, lay in stern corrective measures.

In England, the 'reformation' of criminals was being pursued through a programme of deterrence. The object of imprisonment, du Cane had said, 'is not to make it pleasant'.[3] It was this model which Hume wanted to re-create in New Zealand, and he resolutely pursued his task. By 1892, progress had gone far enough for Hume to comment, 'the English system is the right one and is, as far as practicable, being carried out in New Zealand prisons'.[4] The first goal of incarceration in the colony would be deterrence and prevention. Prisons, Hume believed, should be places to be dreaded.[5]

When Hume arrived there were four major penal establishments in the territory of New Zealand: at Dunedin, Lyttelton, Wellington, and Auckland.

Along with minor lock-ups these accounted for a total of about 700 prisoners. The largest of all the gaols was at Auckland, and in 1878 it was reported to have accommodation for 281.[6] This facility, known as the Stockade, had opened in 1856. Sited beside the extinct volcano of Mt. Eden, it had become the city's major place of confinement when the old City gaol in the centre of the town was pulled down in 1865. A stone wall had been erected around the site between 1862 and 1876, but the wooden buildings were decrepit, and conditions inside them were poor.

The condition of this institution, among others, had been repeatedly criticized since the early 1860s, and in 1877 a Commission of Inquiry into the Stockade had resulted in plans for a new prison on the site.[7] Action on these plans, the 'Mahoney Plans', had been deferred, however, pending arrival of the new Inspector-General whose appointment was about to be advertised.

When Hume took up his duties at the turn of the decade, he was unimpressed with what he saw. His first report of May 1881 was scathing, and one of his recommendations was for the immediate construction of a large central prison in Auckland. The Mahoney draft was scrapped and Hume prevailed upon Pierre Burrows, the Colonial Architect, to draw plans for two large new institutions. One of these, Mt. Cook prison in Wellington, was to be for first offenders and was at the time of its completion the largest building in the colony. The other, at Mt. Eden, would be for hardened convicts and would be constructed in the place of the old Stockade.

Mt. Eden prison, the only building of the two that still stands,[8] clearly reflects the architecture of the period. Built upon the same 'radial' pattern used at places like Trenton (New Jersey, 1836) and Pentonville (UK, 1842),[9] its imposing Gothic structure marks it as unmistakably Victorian. The link between it and the early English gaols is direct. In 1874, the Colonial Architect W. H. Clayton had requested that all available material on prisons be sent to him from the Agent-General in London. Among this information came the Blue Books, which contained the plans of nearly all the institutions in England and Ireland. It was from the Blue Books that New Plymouth prison, Taranaki, had been developed in 1876 and it was to these books that Clayton's successor, Burrows, now referred as he worked on the plans for Mt. Eden. Although Mt. Eden is reputed to have been based on the structure of a prison in Malta — itself built on the English model[10] — Stacpoole believes that the former was probably the result of an amalgam of ideas drawn from the Blue Books.[11] The result was an institution highly reminiscent of those at places like Pentonville and Dartmoor. Burrows' plans for Mt. Eden — then known as Auckland gaol[12] — were ready by 1882, and in the same year, on 20 November, and under the supervision of Captain Hume, labour on the foundations began.

Mt. Eden prison was designed so that as the need arose, extra wings could be added, and with this intent, the project advanced over the next thirty-five years. Built out of bluestone excavated from the adjacent quarry,[13] it was 1888 before the first prisoners were transferred to the new buildings and demolition of the wooden quarters began. The institution's new wings and extensions were added in 1893, 1905, 1906, 1911, 1913, and 1917. In 1917 the gaol was classified as New Zealand's strongest, safest, and chief penal institution.[14]

Mt. Eden prison was planned twenty years after its antecedent at Pentonville and by the time it was completed, forty-five years after it had been designed, it was already out of date. The punitive philosophy in the service of which it had been built died with the retirement of Hume in 1909. By 1910, the Minister of Justice had already declared it obsolete.[15] However, this inert stone bastion, with its solid steel doors and walls over a metre thick, was to serve as the country's only maximum custody facility for more than seventy years.[16]

THE NEW METHOD

From the very beginning, the appointment of Hume had been controversial. His rigorous performance of duty was criticized often during his twenty-nine years in office, and although he mellowed as his colonial experience grew, it was with relief that Hume's resignation was accepted in 1909.

One of the characteristics of penal reform in this country is that a period of stagnation is often followed by one of intensive activity. It is almost as if a succeeding regime behaves in reaction to that which preceded it. Hume's disciplined and orthodox programme was a direct response to the chaos he inherited. The regime of his successors came as a challenge to his rigidity.

When Hume retired, control of the Prisons Department was taken over by the Minister of Justice, and it was the Member for Parnell, Sir John (Dr) Findlay, who now became the Department's guiding figure. Findlay's objective, later known as 'the new method', was to discard the shackles forged by Hume and to create a prison system which, he believed, would be 'as efficient as any existing anywhere in the world'.[17] His principal tool would be the Crimes Amendment Act of 1910. The emphasis of punitive justice would be replaced by correctional care.

The Crimes Amendment Act brought welcome change to the work of the Prisons Department, but no sooner was the legislation passed than Findlay's seat was lost in the general election of 1911. Six months later, a confidence vote led to the dissolution of Parliament. The Liberal administration, led for most of its twenty-two years by Richard J. Seddon, was ousted and William Massey's Reformists took power. With the public service newly reorganized in 1913, responsibiity for prisons passed to Charles E. Matthews, a former Private Secretary to the Minister of Justice.[18] Findlay's brief tenure had prevented his experiment from being fully tested.[19] Matthews, however, energetic and already versed in the workings of the Department, sympathized with his predecessor and eagerly put the new method into effect.

Compared with John Findlay, Matthews was a prudent man and his interpretation of duty reflected the Ministry he served. Massey's Government arose out of rural conservatism, drawing its support from the small North

Island dairy herders. Industrious and practical, they saw purity as springing from the land, and the countryside as a source of virtue. Moral edification could come through productive labour, and it was this reasoning which promoted agriculture as a linchpin of prison industry. As a consequence of these ideals, between 1910 and 1922, tens of thousands of acres were purchased or reclaimed at Waikeria, Turangi, Trentham, and Templeton. Large prison farms were founded on these sites, and by what Matthews called 'the gospel of hard work',[20] the key to reclamation was pursued. Matthews believed that if handled properly, 75 per cent of all prisoners could be managed like ordinary people. Thus, from 1910 to 1923, the number of New Zealand prisoners employed in outside work schemes grew from eight per cent to 70 per cent.[21]

The contribution of Findlay and Matthews was targetted principally at the open institution. However, their interests were not confined here and there were wider applications also. The maximum security gaol was finished under Matthews, and although little could be done to remove the stamp of its creator, 'new method' principles were incorporated here as well as they could be. The ethic of work was pursued at Mt. Eden too, and there was an effort to improve its neglected industries. Activities like tailoring, bootmaking, and concrete-block casting were introduced, and after the end of construction work, metal crushing was reorganized.[22] The latter soon became the gaol's biggest business, so that by 1925, 15.5 per cent of Mt. Eden's labour force was employed in the quarry. Between 1918 and 1925 cash returns for that enterprise relative to others increased from 12.9 to 38.14 per cent.[23] Quarrying continued as a major source of revenue until the early 1960s.

The new stress on reform brought with it a renewed concern for inmate welfare. In 1914, the first schoolteacher was appointed to Mt. Eden, and in 1920 wages were introduced for married inmates. Lighting was improved and lighting hours were extended. Libraries were enlarged. The emphasis on security was relaxed. Broad arrows were removed from prisoners' clothing after 1913, and by 1918 the general issue of arms to prison officers had ceased.[24] A sincere effort was made to replace the regimen of stark containment at Mt. Eden with one where custody and reformation could exist side by side.

In 1919, the Prisons Department was again split from the Justice Department, and henceforth Matthews bore the title of Controller-General. He died five years later, but by then, after almost fifteen years of constant review, reorganization was nearly complete. Findlay had foreseen a programme as efficient as any in the world. Had he still been in politics, he may have considered his goal has been reached.

While it was generally applauded, enthusiasm for the new order was not shared by all, and many thought libertarianism had been taken too far. Among staff, for example, a decision to recognize inmate petitions after 1909 was a source of unease and confusion. Administration was becoming inefficient and it was felt, even at head office, that insufficient regard had been paid to order.[25] Discipline had become unstable and there was criticism from the newspapers as well. For although Matthews had rationalized training by shifting it from humanitarianism to rehabilitation-through-work,[26] his zeal and lack of expertise had led to promotion of unsound schemes.[27] As we shall see, this may have been why an accountant was chosen to replace him after his death.

The Department's feeling that penal progress had gone far enough was matched by electoral sentiment. The decade or so after 1912 was a period of growing conservatism and myopia, which Sinclair suggests was attributable largely to the leadership of Massey.[28] It was a time of change, when economic power shifted from the large landowners of the south to the small farmers of the north; to self-made materialists who valued the practical and the ascetic over the academic and the abstract. Adding to this concern for the mundane was the fact that after 1920 the national economy started to falter. Export prices fell and, although they recovered briefly between 1923 and 1925, they dropped again the following year.[29] Credit became tight, permanent unemployment appeared, and the economy wavered precariously before collapsing after 1929.

By the time of Matthews' death in 1924, finance was already a priority. The nation could ill afford the extravagance of further refurbishment in prisons and in any case, the public was not really interested. As the Commissioners searched for a replacement head for the Prisons Department, philosophical considerations doubtless took a back seat. What was sought now was a practical man who could conform to the conservative mood of the times. In 1924, a temporarily liquid Treasury removed the need for urgency, but by the time an appointee was found, the economy has recommenced its erratic downward slide.

B. L. S. DALLARD, CONTROLLER-GENERAL OF PRISONS

When Matthews died, the control of his Department passed briefly to the inspector, M. Hawkins. The following year, Berkeley Lionel Scudamore Dallard from the Public Service Commission signed on as the Prisons Department's new permanent head.

Born at Rangiora in 1889, a former scoutmaster and a wartime soldier, Dallard's professional experience prior to his Prisons appointment palpably affected his approach to the job. In 1909 he had joined the Public Service as a cadet, becoming audit examiner in 1910. In 1919, after his return from the war, Dallard was appointed Advisory Consultant to the Board of Trade, and in 1924 was promoted to Inspector in the office of the Public Service Commissioner. As Public Service Commission Inspector, Dallard proved an able accountant, whose principal strengths lay in finance organization and management. It was through working with him that the Commissioner became aware of the man's qualities and it was he, personally, who urged Dallard to take the Prisons appointment in 1925.[30]

In these years of uncertainty and disenchantment with penal liberalism, Dallard must have seemed an obvious choice. Here was a manager with a mind for efficiency, who seldom took a step without analysing it from a cost-benefit point of view. He was also an austere and patriotic gentleman and as such, fitted in well with the mood of the day.[31]

Although Dallard's knowledge of correctional theory was scant, he did not allow it to remain so. In 1925, straight after his appointment, he attended the International Penal and Penitentiary Congress in London. The experience of the Congress was important to the new head, and having acquired the written reports from it, he used them as his 'Bible'.[32] He also studied more widely, and familiarized himself with the philosophy of his Department. Having done this he visited every prison in the Dominion to observe how ideas were being put into practice.

The Controller-General was not satisfied with what he saw and immediately began to attend to the question of efficiency. Mechanization in the rural sector was reducing the value of cheap labour, but at the same time superphosphates were making farming more profitable. Convinced that Matthews' programme was working below optimum, Dallard resolved to restore its viability. Thus, in his first Annual Report of 1926, Dallard outlined a revitalized prison industry scheme, the accent of which was to remain on primary production. One of his first moves in this direction was to begin developing the land reclamation camps at Hautu and Rangipo into fully operational farms.[33]

It was Dallard's expanded agricultural programme which proved his most important contribution to penal justice. Although his objective of self-sufficiency was never achieved, by 1933 prisons were producing up to 40 per cent of the cost of their own rations.[34] It was due to the success of this scheme that the economy of the gaols during the depression suffered less than it might have, and conditions in the borstals — where particular interest was taken — were probably better than in many sectors of the free community.

Lock-up prisons, however, particularly maximum security, did not appeal to Dallard and they barely profited from his rule. His first reaction to Mt. Eden was of repugnance. The stone cells and high, slit windows reminded him of the works of Dumas, and he disdained the smell of sweaty, unwashed bodies. He promptly organized carbolic soap to be made up and distributed to all inmates, and he increased the number of showers permitted from one to two each week. He had air vents fitted to the cells, with floorboards installed in some of them. He ordered that bedsheets be distributed to all prisoners and he provided an oil-burning stove for use in the prison kitchen.[35] The bootmaking plant was upgraded, the quarry was improved, and from 1927, all 'sex perverts' (Dallard's term for sex offenders) were transferred to New Plymouth prison, which he had especially set aside.[36]

These concessions were, of course, only minor and reflected mainstream sentiment of the day. Dallard's general lack of empathy with prisoners was reproduced in Parliament as well as the community at large, and this accounts partially for why there was so little pressure for change. In spite of improvements in areas like farming and borstals, the entire Dallard period is remembered for its rigidity, its parsimony, and its want of vision. A principal reason for this was economic. Dallard took over the prison system at a time of financial decline, and the scarcity of public-works funding meant that for most of the thirties any large-scale reform programme was impracticable.[37] The Controller-

General knew this, after all it was his tight-fisted approach to spending that had endeared him to his employers in the first place. After eight years in office his performance was rewarded, and in 1933, at the height of the depression, Dallard was also given charge of the Justice Department.[38]

Another reason for the inertia of the period is political. No sooner had the economy begun to recover than the clouds of another war began to gather over Europe. Once again, attention was drawn away from prisons and towards more critical issues. Prisoners — and the 700 or so military defaulters who later became committed to custody — attracted little political sympathy. After Labour came to power in 1935, the Justice Minister was never particularly interested in prisons,[39] and what little he did attempt was apparently obstructed by his leader.[40] In the first two years after Labour took office, there was a 29 per cent drop in national prison populations and a 40 per cent one in maximum security, but it was not long before numbers began to creep up again. Within the gaols the routine remained unchanged. Even during the tremendous bursts of activity in 1936–1938 and 1945–1946, when Labour created the Welfare State, there was little done for convicted criminals. In contrast to the decades before it, for a quarter of a century after 1925, only one new penal institution was built.[41]

Much about the status of the Prisons Department during this period, therefore, is explicable in terms of the prevailing politico-economic climate. However, if the personality of Dallard was favoured specifically in response to this climate, then it is also unlikely that had he been selected at another time, his approach would have been any different. The Controller-General's personal ideology did not endear him to the notion of penal reform. Dallard's personal writings and speeches[42] are those of a man committed to orthodoxy, to sovereign morality and rule, and to the iron law of super- and sub-ordination. Hard working, dedicated, and guided by inveterate tradition, Dallard was as predictable as he was unimaginative. Precepts he held when he took over the Prisons Department altered little through the passage of the years. As late as 1980, Dallard still advocated sterilization of forensic 'mental defectives',[43] the imprisonment of active homosexuals,[44] and the use of corporal punishment, especially for sex offenders. Many of the latter, he argued, are paedophiles.

Dallard did not believe in corporal punishment as a special deterrent to crime, but he saw it as an appropriate expression of State disapproval towards certain forms of conduct. Likewise he believed in the retention of capital punishment, not because he felt murder rates could be affected by it, but to 'satisfy the claims of justice'.[45] It was necessary, he felt, that society should make a strong statement about behaviour it abhorred. An illustration of how this philosophy was implemented appears in instructions he issued on executions procedure in 1931.

> As the Union Jack is emblematical of British justice, such a flag will be hoisted to the main flagstaff on the morning of an execution. At the time the execution takes place a black flag will be hoisted to a lower flagstaff, the Union Jack in the ascendant, silently proclaiming the suzerainty of the law.[46]

Dallard made frequent claims to 'liberalism', but evidence of professional enlightenment is scant. Asked in 1976 about specific reforms he made, the former chief was vague and non-committal, speaking briefly on the question before changing the topic.[47] Such reformative measures as he did approve were tempered with a good measure of the authoritarian. Despite the shift in prison officers' duties from custody to 'reclamation', for example, Dallard retained the disciplinarian provisions created before his appointment. Throughout his office, prisoners were required to march to and from their places of work, and to stand to attention and salute whenever addressing officers or visitors.[48] Familiarity between staff and inmates was discouraged.[49] When interviewed, Dallard described his administration as one of 'cautious progressiveness'. 'I didn't want to make prison a comfortable place,' he said, 'but there is a certain minimum below which it becomes inhuman. So my aim was to treat prisoners fairly.' Later he remarked:

> I was careful not to give undue publicity to improvements in prison. I didn't want to do anything that would detract from the fear of prison. Nothing annoys me more than to hear prisoners say that they can take it standing on their head. Or that conditions in prison are not at all bad. . . . Well, that's not satisfactory from the point of view of the deterrent factor of prisons.[50]

Yet in spite of his very static set of personal beliefs, and notwithstanding the fact that fourteen of his years were served under a leftist Government, Dallard appears to have got on well with most of his political masters.[51] The Controller-General never voted Labour, but of Rex Mason, the Minister who removed corporal and capital punishment, Dallard speaks with particular affection. During his term in office, and when he retired, Dallard says he received messages of thanks from Mason, as well as the other politicians he had served.[52]

There is no doubt that Dallard earned many of the accolades he received. For whatever his shortcomings, Dallard was a dedicated and faithful servant, committed to discharging duty as he saw it. During his years in the Prisons and Justice Departments, he claimed to have worked late regularly, and regretted that due to his job commitments he saw little of his wife, who died a few months after his retirement. The weight of responsibility he assumed was considerable, and although he operated in the name of his Ministers, most departmental activity was taken upon his own initiative. In 1941, when union secretary Tom Stanley attacked Mason for ignoring the nation's gaols, for instance, a quick response from Mason — prepared by Dallard — outlined the many advances made since Labour had taken office. Dallard asserted, and evidence supports, that he had a major say in the conduct and promotion of the Department. Ministers simply acquiesced, he said, in most of what he wanted to do.[53]

Dallard's approach to his job was meticulous and he took signal pride in his work. Archival files brim with correspondence written by him to explain current policy and frequently, as we shall see, to defend it. Dallard's writings were informative and detailed, and it mattered little to him whether a letter was to a major news agency or merely a worried citizen. He took care in replying to any quarter that expressed an interest in his office.

Initially, then, during the depression and the war years, concern about prisons was sporadic and inconsequential. From 1945 onwards this situation changed and the volume of comment received by the Department swelled. Dallard reacted defensively to criticism, and increasing pressure deepened his resolve. It was not long before he was forced into a pitched battle with forces bent on achieving reform.

THE CRITICS

Pre-War Inertia. Since 1925, the Department of Prisons had seen little in the way of formal publicity. Dallard did not favour profiling his penal establishments, because he felt that advertising their comforts might erode their deterrent value. For some time media interest was restricted to sensational coverage, and this added pitch to the prisons' mystique. However, although they were heavily publicized, escapes at this time were quite rare. In the five years before 1940, for example, not a single break out had occurred from the security gaol at Auckland. Disturbances too, were infrequent. Inmates apparently accepted their situation and sporadic challenges were easily repressed.

Some of the reasons for this will be considered later, but one of them was that information about incidents was strictly regulated to prevent public reaction. A good illustration of such control appeared on Anzac day, 1936, when a united refusal by prisoners to return from the Mt. Eden yard led to police reinforcements being called. Declining to comment after the men had given in, the superintendent said the incident was a 'domestic matter', of no concern to the public.[54] Press access was denied, and calls for an inquiry were refused. The precise reasons for the incident were never disclosed.

The Department's monopoly on official information led to a vacuum in sympathy for rebellious convicts, and part of their reluctance to demonstrate can be understood in this light. Conditions in Mt. Eden were poor, but few released men ever spoke out about them. Those who did so were vigorously rebutted. However, despite the success of such tactics, Dallard's claims left a residue of doubt. At first this was small, but a succession of challenges caused scepticism to mount. For years, official explanations had served as effective sops for concern. However, times changed quickly in the 1940s, and with them, so did the face of prisons.

The War and After. In 1939 war broke out in Europe and attention became focused on what soon became a life-and-death struggle against Hitler's Third Reich. New Zealand declared war in September 1939, and in 1940 the first contingent of volunteers was dispatched to England. Due to the depression, National Service had been abolished in 1930, but, as hostilities intensified, conscription was reintroduced by invocation of the National Service Emergency Regulations in July 1940.

During the First World War, detained military defaulters, numbering 212 in 1917, had been held in institutions administered by the Prisons Department,[55] although an attempt was made to keep them apart from ordinary criminals.[56] The year after the Second World War began, a different policy took effect and camps for conscientious objectors were established by the National Service Department. While this Department remained entirely separate from the Prisons and Justice Department,[57] a number of men who refused to comply with regulations in the defaulters' camps were charged in civilian courts and sentenced to confinement in the regular gaols. By March 1945, there were about 650 objectors incarcerated in New Zealand, forty-five of whom were held in prisons. Although their activities had been restrained during the period of hostilities, the impending defeat of Germany and the return of the first servicemen in April 1944 caused a growing number of defaulters to begin clamouring for freedom.

Men who refused to go to war had mostly done so on moral, religious, or political grounds, and many of them were people of determination and awareness. As such they posed a potential threat to any autocratic system and the most dangerous in this regard were the recalcitrants — the men who had been transferred to the prisons. It was they who now kindled the movement which brought penology (the study of punishment of crime) out of the recesses and back into the public spotlight.

Dallard, of course, had no time for what he called the 'illusory philosophies' of pacifists and, supported by Mason, he refused to listen when the liberation demands began. It soon became clear to prisoners that no early release was contemplated and a general strike was organized from 22 January 1945. It was this tactic, consisting of a hunger strike and defiance of authority, which drew comment from the media and returned attention to the prisons. Information about gaol conditions was smuggled out, forcing Mason into the fray. In concert with Dallard, the Minister attacked the men's motives and denied their claims of mistreatment, describing them as 'exaggerated falsehoods', disseminated as part of the pacifist plan.[58] In the meantime, protest leaders were transferred to Mt. Eden. Although appeals for release were rejected by the courts, gradual discharge of defaulters began several months later, and continued into 1946.

It was the issue of the military objectors which first revived an interest in prisons, but theirs was only one of several disputes which emerged as hostilities ended. On 2 February 1945, in the middle of the defaulters controversy, Dunkirk veteran P. J. O'Brien received a month's imprisonment for ship desertion. For refusing to work, he was given a total of nine days on bread and water, and through a series of cumulative punishments, spent the whole of his sentence in solitary confinement.[59] Upon release, O'Brien's case was taken up by *The Auckland Star*[60] and the Communist Party paper *The People's Voice*,[61] to initiate a lengthy press battle. The comment it raised drew into question not only the legality of consecutive punishment, but the condition of the entire penal system. The fact that O'Brien was a war veteran did not go unnoticed either, and newspaper correspondents were almost entirely favourable to his plea.

It was the obstructive arrogance of officials dealing with these matters that provoked the rancour of the press. Mason, who seemed disinterested in explaining or justifying anything, was lampooned in a cartoon in *The Auckland Star* captioned 'Above the Law', depicting him and an unnamed bureaucrat sitting atop a locked copy of the Prisons Act.[81] Although Mason never admitted it publicly, the legality of cumulative sentencing (which had in fact been recommended by Dallard) was queried by Crown solicitors, and withdrawn before the year's end. Rather than answering critics directly, in other words, the Department had opted for the tactic of bland denial, followed by a hidden remedy.

By October, the interest in O'Brien had begun to wane, but the issues he had raised endured. Large numbers of otherwise law abiding men had been imprisoned during hostilities, and it was they who continued to protest. Many of them were extremists, with a credibility that was weak; others were inconspicuous people whose voices would never be heard. However, among a minority of both thoughtful and articulate detainees was one who in September 1945 threw down a challenge which became the most devastating controversy Dallard ever faced.

Born in Auckland in 1893, Ormond Edward Burton, a schoolteacher, had joined the army in 1914, serving overseas at Gallipoli and in France. Wounded twice in action and awarded the Military Medal for bravery as well as the French *Medaille d'Honneur*, Burton returned to New Zealand as a Second Lieutenant. Shocked by the horror of war, in 1921 he turned to pacifism and, rejected by his Presbyterian Church, became a Methodist minister in 1934. A personal friend of Prime Minister Peter Fraser (who also had been imprisoned as an opponent to conscription in World War I), Burton became a prominent anti-war campaigner after 1939. In addition to being expelled from the ministry, from teaching, and from the Labour party for his views, he was arrested and gaoled five times during the Second World War, serving a total of thirty-two months in custody.

Ormond Burton already had three books and two published plays to his credit before *In Prison*, his challenge to Dallard, was published in 1945. Although he never served time at Mt. Eden, Burton had experienced Mt. Crawford and Napier prisons and was able to portray, in often witty detail, the pointless, menial, and monotonous routine of a hard labour sentence. Although fair, even to a fault,[62] the book was a candid and revealing critique of New Zealand gaols in the 1940s. Media response was favourable, but had the claims been humoured it is unlikely that more would have come of them. Dallard, however, felt personally impugned by the attack and once more felt obliged to respond.

At a luncheon talk in the Wellington Rotary Club on 18 June 1946, Dallard launched his defence. Accusing Burton of a 'pro-criminal and anti-authoritarian bias', Dallard slated law breakers, comparing some selected cases of his own against Burton's mild picture of the 'average' offender.[63] Reaction to the speech was even greater than Dallard might have anticipated. The Rotary address was covered in detail by the major tabloids, accompanied by letters and supporting editorials. Responses from persons to whom Dallard sent copies of the talk were almost totally in his favour. Messages of support came from prominent law offices and other organizations as well.

As media focus sharpened, the profile of the dispute increased. Book sales soared and the edition sold out. Throughout the dominion, newspapers and their correspondents began taking sides. Many of them supported Dallard but an equal number was opposed. The most virulent of these was *The Press* of Christchurch, which maintained a vigorous dialogue with him over the ensuing weeks. Other South Island papers rallied behind it, and opinion was soon fixed in a deadlock. The most forceful reaction came from reader-correspondents, and it was the weight of their input which eventually broke the stalemate. Up and down the country letters began flowing into press offices, nearly all in favour of Burton. The Christchurch branch of the New Zealand Howard League for Penal Reform called for a Royal Commission and was backed by its Dominion Executive. The clergy entered the fray and chorused for an investigation. Attacked from all sides and steadily losing ground, Dallard showed signs of uncertainty. From October 1946, he refused to answer any more critics. Publicly invited to a debate with Burton by the Christchurch Howard League that same month, Dallard, uncharacteristically, declined.

In spite of Dallard's reticence the question was far from closed. Burton's allegations were noted in Parliament and while Dallard fought his solitary battle with the media, he was also facing questions from above. There was an election in 1946 and the popularity which Labour had enjoyed for over a decade was now in decline. Political interest in Burton had been recorded in December 1945,[64] followed by calls for Mt. Eden to be demolished. When the matter was referred to Dallard he advised his Minister that the deficiencies outlined no longer existed, or never had existed, or were irremediable. In short, no action was needed.

Apart from the damage it did to public relations the Ministry's refusal to recognize problems, much less rectify them, had a number of internal consequences. For many months, the Public Service Association had been complaining of shortages in staff. In October 1945 the sixty members of staff at Mt. Eden had dropped to forty-five and remained that way for some weeks. Insufficient supervisory personnel forced the closure of the quarry and the risk of escape increased. Little relief was offered and, by May 1946, staff was down to half the establishment strength. The Association claimed that sick leave was abnormally high because of overwork, and there had been six nervous breakdowns over a period of twelve months. The quarry remained shut and a need for staff relief led to a new salary scale from 1 April, and from 1 October 1946 a forty-hour week was introduced. In spite of these incentives, however, recruitment was slow, and it was August 1947 before the quarry could be brought back into service.

Unemployment arising from supervision shortages and awareness of the reform debate heightened inmate perceptions of their own poverty. Where unrest among convicts had been rare, after 1947 rebellion suddenly became noticeable. In February 1948, in protest at the food, medical facilities, and general conditions, male prisoners joined those in the female division with a noisy demonstration. Local residents claimed it was the worst fracas they had known at the gaol.⁶⁵ This incident was followed by others, and the seed of sympathy which had been sown in 1945 started to take its root. When in April 1948 a Waikune prison escapee complained in court about conditions at the institution, a magistrate ordered a report to be supplied to the Controller-General.⁶⁶ Six months later, another inmate absconded, this time protesting at injustices at Mt. Eden. Commenting on the charge of escape, the Magistrate's Court said it felt the protest was justified, and discharged the offender with a suspended sentence.⁶⁷

Rising fractiousness among inmates and a belief that central administration could no longer cope caused staff morale to drop further. Within eleven days of the February 1 incident, six staff resigned and it was revealed that two others had left just before.⁶⁸ Dallard accused them of 'windiness' and proclaimed the Department 'well quit of people who conjure up grievances to which they had not hitherto given expression'.⁶⁹ Reproved for the remark by the Public Servants Association, Dallard disclaimed any defamatory intent then repeated himself in different terms.⁷⁰

In the meantime, resignations continued. By 17 February 1948, staff shortages at Mt. Eden were so serious that the police had to be called to replace prison officers guarding inmates at the Auckland Supreme Court. In an attempt to stop the exodus, the Public Servants Association investigated and made a number of recommendations of its own. Following these, a wide range of staff amenities was upgraded and a formal training programme for prison officers was promised. As a result, by the middle of 1948, the staffing situation had gradually begun to improve.

The year 1948 had been climactic for the post-war reform movement, and by the end of that year decreptitude in the gaols was an accepted fact. The public, sick of revelations about neglected prisoners, was impatient for positive action. After 1945, only the intransigence of Dallard had stood in the way of progress and now he, tired and dispirited, was preparing for his retirement. A general election was soon to follow in which one other issue remained undecided. This was the question of the death sentence, which Labour had removed as the penalty for murder in 1941.

As the year of 1949 began, the future of penal policy thus rested on two important outcomes. Firstly, if Dallard's example was anything to go by, his replacement would be an important steersman in the course of the Department. Secondly, Government attitudes towards punishment — particularly that of hanging — would depend on the result of the election. Labour was opposed to the gallows, but, as we have seen, its record in the prisons was poor. Given the strength of the reformist lobby, the best chances seemed to lie with National. However, this party was also committed to the return of the rope, and whether it could reconcile that with a policy of progressive change remained to be seen. When Dallard retired in August 1949, therefore, the announcement of his replacement left many matters unsealed. With the existing administration being occupied with the demands of its campaign, there was little to do for the moment, but await the return of the polls.

NOTES

1. Mayhew (1959) 38.
2. Some authorities assert that Hume served as governor at Wormwood Scrubs. Mayhew (ibid.) 39-40 finds no justification in the claim. More detailed discussions of New Zealand's early system of justice can be found in Lingard (1936); Matthews (1923); Mayhew (ibid.); Missen (1971); *Report of the Penal Policy Review Committee 1981*, 13-21; Ritchie (1984); Stacpoole (1976) 138-9 and Webb (1982).
3. Quoted in Walker (1971) 13.
4. Quoted in Ritchie (1984) 31.
5. Webb (1982) 83.
6. *Report of the Gaols Committee 1878*, 55.
7. Ibid.
8. Mt. Cook prison was only occupied for a short while before being handed over to the Education Department. It was demolished in 1925 (Mayhew, 1959, 65).
9. The English 'radial' design, which involves a series of wings emanating from a central hub like the spokes of a wheel, was first seen in such early US prisons as Eastern Penitentiary (Philadelphia, 1829) and Trenton (New Jersey, 1836). Interesting, however, is the fact that the idea apparently came from England. John Haviland, who designed America's first radial institution at Eastern, was himself an English immigrant and was probably influenced by the design of the early English institutions (Johnston 1973, 29-30).
10. Johnston (1973) 34.
11. Stacpoole (1976) 139.
12. By the time Auckland gaol was completed in 1917 its name had been changed again, to Auckland prison. It remained so until the opening of Paremoremo in 1968. In 1968 Paremoremo became Auckland prison and the old institution became Mt. Eden prison. In order to avoid confusion, I shall keep to popular usage and refer throughout to Mt. Eden prison and Paremoremo prison, respectively.
13. In 1912, three-quarters of Mt. Eden's inmates were involved in constructing it (Edridge 1964, 119).
14. Ritchie (1984) 98.
15. NZPD vol. 150, (1910) 352.
16. General information about the history and architecture of Mt. Eden and other radial prisons is contained in a variety of literature. Of particular value are, Anon (1978); Edridge (1964) 96-122, 105-113; Fields & Stacpoole (1973); Johnston (1973) 10-41; Mayhew (1959) 22-25; National Archives (1984) 5; Newbold (1978) 89-102; *Report of Royal Commission on Prisons 1868*; *Report of Auckland Gaol Commissioners* (1877); Ritchie (1984) 31-111; Stacpoole (1976) 138-9.
17. Missen (1971) 10.
18. Matthews had relinquished the post of Private Secretary to commence duty as Deputy Inspector of Prisons in 1912. He became Inspector the following year, replacing Hume's successor F. Hay. Although Hay's job had been mainly supervisory, with the Minister retaining executive power, the Public Service Act brought prisons back under public service control after 1912.
19. For example, although Findlay quickly put Hokitika and Wanganui out of service, his intention in 1910 also to close Napier and Dunedin Gaols 'as soon as possible' (NZPD, 1910, vol. 150, 313), was forestalled.

    After he was defeated, Findlay's office was occupied briefly by J. A. Hanan, uncle of J. R. Hanan. The latter was to become a powerful figure in the Holyoake cabinets of the 1960s.
20. Matthews (1923) 8.
21. Ibid. 9.
22. Ibid. 7-8.

23 Ritchie (1984) 112.
24 Matthews (1923). Apparently the arrows remained on the older clothing. In 1917, pacifist Archibald Baxter reports the arrows were still present on the clothing of prisoners at the Terrace gaol, Wellington (Baxter 1939) 29.
25 Mayhew (1959) 109-11.
26 Young (1981) 15.
27 Dallard (1980) 46-51.
28 William Massey, an Ulsterman and British Israelite, remained Premier until his death in 1925 (Sinclair 1980, 239, 247).
29 Sinclair (1980) 244-5.
30 Dallard (1976).
31 Dallard's patriotism was such that in 1940 he found himself compelled to report the Minister of Marine D. G. McMillan to the Prime Minister for 'disloyalty'. The aberrant Minister had made a derogatory comment about Queen Victoria (Dallard, 1976).
32 Dallard (1976).
33 This objective was not in fact achieved until after 1935.
34 Ritchie (1984).
35 Dallard (1976).
36 This represented a refurbishment of the neglected classification programme: sex offenders would be sent to New Plymouth; incidental offenders would go to medium security prisons and camps; the old and senile would serve their time at Wanganui; and Mt. Eden was set aside for long-term, dangerous, and habitual criminals.
37 Dallard mentions in his 1976 interview that due to his acquaintance with the Public Service Commissioner he could have had what he wanted in prisons, but for the question of finance. During the depression years and later, the war years, he says, Prisons appropriations were subject to constant pruning.
38 Although not incorporated formally until 1954, the Prisons and Justice Departments operated together from this point.
39 M. Finlay (pers. comm.).
40 Mason (1968).
41 This was the women's borstal at Arohata, Wellington. Construction of Arohata was forced when a wartime amunition dump near the old institution of Pt. Halswell made occupation of that site unacceptable (Dallard 1976).
42 Dallard died in 1983 but his 1976 taped interviews are instructive.
43 Dallard (1980, 75).
44 Ibid. ch. 11.
45 Letter from Dallard to O. V. Church 23 June 1932.
46 Memo from Dallard to superintendent, Wellington Prison, 5 December 1931.
47 Dallard (1976).
48 Prison Regulations 1925 ss. 252, 253.
49 It was because of their supposed propensity in this direction that Dallard had not favoured recruiting Māoris as prison wardens (Dallard 1976).
50 Dallard (1976).
51 The single exception he instances was D. G. McMillan. McMillan was never Minister of Justice, but may have served in acting capacity in 1940-1941.
52 Some of these he reads out in his 1976 interview. Others are quoted in Dallard (1980) 148-51.
53 Dallard (1976).
54 *NZH* 27 April 1936.
55 Webb (1982) 34, for example, mentions the road-making camps at Rotoaira and Erua.
56 Mayhew (1959) 112.
57 For an ethnographic study by an inmate of such a camp, see Scott (1949).
58 Letter from H. G. R. Mason to S. N. Zimon 24 September 1945.
59 This policy of charging inmates with a succession of minor offences and awarding

cumulative penalties of solitary confinement was conceived by Dallard himself. Its aim was to stop the publicity that pacifists got when they were released from long sentences of solitary confinement with bread and water, after aggravated charges. In a memo to all controlling officers in September 1945, Dallard recommended pacifists be silenced by keeping them in solitary wherever possible.

[60] *AS* 29 March 1945.
[61] *The People's Voice* 10 April 1945.
[62] Hamilton (1953) 155, for example, takes Burton to task for erring too far on the side of administration.
[63] Dallard (1946).
[64] NZPD vol. 272 (1945) 312–3.
[65] *NZH* 3 February 1948.
[66] *NZH* 6 April 1948.
[67] *NZH* 30 September 1948.
[68] *NZH* 12 February 1948.
[69] Ibid.
[70] *NZH* 23 February 1948.

2: THE ELEMENTS OF CHANGE 1949

The New Regime

The ground upon which the 1949 election was fought was of politico-economic, not penological, substance. The Labour party, at its peak in 1938 when it took more than half of the votes cast, had lost four seats in 1943 and three more in 1946. In the final years of the decade, the ranks of the Labour party, split by an oligarchical leadership, had fallen into disarray and conflict. In 1947 there was industrial trouble over prices and wages, followed by the deregistration of several unions. Displeasure arising from these actions was compounded by former pacifist Peter Fraser who, keen to restore peacetime conscription against party opinion, held a referendum on the matter in 1949 to gain support for his stand. Absences of Fraser and his Finance Minister Walter Nash through illness, and frequent overseas trips in the post-war period, aggravated dissatisfaction and caused the Government to lose touch with the electorate. These factors, according to Sinclair, were significant in the decline of New Zealand's first Labour Parliament.[1]

In contrast, the National party, led by the purposeful Sidney Holland, was gaining in confidence and strength. Commodity shortages, caused in part by a retention of wartime controls after 1945, were adding impetus to calls for a return to free enterprise. Campaigning under the slogan 'Make the Pound Go Further', National promised an end to economic stabilization and the liberation of market forces. Although Fraser, one of the first of the Cold War defenders, was strongly opposed to Russian Communism, both he and Nash were effectively denounced as Socialists responsible for the fostering of class antagonism. By November 1949, these factors had caused support for Labour to fall. In a decisive victory that month, the conservatives were swept to power with a majority of twelve parliamentary votes.[2]

T. C. Webb, Minister of Justice

The National Government took power at a time of a conservative swing in political and social attitudes. As far as prisons were concerned, however, this was a period of considerable sympathy for progressive change. Adding complexity to that area, however, was a vein of intense pressure for the return of capital punishment for murder.

Allocation of portfolios after a general election is a task for the Prime Minister, who assesses a Member's aptness for a Cabinet position partially in terms of the prevailing attitudes of the day. The reform of gaols had taken a fairly low profile in the 1949 campaign, but the issue of hanging formed a component of National's platform. If penal philosophy had been of any importance in the choice of Justice Minister, therefore, it is unlikely that broader penal issues weighed very heavily. When Holland appointed his cabinet, the Justice option reflected this consideration. The new Minister of Justice was a man personally committed to the return of the death penalty, whose interest in penal reform was subsidiary.[3]

The son of a local doctor, chemist, and dentist, Thomas Clifton Webb was born at Te Kopuru, near Dargaville, in 1889. Educated initially at a local school, the young Webb won a junior scholarship in 1902 and gained entry to Auckland Grammar School. He attended Auckland University, was admitted as a solicitor in 1911, and began working in a law office. He then left to serve overseas with the army between 1917 and 1919. Upon his return, Webb entered legal practice in Dargaville, becoming a member of the local borough council for two years after 1921. Webb's career in national politics began relatively late in his life and it was not until 1943, at the age of fifty-four, that he entered Parliament as MP for the Kaipara (later Rodney) electorate. In 1949, in addition to the Justice portfolio, Webb was also given control of Prisons and Patents and was made Attorney-General.

Although Webb's tenure in Justice was not a long one, his period, as part of the new administration, was productive. Webb contributed to a wide range of legislative activity, in areas as diverse as matrimonial property, legal jurisdiction, and parliamentary procedure. As Attorney-General it was also Webb's frequent duty to justify and elucidate proposed Government action, a role which earned him the opposition label of 'Official Explainer'. His successor, John Marshall, describes Webb as having been possessed of common sense, clarity of mind, and integrity. He was Marshall's mentor and, according to Marshall, it was these fine abilities that made Webb one of the most influential men in Holland's first Cabinets.[4]

One subject which Webb did not spend much time on was that of penal policy, and except in relation to hanging, his personal input was small. As Justice Minister, Webb was concerned mainly with the punitive aspects of imprisonment, and he had little sympathy for reform. According to Dr Martyn Finlay, a later Justice Minister and one of Webb's electoral neighbours, Webb was not interested in corrections and seemed to leave the Prisons Department to run itself. Webb's daughter Patricia — a former Chief Legal Adviser to the Justice Department — concedes that her father's attitude towards punishment was 'a fairly tough one', but points out that, none the less, he went along with the significant changes that occurred while he held office. She also remarks that he was not the sort of man to be unaware of what went on in his Department. Marshall agrees, and has no doubt that a 'proper' relationship would have existed between Webb and his permanent head.[5] This means that all proposed administrative activity would have been discussed with the Justice Minister and that before being put into effect it would have needed his approval. So while he may not actively have initiated major reforms, Webb apparently nodded to the developments that did take place.

The early 1950s was thus a period in which liberalization in prison routine was possible. Most changes, as we shall see, were administrative in nature and occurred without the need of legal sanction. We can conclude that it was not from the new Government that energy for the changes was drawn. The source of activity in the early 1950s was the Justice office itself, and in particular the Department's new permanent head, who had taken control in September 1949.

S. T. BARNETT, SECRETARY FOR JUSTICE

The successful applicant for the control of Justice and Prisons in 1949 was a controversial figure who, at the time of his appointment, had served the Government for more than thirty of his forty-eight years. Born at Dunedin in 1901, Samuel Thompson Barnett was educated in Southland and at Canterbury University. Graduating with a Bachelor of Laws, he was admitted as a solicitor of the Supreme Court in 1930. By this time, Barnett was already rising in rank in the New Zealand civil service. He had joined the Lands and Survey Department in Dunedin in 1918. From here he had transferred to Christchurch, and rose to the rank of Assistant Under-Secretary before eventually transferring to the Education Department as Assistant Director (Administration) in 1947. The following year he transferred again, this time to the Justice Department as Deputy Under-Secretary and also Deputy Controller-General of Prisons. He had only been there a year before he assumed their leadership.

Mason was opposed to Barnett's final appointment and it is a credit to the independence of the Public Service, as well as to the latter's sheer ability, that he won selection in the face of ministerial disapproval. Barnett's public service record and the fact that he was a lawyer made him an attractive contender for the top job, and it is probable that his promotion the previous year was specifically contrived to prepare him for it. Barnett's succession to Dallard was in fact confirmed six months before the chair even became vacant. So anticipation of the input that Barnett might provide was certainly established by the time he took control. Finlay explains the circumstances of the appointment in these terms:

> How [Barnett] came to be appointed I think was simply because the chauvinism, the conservatism of Dallard, sparked in the appointing authorities a desire for somebody who was a bit out from the general mould, an experimenter, a radicalist. And he showed that in every position he had, in Education, in Police and on the Licensing Control Commission.
>
> He was put in simply to break the mould that had been established so firmly by Dallard and, having done so, he opened the way for further experimentation. Barnett was, and always had a reputation for being, an innovator. An irritant too, an irritant innovator, he annoyed all the people he worked with because he sparked them along so much. He was quite unconventional. So I think his having penetrated into the Justice Department left a crack.

With his long record of service, the Public Service commissioners can have been in little doubt about the type of impact Barnett was likely to make. Knowledge of his character, well established in Government circles by 1948, soon became general. John Marshall, under whom Barnett was later to work, spoke particularly highly of him:

> Sam Barnett was a strong, positive personality who prided himself, with justification, on the efficiency of his administration. He did not suffer fools gladly and sometimes his definition of a fool extended to sensible citizens who disagreed with him. . . . His tendency to arrogance and rudeness was softened by a Christian spirit of concealed goodwill and silent reconciliation.[6]

Progressive, realistic, and thorough, he was, Marshall said, 'among the greatest of our public servants'.[7]

Samuel Barnett was a self-assured man with a powerful personality, and it was his individual qualities as much as anything which lay behind his success. Although the 1950s was a period of social restraint, the Prisons head was able to exploit post-war sympathies to their fullest. Of Barnett's administrative manner, a later Justice secretary told me:

> He was a very competent man. He was autocratic in a lot of ways. He had an excellent mind. Got things done. And he didn't mind who he trod on in getting things done. He could be a very kind-hearted person or he could be the most thoroughly rude person.

Patricia Webb, who joined the Department in 1951, remembers her boss's huge office — reminding her of the town hall — with Barnett sitting at the far end of it. '. . . when it came to leave you thought you ought to back out', she said. 'He wasn't exactly the sort of person to make you feel terribly at ease.'[8] Another junior, who later became Deputy Secretary for Justice, was impressed in a similar way.

> I think at that stage I found him fairly formidable! . . . And I certainly realized he had a very considerable 'manipulative ability' . . . it was also my impression that he was fairly adroit in persuading people to accept the line that he thought should be followed.[9]

However, although the Justice Secretary was able to achieve much in the face of political indifference, the association between him and his Minister was strained. The permanent head did not always get his way, and insoluble disagreements between the two soon eroded their relationship.

One region of conflict was penal reform, and immediately after the change of Government Barnett wrote to Webb, welcoming his leadership and outlining the Department's principal areas of deficiency. The politician was not merely unmoved by some of the proposals, he made it plain from the start that he did not support vigorous change.[10] At once, the Secretary's aspirations fell. Another area of conflict, and one in which the pair were more deeply incompatible, was that of capital punishment. The death penalty for murder was restored in 1950, and in 1952 executions recommenced. Barnett was profoundly opposed to the principle of hanging and he loathed a practice which he was forced to administer.[11] Webb's important role in drafting the Capital Punishment Bill, and his powerful influence over decisions to carry out death sentences, gave Barnett's duty a highly personal flavour. Organizing executions was a task which the permanent head accepted, but which he performed with increasing distaste.

The Capital Punishment Issue

Clifton Webb observed that where national politics were concerned, there were two areas on which New Zealand's major parties were distinct: one involved the part the State should play in economic management; the other was capital punishment. Webb had for years favoured the latter, and when his party was sworn into Government in December 1949, one of his first moves had been to restore the death penalty for murder.

It is tempting to look at the matter of capital punishment simply in terms of party politics. National always defended the death penalty for murder and its caucus had lambasted Labour for removing it in 1941. Likewise, just as Webb was a long-time advocate of hanging, so was Mason opposed to it. However, to restrict analysis to party philosophy is to ignore a fundamental of politics. For although Governments may bring in laws in defiance of public mandate, it is to the electorate that they must finally answer. So it was with hanging. The repealing legislation of 1941 had passed through the House unsupported by public opinion and since then, opposition to repeal had risen. Labour had continued to favour abolition, but in 1949 the National party had restoration written into its manifesto. Although not a major election issue, when the country went to the polls that November, it knew that a vote for Labour would be a vote for the status quo. A vote for National would be one for the gallows. We have seen that the election result was an overwhelming victory for National.

Death by hanging — following British tradition — had always been the favoured penalty for murder in the colonies, and it officially commenced in New Zealand in 1842.[12] Since then there have been forty-four recorded executions. Between 1900 and 1935, fifty-six men were sentenced to death, of whom nineteen were executed.[13]

When Labour took power in 1935, it had brought with it a promise to abolish capital punishment. Although legislation had for the moment been deferred, from this time all sentences of death were commuted to life imprisonment. Slender support for the Government's stand made its leader hesitant to enact abolition and it was not until 1941, with the Prime Minister overseas, that an uproar over flogging allowed both corporal and capital punishment to be formally struck out.[14]

While the move had never been a popular one,[15] distraction over events in wartime Europe and the low incidence of murder had eased the acceptance of abolition. However, after 1941, a dramatic increase in murder convictions — with the 1935–1941 total of six increasing to twenty in the next six years — and a series of unsolved homicides, had brought the death penalty back into focus. It was easy for those who supported retention to blame the rise on the lack of an effective deterrent. Thus, at the same time as the defaulters' sympathizers were successfully calling for prison reform, an equally vociferous group began clamouring for return of the noose.

If public opinion was not already resolved on the matter by 1948, an unprecedented six murder convictions that year and two sensational cases towards the end of it gave the capital punishment lobby the very fuel it needed.

The first of these, the brutal sex-slaying of a middle-aged widow at Wellington in September (the Kitty Cranston murder), sent shock waves through the nation and caused an Imprest Supply Bill in Parliament to be interrupted by a spontaneous debate on hanging.[16] Wide media coverage followed and the opposition pledged that if elected, it would 'not flinch in carrying out its duties to give the people the protection they are entitled to'.[17]

The second incident occurred two days before the end of the year. On this occasion a mentally unstable man called Tume battered his son, his mistress, and her mother to death with an iron bar. During his trial, Tume displayed no remorse, and evidence was given that he had openly contemplated the killing for some months prior to carrying it out. His words before the crime, 'Even if you do murder these days you only do eight years, that's because of the good Government we've got,' achieved wide notice,[18] as did the convicting jury's rider calling for re-introduction of the capital penalty.

The Cranston murder and the trial of Tume in April 1949 ensured that the issue of hanging had no chance to cool in the vital months leading up to the election. As soon as the new Government was sworn in, the Justice Minister asked Barnett to prepare a full memorandum on it. As a result, when Parliament resumed in June 1950, capital punishment was one of the first items on its agenda. Debate over the final reading ended in late November, and despite a free vote on the matter, only one National Member, J. R. Hanan, crossed the floor to vote with a unified opposition.[19] The Capital Punishment Bill was passed by a majority of nine and from 1 December 1950, New Zealand again became a retentionist State.

The commencement of 1950 saw the penal system of New Zealand at an ill-marked crossroad. Imminent change was assured, but its course and destination were uncertain. From the time Mt. Eden had been completed, almost no alteration in routine or policy had occurred. Because of official secrecy, the maximum security prison had been obscured from the public. As the war ended and the military defaulters anticipated release, this condition was transformed. Although exact prospects were still unclear, by the end of the 1940s radical and significant change were expected. Events of the next decade were to prove among the most intensive and far-reaching of any in the nation's penal history. In order to appreciate the extent of these developments, it is necessary that the situation which Barnett inherited at Mt. Eden be described.

Pre-1950 Conditions

Between 1949 and 1950, Mt. Eden prison employed sixty-five members of staff, who were responsible for the supervision of about 300 prisoners. Housed in a separate section of the gaol, about a dozen of these inmates were women.[20] Of the total muster, approximately ten were serving life sentences for murder. Although no racial breakdown for Mt. Eden is available, about 22 per cent of the country's inmate population in 1949 was Māori.[21]

At the time, adults not subject to indefinite penalties could be given imprisonment with or without hard labour[22] (usually the former), given reformative detention,[23] or be declared habitual criminals. This latter status was a controversial one that allowed repeated offenders to serve indefinite terms of imprisonment, subject to release by a review authority known as the Prisons Board. Such boards had existed since 1910. At Mt. Eden gaol, inmates were roughly classified according to the type and length of their sentences. Lifers and hard labour men, numbering over 100, were held mainly in the north wing; the thirty or so habituals in the south. Remand prisoners occupied the west wing. In the east wing lived the seventy-odd reformative detainees, while about forty short-termers were kept in the east wing basement, and in the part of the north wing extension not occupied by women. Other than this, classification and specialized treatment for particular categories of offender were absent. Except that the latter could appear before the Prisons Board, there were no significant differences between the routines of hard labour convicts, and those doing reformative detention.[24] Most young and physically able men were assigned to work in the quarry. The majority of lifers, and a number whose possible escapes might create a hazard, were employed in one of the shop industries located within the institution. These included tailoring, boot manufacturing and mailbag making. Others worked at domestic and maintenance tasks.

In 1925, the Mt. Eden quarry had engaged 15.5 per cent of the labour force when, accounting for 38 per cent of industrial returns, it was by far the gaol's biggest earner. By the late 1940s, the population working in the quarry had risen to eighty — that is, to over a quarter of the total muster — with three officers assisting. However, due to inefficient working and competition from private industry, quarry returns had dropped to about 25 per cent of productivity by 1936.[25] Although profitability continued to fall, the quarry remained the principal employer of security prisoners throughout the 1950s.

Routine before the fifties was simple, and it varied little from year to year. A prisoner's day started at 6.30 a.m. with the sounding of a rising bell. Doors were opened and two inmates, carrying a half-44 gallon drum set between two poles, visited each cell, collecting 'slops' — the contents of the night pots. Directly after slops came the breakfast detail carrying another can, this one full of porridge. A ladle of porridge would be served to each inmate, together with a cup of tea, before doors were again locked shut. At 8.00 a.m. doors re-opened for 'second unlock'. Prisoners would now be released from their cells and frisked before being marched out to the yard for work parade. They laboured for about three hours and came in at 11.30 a.m. for lunch.

From before the 1930s through to the beginning of the fifties, the prison's noon meal was the main meal of the day. Food was served in dixies, with vegetables in a top compartment and meat or stew in the base. Dixies were made up in the kitchen and transported to the wings in a tray filled with hot water. Inmates, lined up beside a number corresponding to that of their cell, would be given a dixie and a cup of tea as they filed back to their 'slots' for lock-up. At 1.00 p.m. prisoners were released again for work and they returned in the afternoon at 4.30 p.m. to receive their evening meal. Before 1950, this consisted of a ration of bread, milk, and sugar, a mug of tea, and a dixie of either porridge, semolina, or rice. The men were secured for the evening at 5.00 p.m.

Each Friday, prisoners whose work was satisfactory were issued one ounce of tobacco, but no cigarette papers.[26] Every day a single match was given to each man, although Dallard later increased this to two. Ingenious means were devised to evade the restrictions which match rationing produced.[27] The weekend routine at Mt. Eden was little different, except that morning unlock was an hour later. Saturday was a half-holiday and on Saturday afternoons prisoners would be allowed a shave, shower, and kit-change before being put into the main yard. Little in the way of recreation was available, although men would often gamble clandestinely with dominoes and cards, or bet on horse races which could be listened to on (illegal) home-made crystal sets.[28] Sunday was a complete rest-day and Church was compulsory for those claiming religious adherence. Sometimes the Salvation Army band would play in the prison compound, with band members roped off from inmates and forbidden to communicate with them.[29]

In spite of the fact that many were engaged in sweaty, dirty work, the number of showers permitted remained at two throughout the Dallard period. Most inmates could shave — or be shaved — twice a week, and clothing was changed once a week. Standard convict kit of the day consisted of boots, white or brown moleskin trousers, moleskin cap, red, white, and blue fine-checked shirt, and a grey jacket. From 1947, prisoners over the age of thirty-five, and others on recommendation of the medical officer, were issued with pullovers. Medical inspections were brief and perfunctory, and treatment was often crude. A former officer who served at Mt. Eden during this period described medical parade to me in these terms:

> They fell into the big yard. There was a medical officer out there with a bottle of aspirins and a bottle of cough mixture and they marched off to work. If a bloke wanted an aspirin he got one and if he wanted cough mixture they'd give him some, but he kept on walking.

The regimen of the old institution at the time Barnett took it over was thus highly restrictive. Prisoners spent at least sixteen hours of every day locked up and used to reckon that during an ordinary twelve month stretch, eight would be spent in their cells. Conditions in cells, like the rest of the gaol, were spartan. Although quite large,[30] units contained only the barest of furniture, and prisoners slept on canvas hammocks or straw palliasses. Hobbies were not permitted, and all newspapers, other than *The Weekly News*, were banned. Apart from approved educational reading the only books permitted were those held in the prison library. Most of these were old, tattered, and out of date. Swapping of books among inmates was forbidden.[31] Some long-termers were allowed a writing desk and each cell was equipped with radio headphones. The frequencies of these were controlled by administration.

The result of such bridled management was a system in which staff-inmate relationships became simple and clear-cut. The job of the warder was the locking up and unlocking of prisoners, supervision of routine activities, and the prevention of disturbances and escapes. To this latter purpose, all night and sentry personnel were armed. Communication between convicts and officials was restricted and formalized. Recruits were kept under supervision to prevent them from becoming too intimate with the prisoners. A former warder summed up the relationship with inmates in the following words:

> You couldn't give them anything because there was nothing to give. They didn't expect anything because there was nothing to expect. Except their meals and their ounce of tobacco a week. . . . A bloke did as he was told, sure, but the type of man you had, he *knew*. He knew the routine so you virtually didn't have to tell him anything.

The type of inmate society which emerged beneath this regime contrasted starkly with those of later years and was a direct reflection of management. Monitored communication,[32] long hours of confinement in single cells, and an absence of special privileges affected both the opportunity and the capacity for rebellion. Not only did inmates lack a strong sense of cohesive identity, they were also without any firm social structure. It is in the light of prisoners' inability to interact that the scarcity of rebellions must be viewed.

In 1950, Barnett travelled to study correctional institutions abroad, and it was then that his interest in convict therapy matured. By the time he returned to New Zealand later in the year, Barnett's zeal for reform had been well nourished, and thus it was that from 1950 onwards, managerial policy at Mt. Eden prison began a sudden, rapid shift. The new permanent head brought a fresh concept to penology, familiar in its rudiments to that of John Findlay. Once more, the retributive aspects of imprisonment would be relaxed. No longer were offenders merely to be punished; a revised doctrine demanded that they be treated, resocialized, rehabilitated, and reformed. The 'new penology', which the United States was developing in parallel contour,[33] saw the prison as a kind of deviance clinic, where 'treatment' programmes could be deployed to cure the criminal of his disease. A substantially revised system of penal treatment became the permanent head's overriding objective, and it was this goal that underlay the prospectus which he now began to draft.

NOTES

[1] Sinclair (1976) 265.
[2] Authoritative accounts of this election appear in Sinclair (1976) 265-78; (1980) 285-7; and Chapman (1981) 351-6.
[3] Another pertinent factor is probably Webb's seniority, and the fact that he was one of the only legally qualified men in the National caucus.
[4] J. Marshall (pers. comm.); (1983) 213.
[5] J. Marshall (pers. comm.).
[6] Marshall (1983) 216.
[7] *NZH* 1 March 1972.
[8] P. Webb (pers. comm.). Patricia Webb is a daughter of Clif Webb. She joined the Department of Justice as librarian and Law Research Assistant, rising to the

Assistant Secretary position of Chief Legal Advisor by the time of her retirement in 1976.
9 J. Cameron (pers. comm.). Cameron joined the Department in the early 1950s and when spoken to held the position of Deputy Secretary.
10 Robson (1972) 18.
11 J. Cameron (pers. comm.); Engel (1977); Robson (1971) 25.
12 This was the case of *Maketu*, executed at Kororareka in February 1842.
13 All of those executed in this period were men. The only woman to have been hanged in New Zealand was Minnie Dean, in 1895.
14 The story of abolition is a complex one and aspects of it are covered in detail in Engel (1977) ch. 3; Sinclair (1976) 204 and in a memo from J. Gifford (acting Under-Secretary for Justice) to T. C. Webb (Minister of Justice) 29 May 1950.

In October 1940 a reprieved murderer, R. R. D. Smith, and three accomplices escaped from Mt. Eden prison, permanently disabling a warder in the process. Quickly recaptured, the quartet was sentenced in February 1941 to twelve years' hard labour each, and to twenty strokes each of the cat-o'-nine-tails.

When the penalties were upheld by the Court of Appeal the Government, embarrassed about having to conduct a procedure that it had for years promised to abolish, desired to have the floggings commuted. The Governor-General, Sir Cyril Newall, was himself an avid abolitionist, however, and he refused to sign the Government's remittal recommendation unless an undertaking was made to abolish flogging altogether, along with capital punishment.

It was Fraser's opposition to the repeal of either that was holding legislation up, but now, with an election pending, neither he nor his deputy Walter Nash wanted a viceregal confrontation. It was largely for this reason that, when Fraser went abroad to visit beleaguered New Zealand troops in Crete in May 1941, Nash seized the opportunity to have legislation put before the House before Fraser returned, and before the nation went to the polls in September 1943.
15 Public opinion polls as late as the 1950s continued, almost invariably, to support capital punishment (Engel 1977, 71).
16 NZPD vol. 283, (1948) 2629-53.
17 NZPD vol. 238, (1948) 3270. The speaker was D. C. Kidd (Waimate).
18 NZPD vol. 285, (1949) 494-5; Engel (1977) 39.
19 Hanan's only party sympathizer, E. P. Aderman (New Plymouth), was absent.
20 In the 1950s Mt. Eden's women prisoners were held in the basement of the north wing extension and in an old wooden building within the walls where the separates division is now located. In 1958 women were moved out of the north extension and the remainder vacated the old wooden structure early in 1965. This was demolished that same year and the separates division opened in 1966.
21 Māori at the time constituted around 6 per cent of the total New Zealand population.
22 Criminal Code Act 1893.
23 The Crimes Amendment Act 1910 introduced this sentence, which Magistrates could fix for a period of up to three years; Judges for up to ten years.
24 This fact was well recognized by C. L. Carr (Labour, Timaru) (NZPD vol. 270, 1945, 313); and conceded by Mason (NZPD vol. 270, 1945, 315). It was also commented on by J. Watts (National, St. Albans), (NZPD vol. 278, 690); and had been remarked upon by the Minister in Charge of Prisons as early as 1924 (*App.J.* H-20, 1953, 6).
25 Lingard (1936) 64.
26 Cigarette papers were not issued until December 1949, after Barnett took over. Prior to this, toilet paper was used (Anon. 1936).
27 One method was to mix the phosphorous head of a match with water in a spoon, then split the match into pieces and make new matches from the paste in the spoon. Another way of overcoming the match shortage was by means of what was known as 'the gear'. Tinder would be made by burning a piece of rag and slapping it shut in a book. The tinder would then be kept in a small box or tin and ignited

by striking a piece of flint or quartz from the quarry against a piece of steel to produce a spark. Possession of tinder was illegal and detection could mean forfeiture of tobacco (I.S. pers. comm.); *NZH*, 21 January 1931.

28 Crystal sets could be made from a pencil lead or a piece of razor blade and some copper wire, attached to a set of headphones. The sets were illegal but one lifer told me he had seen one so small it could be hidden inside a walnut shell, with two pins sticking out of it for aerials (I.S. pers. comm.).

29 W. Limpus (pers. comm.). Bill Limpus, who played in the Salvation Army and in the late 1940s became involved with Auckland prisons for years as the city's representative of the Prisoners' Aid and Rehabilitation Society.

30 Cells at Mt. Eden measure 3.3 m × 2.3 m × 3 m in height (Edridge 1964; 108).

31 Anon. (1936). This situation improved after publicity in 1945 (*NZH* 5 April 1945). The National Library Service was extended to prisons in 1947.

32 Evidence of this communication restriction is visible today among the older criminals, some of whom still habitually talk out of the sides of their mouths, without moving their lips. The habit is said to have originated in the days when prisoners were often forbidden to speak to one another. Today, to describe a person in gaol as one who 'talks out of the side of his mouth' is to say that the person is an 'old lag' or an 'old boobhead'.

33 See for example Mitford (1977) ch. 7.

3: THE REFORMIST'S ZEAL 1949-1954

By the time Clif Webb took office, Sam Barnett, already a seasoned administrator, had spent over a year as a top-ranking Justice official. Although by no means expert in penal matters — as he readily acknowledged — Barnett worked dutifully to acquaint himself with the running of his Department. In spite of his Minister's reserve, from an early stage he had firm notions about the direction he would like to take.

Early Ideas

A message sent to Webb in December 1949 was Barnett's first statement of intention. To begin with, he criticized the Prisons Act of 1908, which he found archaic, confused and in need of review. The Prison Regulations, he felt, were deficient also, and he sent for a copy of the latest English regulations, hoping to adapt them for application in New Zealand. An important concern was classification. The previous Controller-General had been distrustful of psychologists and had taken charge of classification personally. Barnett was more ambitious and envisaged a diagnostic committee at every prison, with continuous monitoring of inmates from reception to release. He attacked the habitual criminals legislation of 1906, as well as that of reformative detention, 1910. The first, he said, was not being implemented properly; the second, with a maximum of either three or ten years, was proving too short. He also wanted brief prior terms reduced to a minimum and recommended that the whole question of sentencing be addressed by the Justice Minister in the coming parliamentary session.

Barnett was unhappy with Mt. Eden, which, he felt, was too old to be fully secure in an urban area. His belief that it should be removed to the country was one which would dominate his thinking for much of the coming decade. Another problem was personnel. Prisons were still undermanned and the low quality of certain senior staff, together with the poor standard of recruits, had adversely affected morale. In the twelve months before March 1950, resignations had exceeded enlistments. Barnett hoped to establish a training centre at Wellington and to upgrade employment conditions and pay. He hoped that these provisions would boost flagging morale and improve staffing quality.

Barnett's memo of 1949 was a long and involved one. Occupying a total of twenty-two typed pages, its object was to introduce the incoming Minister to his Department and to suggest areas which might benefit from attention. In closing, Barnett promised to do his 'level best' to make Webb's administration a success. Ministerial reserve notwithstanding, several of the Barnett proposals were ready to bear fruit by the time Webb stepped out of politics.

Early Action

Although strongly at variance with his predecessor, now only three months out of office, Barnett was careful not to disparage him. However, excitement over Dallard's departure was hard to conceal, and the Department was lively with optimism.

The new boss went abroad in 1950 and spent eight months studying the penal systems of the United States and Europe. However, before these travels had even begun, work was already under way. By the end of March 1950, facilities for staff were improving. Changes in uniforms had been proposed and the old style of kit was being phased out. The re-designed dress was navy blue and the old khaki 'bandsmen's tunic'[1] became an open-fronted jacket, with blue shirt and black tie. Accompanying the modern attire was a palatable new title. Custodial personnel would no longer be referred to as 'warders', they would now become 'officers'. To underline the new professionalism, a staff training school was established and would be attended by all recruits, for a period of four weeks.

Conditions for prisoners also got better. At Mt. Eden, a part-time welfare officer, who had been working there since February 1949, was signed on for full-time employment. His functions would be broad and include mail censorship, escorting prisoners to the infirmary, and talking with them around the institution. The man who took this position was Stanley Robert Banyard, a captain in the (Anglican) Church Army. An Englishman who had migrated to New Zealand in 1935, Banyard remained at the prison until his retirement in 1960.[2]

General welfare services received attention. In December 1949, Barnett issued a memo to all controlling officers,[3] directing that inmates considered likely to benefit from tuition should be encouraged to apply for enrolment. All reasonable fees would be borne by the Department. In March the following year the recently retired Assistant Director of Education, Mr A. F. McMurtrie, was appointed Acting-Supervisor of prison education, with the task of reviewing it regularly and reporting on progress.

As a step towards better recreation facilities, regular entertainment was envisaged. At Mt. Eden, for example, concerts would be given. Monthly band recitals were planned, together with shows of some other type every two months. A movie would also be screened every six weeks, so that each fortnight, maximum security inmates would receive some form of organized amusement. In addition to this, basketball courts and a bowling green were marked out in the main exercise yard. Like education, equipment associated with these activities was supplied by the Government.

EARLY RESPONSES

Initial reaction to the changes was enthusiastic, and was matched by terse criticism of previous management. In the House of Representatives, Labour MP the Reverend C. L. Carr said that New Zealand was now well on the way to achieving the reforms which parliamentarians had advocated for years. Lack of progress had been due to lethargic and indifferent prison authorities.[4]

The new attitude, visible in Barnett's first report of 1950, came like the 'winds of Heaven', Mr Carr said, compared with a 'valley of dry bones' in recent years.[5] Of specific concern was the agricultural programme, the development of which Dallard had been so proud. This enterprise, P. Kearins (Labour, Waimarino) now revealed, had been suffering for years as a result of neglect and inadequate supplies.[6] Because of failed supervision, the rabbit population at Hautu and Rangipo prisons was exploding and quantities of fertilizer, essential to productivity, had been promised but never arrived. Having been let down by the previous administration, said Mr Kearins, he was pleased to see a new Controller-General appointed and he looked forward to a more positive policy in the future.[7]

Parliamentary comment on Justice and the condition of prisons was greater in 1950 than it had ever been, and nearly all of it referred to Barnett's reforms. Praise for the new head was unanimous and in recognition of the work being done, the prisons vote was increased by 28 per cent, to almost nine hundred thousand pounds. Even the reluctant Webb appeared moved. Having visited Mt. Eden with Barnett, in March he spoke of a 'pressing need' to pull the institution down.[8] Eight months later the Minister announced he was prepared to 'throw all his weight' behind a proposal to shift the institution to a new location.[9]

Barnett returned from his overseas tour in November 1950 and by the time he did, numerous innovations were in place. The Department of Lands and Survey had been co-opted to advise on the regeneration of farming. Increased rates of pay for prisoners had been approved, rising from eight pence to ten pence a day, to between one shilling and two-and-sixpence a day. The weekly tobacco ration had been scotched, and canteens introduced in which inmates could purchase tobacco and a number of other items as well.[10] In all prisons, routine hours of lock-up had been cut to allow better use of recreation.

While Justice was proud of the advances it made, publicizing progress was at first rather difficult. Strikes on the Auckland waterfront,[11] the Korean War,[12] and the capital punishment debate upstaged interest in gaols, and not until November did *The New Zealand Herald* accept an invitation to inspect Mt. Eden again. The paper which had seen virtue in a prison stagnating, now trumpeted praise for transition. Three articles followed the November tour and without referring directly to those of the past, existing conditions were generously aired.[13] Personal effects had been seen in the cells and it was reported that their walls now displayed pictures. Evening lock-up was an hour later and prisoners were able to receive extra food, sent in by friends outside. Although the 294 prisoners[14] still ate only cereal at night, the midday meal was much improved, with soup as an addition on some days and dessert twice a week. Concluding its reports *The New Zealand Herald* remarked:

> Old fashioned conceptions of prisons are swept aside when visiting the Auckland Prison. The lasting impression is that there is little more that can be done for the prisoners' physical needs. They are extremely well treated. . . .[15]

Hyperbole aside, there is no doubt that conditions had improved, and despite the paper's impression, material progress continued. In August 1951, rations again received attention. A bacon issue was then introduced and eventually the main meal was transferred to the evening. Superintendents commented favourably upon the new scale and at one institution, Waikeria, complaints about food were reported to have disappeared entirely.

While media interest remained vital, central to the new method was the fact that judgement was not confined to officialdom. For the first time, inmate opinions were consulted. Visiting executives could assess these opinions either through official interviews or through impromptu conversations with prisoners at work. Inmate Archie Banks,[16] confined at Mt. Eden during this period, remembers requesting an audience with Barnett when he visited in the early fifties. Having waited all day in the 'Dome'[17] to see the Controller-General, he was taken back to his cell at 4.30 p.m. and locked up. Believing he had been overlooked, Banks managed, through a Church minister, to get word to the head that an inmate was still wishing to see him.

> Much to my surprise, about 7.00 at night — and this was unheard of, *no one* ever got out of their cells once they were locked up, half-past four you were *all* locked up and that's where you stayed — the door swung open, I was in the south wing ... and two screws, both armed with pistols said 'You're wanted down the superintendent's office.' And they both marched me down there, took me to the superintendent and there was Barnett, a larrikin of the first order then, he was sitting back [at the superintendent's desk] and he was smoking a cigar. He was one foot up on the desk and he asked me what I wanted.

Barnett's approach to prisons and convicts, like his other duties, was of sanguine self-assurance. He talked easily with inmates, but noted their replies, drafting directives from the knowledge he gained. One lifer I spoke to remembers Barnett interviewing numerous prisoners on his visits to Mt. Eden. He told me that he personally had complained to the Controller-General about the embargo on newspapers, the restrictions on matches, the poor quality of rations, and inadequate exercise for prison cooks. All of these matters were soon attended to.

When Barnett took over it had been rumoured, even inside the gaols, that extensive changes were pending. Inmates, cynical about the promises of gaolers, had regarded reports with disbelief. Soon it became clear that improvements really were being planned, and when, by mid-1950, the extent of the activity was confirmed, doubtfulness turned to surprise. Banks' description probably indicates the reaction generally:

> Oh yes, there were great changes when Barnett took over. Food changed. Food changed *dramatically*. They started upgrading the kitchen, putting in stainless steel — this all took time — changed the cooks. Gas cooker — they've got electricity now, but they had a gas cooker instead of the old wooden stove. Hygiene started to improve, food started to improve, with sweets every day which was unheard of. 'Course, when canteen came in — I heard there was going to be a canteen — they all laughed. 'Ah! Fairy story!' But it came.
>   Porridge was off. You could have a *choice* of porridge, and you got it on a plate. Plenty of milk — well, not too much — but a lot more than you got

before and they took great pains to see that the cooks didn't water it down. You — in fact you got a small half-pint bottle every day. Sealed. From the milk department. Which was better than the cooks filling one up about one-third full and watering it down two-thirds. No, no, it changed dramatically.

Food — they went overboard again. Good food. Bacon and eggs for them twice a week, for example. No choice of meal but reasonably good tucker. On a plate. Dished out from a hot press. They did their best.

By the end of 1958 the framework of Barnett's prospectus was in place. Although its objectives were now well defined, there was still one question unanswered. Mt. Eden's superintendent was due to retire. His power over policy translation meant that a great deal depended on the choice of successor.

J. J. H. LAUDER, SUPERINTENDENT

Between April 1946 and April 1951 the superintendent at Mt. Eden was John James Henry Lauder, a hard-nosed disciplinarian who had spent most of his life working for the Prisons Department. Joining the service in 1910, aged twenty-one, he had remained there — apart from fighting overseas in the First War — until his retirement. 'Baby-Face' Lauder began duty at Lyttelton[18] then served at Dunedin and Tairoa Heads. He worked on tree-planting operations at Hanmer, Waipa, and Kaingaroa, before transferring to Mt. Eden prison and rising to the rank of principal warder.[19] After seventeen years as an ordinary officer — and two years after Dallard took over — Lauder was transferred to Mt. Crawford as a chief. By the time he returned to take charge of Mt. Eden, Lauder had seen thirty-six years of service.

There are few people about who can still recall Lauder and because of the system he worked under, no one I spoke to remembered him well. One officer, who had joined Prisons in 1939, said of him:

> Oh well, he was one of the old school, a really tough disciplinarian. There was a set of rules, for both sides of the line, inmate and officer, and they were set down before you and that was it. There was no give, no take. He wasn't flexible, but on the whole he was a fair man.

Dan Cavanagh, who had entered the service from the police after the Second World War, said that Lauder had welcomed him 'with open arms'. Both had been wartime artillerymen and that fact, together with Cavanagh's police background, had bonded the two. 'A nice man. A gentleman. . . . I had a lot of time for him and he treated me well,' said Cavanagh. 'But again, he had been a product of what went before. The change hadn't taken place in his time.'

Inmates who knew Lauder were agreed upon his method, but their comments about it were less charitable. Les 'Foxy' Townes, who first went to gaol in 1942, said that Lauder, like his antecedent W. T. Leggett, was a strict disciplinarian who ran the prison from his office. He was seldom seen around the institution and never spoke to inmates. Archie Banks, who remembered Lauder from the early forties, held a similar view:

> The superintendent then was a fellow called Lauder. A *big pig*, he used to be in Wellington. He was chief in Wellington in my 1942 days, and then he was transferred to Auckland as a superintendent. But anyway, I landed with him and you couldn't get anywhere with him.

A third inmate, in prison in 1948, remarked:

> Typical bloody gaoler, yeah. Wouldn't talk to you. And if you got put on the mat, he'd never talk to you civilly. It was always a *roast*. Always a roast. Then down the dummy.[20] Old Lauder.

As a remnant of an authoritarian past, the mould from which Lauder was cast did not lend itself readily to change. Old officers like him could adapt, but only with difficulty, to the fifties, and it was reluctantly that many of them tried. Without serious misconduct, however, they remained secure in their jobs. For example, when Paparua's superintendent caused inmates to strike by denouncing the new method at a prison function in 1953, no direct action could be taken against him. Although an early retirement was this time allowed, Barnett could mostly do little but wait for the disciples of Dallard to run their course of tenure.

On 6 April 1951, Lauder stood down and for a number of months his position was vacant. The Prisons Department, in an anxious search, worked hard in its quest for a replacement. The order to be filled was not a short one. Humanity, imagination, and persuasiveness would be essential in an appointee. As well as these traits, a person was needed who could be dispassionate and tough. Because the Government had just recalled capital punishment, and unlike before, Mt. Eden was now the country's only hanging gaol. From 1950, all executions would take place at the maximum security prison. Although in 1950–1951 no one was executed, by the time Lauder left, one inmate had already been condemned.[21] He had been reprieved in April 1951, but nothing was surer than that sooner or later a death sentence would be confirmed. When this happened, the superintendent would become master of ceremonies.[22]

H. V. HAYWOOD, SUPERINTENDENT

On 3 May 1951, almost a month after the demission of Lauder, a small advertisement appeared in *The New Zealand Herald*. Applications were invited for the position of superintendent at the maximum security prison at Auckland. It was requested that applicants have administrative experience, as well as a knowledge of modern penal practice. If publishing such a position in a daily gazette was unusual, then more curious was an entry of several days later. The vacancy had been announced not only within the dominion, but also throughout the English-speaking world. The reason for the anomaly was simple: the Public Services Commission felt it necessary to cast wide for a suitable adjutant.

For seven months the vacancy at Mt. Eden remained, and not until September was an appointment made. Despite its international quest, the Commission's choice was a local one. It was also a little perplexing. The superintendency of Mt. Eden went to a man who, until recently, had driven the loader in the Paparua prison quarry. Horace Victor Haywood was born in 1900 and had joined Prisons as a warder twenty-nine years later. He was seventeen years in the service before gaining his first promotion in 1946, to deputy principal warder. Only three years later, he became a principal warder at Paparua, where he was put in charge of the quarry. This odd service record became even more unusual when, less than four and a half months after his rise to principal, Haywood skipped a grade — that of chief — and was transferred to Mt. Eden as a deputy superintendent. A year later, when Lauder retired, Haywood took over as acting superintendent. On 20 September 1951 Haywood gained permanent control of the security gaol. Thus, a man who took seventeen years to win his first promotion rose from there to the highest prison appointment in the country in a little over four years.

We can only speculate as to why this might have been, but certainly the new executive had much to do with it. From the time Barnett took over, it seems, Haywood was marked for greater things. His rise to principal warder had come within four months of Dallard's retirement. He was top-grade superintendent twenty-one months later. Whether he was already in line for it when he became Lauder's deputy is something we can never know. What is apparent is that, regardless of what applications may have been received, in September 1951 it was the existing deputy who was considered best for the job.

Like the man who had favoured him, Haywood was a character who stood out from the crowd. Forthright, innovative, and unorthodox, Haywood was respected by the people who worked for him. His duties he carried out with honest resolve, and the executions took place with solemnity. Although mishap was to mar his later years, for the moment the pathway was smooth.

The New Penal Policy Under Way

When Haywood took over Mt. Eden prison, the brave new experiment was well on its course. Haywood, little-known though he may have been, now became as much a component of the revision schedule as an agent for its exercise. His early commitment to the project fostered hopes of its success. Thus it was that in 1951, Barnett, confident of political endorsement and reinforced by support from beneath, shifted into higher gear. There was much work to be done in that year and for the first time he disparaged the previous regime. The defects which Barnett identified were several, and they arose from neglect in three major areas: firstly, the Department had failed to revise the law relating to criminal punishments; secondly, it had failed to provide a proper system of classification; and thirdly, adequate trade training and education for prisoners had been ignored.[23]

Barnett wasted no time, and already the situation was being remedied. Consideration was being given to the revision of non-determinate penalties, following the British Criminal Justice legislation of 1948. The first experimental classification committee had recently been set up at Mt. Eden, and education and trade training had improved. In 1951 full-time schoolteachers were appointed to Mt. Eden prison and Invercargill borstal.[24] Daily newspapers were now permitted and inmate discussion groups had been arranged. An experimental programme of unofficial visitors had begun at Mt. Eden, recreation committees had been established, and the Junior Chamber of Commerce had organized the prison's first debating club.

The nettle of staff dissatisfaction was also grasped. Wages had been readjusted from 30 March and working conditions were improving. Although some believed the change was occurring too quickly[25] many, especially younger staff, were pleased with it. Officers' duties were expanding and the revised routine added challenge and variety to the job. Diversity was assisted by the involvement of outside organizations like the Church, which from 1951 had been invited to provide a full-time (Protestant) chaplain to every penal institution in the country.[26]

Finally, and one of the most significant announcements of 1951, concerned the fate of Mt. Eden prison. In November 1950, Webb had conceded that Mt. Eden needed replacement, and pledged to assist in its removal. Although details of plans were then unavailable, their development had gone forward rapidly. The Annual Report of 1951 announced that Mt. Eden would be demolished and possibly rebuilt as part of a National Penal Centre at Waikeria.[27] Although immediate finance was denied, the replacement of Mt. Eden and the Waikeria project remained firmly fixed in Barnett's mind. The Penal Centre was Barnett's brainchild and it soon became axial to everything he planned. We shall see that the factors which prevented its success were the same as those that plagued his entire programme.

In 1951 the Justice Department, along with the rest of the country, enjoyed a time of rising prosperity. The performance of Government met high approval and confidence in its policies grew. In September, when Holland ordered a snap election over striking/locked-out watersiders, support for his handling of them was overwhelming. Fifty-four per cent of the population voted for the Government. Labour lost three more seats, and National was returned with a majority of twenty.[28] The security that was being felt in Government and Treasury spilled out across the nation. At Mt. Eden, Haywood was comfortable in his position and he operated with laconic ease. Early in 1952, when Barnett enquired about the new rations scale, Haywood suggested even further improvement. Later, when head office refused to increase his dripping supply, Haywood pestered until finally he got his way. Reprimanded later for issuing rations without the proper authority, Haywood simply responded that he had used them to provide supper for visiting groups.

Generally, however, the Department approved of the governor's expansiveness. Initiative was to be encouraged. In 1951 Mt. Eden's new welfare officer suggested the idea of setting up a prisoners' council. Councils had been tried (with brief success) in the United States in the early twenties, and had been favourably described by criminologist Max Grünhut in 1948.[29] Their purpose was to give prisoners a small say in the running of institutions and to encourage a sense of responsibility. Councillors would liaise between inmates and management, make recommendations for change, and represent prisoners at official hearings. The response of head office to the proposal was equivocal, and although Barnett liked the idea, it was postponed after opposition from senior head office staff. Some doubted that prisoners were ready for such freedom, and feared councils would upset routine. When several months later a second application was received, however, opposition was mute, and the idea was provisionally accepted by Barnett in February 1952.

Charged with enthusiasm as it was, the prisoners' council at Mt. Eden got off to a hopeful start. In March it requested that canteen payments be permitted for use in the entertainment of visitors, and subcommittees were later formed to advise on education and welfare. A request was made for an increase in visiting time and despite objection from Haywood,[30] Barnett approved hour-long visits for those travelling some distance to the prison. The group also asked for earlier notification of release dates and freer access to gaol regulations. Both were quickly approved and acted upon by the Controller-General himself.

Early in 1952, one of the council's more far-reaching recommendations was for the formation of an inmate concert party. The matter was discussed with local administration and two weeks later it was announced that Mrs Haywood, wife of the superintendent, had agreed to assist. The first assembly of the party's twelve members took place at a special meeting on Saturday, 1 March 1952. The formation of the prisoners' band was one of the most important events in Mt. Eden at the time and it found a strange place in the gaol's internal structure. Feeling about Mrs Haywood's involvement was mixed from the start and some staff soon found it vexing. In the end, as we shall see, it was the band that provided one of the most scandalous incidents in the careers of both Haywood and Barnett.

The tendency of democratic inmate councils, especially in security gaols, is that following a brief period of success, they become corrupt. As the powerful and the self-interested attain positions on them, groups lose the objectives to which they are originally committed.[31] Although precisely what happened at Mt. Eden is unclear, it was not long before it encountered this same sort of trouble. From the beginning, although the work of the Mt. Eden inmate council was encouraging, confidence in it was shallow. As early as August 1952 the Controller-General was prompted to write to Haywood, regretting the prominence on the council of members with long criminal records. The superintendent sympathized, and noted that since undesirable elements had gained power on the council it had become the centre of institution politics. The more powerful men were now dominant and Haywood ordered that they should not interfere in disciplinary matters, nor identify officers in council reports. From that point on, only 'constructive' criticism, and that which was 'in the inmates' interests', would be tolerated.

At head office there was concern for Haywood, and the rising dominance of criminal leaders. By the end of 1952 the council had become a hegemony, whose interests had swung from communal to élitist. The legitimacy of the council had failed and from this point its future was doomed. Early in 1953 an inmate newspaper, *The Council News*, was established, but its life was short and by the end of that year Mt. Eden's inmate councils had ceased to exist.

Involvement of inmates in the affairs of management and the formation of the prisoners' band were only two of the significant experiments tried at Mt. Eden in 1952. Barnett's third year saw the beginning of another scheme that became important in the security prison's future: on 20 March for the first time ever, a group of nine inmates left Mt. Eden to play basketball with an outside team. The idea of allowing long-term prisoners out to play sport with community groups was Barnett's own. The conduct of the first envoys was watched keenly at head office, and their positive reception cleared the way for further activities. In July 1952, inmates were allowed to compete for the first time in a chess tournament held at the gaol, and the following month the prison bowls team took its maiden excursion beyond the walls. In July, Auckland Prison took part in the first of a series of competition debates for the Auckland B. grade trophy, the Robinson Cup. On 28 November 1952, a prison team won the trophy for the first time. The next year prisoners gained the A. grade trophy. The (Athenaeum) cup final, held on 18 November 1953, marked the first time that inmates had been allowed to debate outside the gaol. Interestingly, the second speaker for the defeated junior National party team on that occasion was a future Prime Minister, Robert D. Muldoon.

The object of the many concessions granted in the early fifties was rehabilitation, to be achieved through a programme of training. This process, it was hoped, would take place at work as well as at leisure and be assisted by frequent contact with the outside world. While sounding easy in theory, the Department's objectives faced complex problems. To begin with, the rural location of many prisons made regular outside contact difficult. Additionally, the fact that 90 per cent of gaol receptions were urban meant that the farming emphasis was of doubtful rehabilitative worth. The major industry of city institutions like Mt. Eden — quarrying — was of no great vocational value either, and it was hard to see how competence in rock-breaking could be of any service to virtue.

Barnett knew the farms and quarries had to stay, but he hoped to amend their deficiencies by offering trade training to interested inmates. He also thought the introduction of a home leave scheme might relieve the burden of rural isolation. Although access was easy in urban areas, the old city gaols faced an added drawback, the solution to which was more involved. This was the archaic structure of the institutions themselves. As we know, Mt. Eden had been built for du Canian purposes and its forbidding gothic architecture reflected its outmodedness. This elderly eyesore was set near the centre of the nation's largest city, where it was symbolic of Victorian intolerance. As a vehicle for liberal experiment, the gaol was plainly unsuitable. Its heavy stone construction made it inflexible to changing policies and its conspicuous antiquity was embarrassing. Reform seemed doomed to truncation as long as Mt. Eden stood. The interests of progress, therefore, made its demolition vital.

In 1950 Clif Webb had agreed to remove Mt. Eden and had stated the following year that it would be re-sited at Waikeria. Ideas were then developed and by March 1952 the Controller-General was able to announce that the proposed National Penal Centre had been approved in principle. A general layout plan was being prepared. Although certain other institutions would remain, their utility would gradually reduce. Penal services in New Zealand would become specialized, and concentrated in the large new complex at Waikeria. At the current rate of progress, it seemed, New Zealand might anticipate a restructured correctional system by the end of the 1950s.

So in mid-1952 the target of Barnett's efforts was closing rapidly. The replacement of Mt. Eden seemed assured, although commencement of work was still uncertain. In the meantime, the institution's 276 male inmates[32] had had a wide range of activities made available to them. A bee club had been established, as well as a bridge club and a weekly art class. One prisoner was reported to be studying radio electronics and another was half-way through a BA degree.[33] A bowling green in the yard had been completed and games of bowls with outdoor teams were scheduled once a month. The prison basketball team played every fortnight, alternating 'home' with 'away' games. The competence of the prison band was growing and interest in it was booming. This was stimulated by regular visits from outside concert and drama groups. 'Two or three years ago', an observer in *The New Zealand Herald* wrote, 'privileges the men now enjoy would have shocked the authorities. Today they have proved their worth in furthering the true purpose of a prison — the rehabilitation of criminals.'[34]

Some aspects of the old routine persisted. The main meal was still served at noon, for example, and inmates slept on beds of canvas and straw. Moleskins had not yet been replaced[35] and, although lock-up hours had been reduced, lights were out at 9.00 p.m. Recreation facilities, perhaps more impressive in 1954 than ever before or since, still, and always, remained unavailable to short termers.[36]

These deficiencies were minor, and most would be remedied in time. In his annual report of 1953 the Controller-General noted proudly that nearly all of the programmes he had indicated in 1951 were now in operation. A new Criminal Justice Act was also being planned, along lines suggested two years earlier. To be drafted with reference to the English law of 1948, the legislation would provide for detention centres, corrective training centres, and a new form of indeterminate sentence for the treatment of persistent offenders.[37]

1953 was the last year in which Barnett enjoyed the unfettered freedoms that had been his since the change of Government. Four years of progress had continued unabated and the effect it had produced was impressive. At Mt. Eden, in particular, enormous advances had been made. Barnett had revolutionized the regime for security inmates, and it was difficult to see that much more could be done. Without great financial input, Mt. Eden was approaching its limits of adaptability. After March 1953 the solution to the problem, in the form of a new complex, seemed imminent. Nine months later the prospect suddenly faded. For in October that year, the Waikeria proposal received the first of what was to become a fatal series of set-backs: Webb announced that owing to a lack of finance, plans for removing Mt. Eden were being postponed.[38]

Webb's statement to the House in October 1953 marked the beginning of the end of Barnett's halcyon days. From 1954, the early fifties' bright horizons became increasingly bleak. Prison musters, which in 1949 had dropped to a mean of 1024, had begun thereafter a slow and unremitting climb. While the problems associated with these proved minor in retrospect, they seemed serious enough at the time. Overcrowding was acute, Barnett warned, and a crisis point was approaching.[39]

Between 1949 and 1954 daily average prison populations rose by 17 per cent, and the maximum of 1449 recorded in 1954 was 19 per cent higher than on any day five years earlier. That situation was worsening, and its development was aggravated by two prevailing factors. One of these was the courts. In spite of Barnett's clear rejection of short sentences, 68 per cent of those sent to prison in 1953-1954 had received terms of six months or less. In Barnett's view, those deserving minor sentences should not have been incarcerated at all but lacking any voice in the sentencing process, he was powerless to interfere.

The second element affecting intakes was more complicated, but was related to social, demographic, and economic factors. This concerned a rise in overall offending, particularly among youthful Māori. In 1950 Māori had constituted only 20 per cent of prison musters. By 1954 Māori, comprising only 6 per cent of the total population, were represented in 21 per cent of the prison community and in 38 per cent of the borstals.[40] Mt. Eden prison reflected this trend, with sixty-four Māori prisoners, or 22 per cent of its inmates.

In early 1954 Barnett hoped that increasing publicity would alert the courts to the problems he faced. With this in mind he was ready to release a policy pamphlet, which outlined the principal philosophies, methods, and objectives of his Department.[41] The problem of Māori crime on the other hand was only the vanguard of what was clearly an emerging nationwide problem. This would be attacked along a broad front, starting with the passing of the Criminal Justice Bill. Draft legislation was soon to go before Parliament, and its provisions would eventually be implemented at the National Penal Centre. Barnett trusted that the postponement of this facility would be short.

So apart from some cautionary reserve, a spirit of hope remained at head office in 1954. Optimism was high, suspended by Barnett's belief that, having been forewarned of incipient troubles, forces could be mobilized to prevent their expansion. What the head could not know at this point was that the difficulties he had perceived in population growth so far were only a foretaste of what was to come. Just as his efforts to remedy their causes had proved barely effective in the past, so were they to prove in the future. The burdens of Justice would grow heavier before any relief would be found.

Notes

1. The old attire had for years consisted of a khaki tunic, buttoned to the collar, with khaki cap and trousers, and heavy black boots.
2. Captain Banyard was awarded the MBE for his services in 1959. He retired in 1960 and died in 1982.
3. The controlling officer is the person in charge of an institution. This person usually does, but may not, hold the rank of superintendent.
4. NZPD vol. 292 (1950), 3901.
5. Ibid., 3896.
6. Ibid., 3895, 3900.
7. Ibid., 3900.
8. *NZH* 27 March 1950.
9. *NZH* 2 November 1950.
10. The pay scheme was organized on a credit basis with no actual cash involvement. Forty per cent of the weekly wage could be spent at the canteen; the other 60 per cent went into a personal account for use by the inmate on his/her release.
11. These strikes occurred in August and September 1950.
12. The Korean War began in June 1950. New Zealand volunteers (the 'K Force') sailed for Korea on December 10.
13. *NZH* 11, 14, 15 November 1950.
14. The prison at the time held 280 men and fourteen women, supervized by sixty-five male staff and three female matrons (*NZH* 11 November 1950).
15. *NZH* 11 November 1950.
16. Archie Banks was father of the current National party MP (Whangarei) and prisons hard-liner, John Banks. Archie Banks died in 1984.
17. The Dome is the 'hub' of Mt. Eden, off which the wings emanate.
18. Lyttelton Gaol, one of the first of New Zealand's major prisons, had been built in 1860 and was closed in 1920.
19. Prior to 1960 the promotion scale for penal personnel proceeded thus: temporary warder (a mandatory probation status), warder, deputy principal warder, principal warder, chief warder, deputy superintendent, superintendent.
20. Solitary confinement.
21. This was Gordon Malcolm McSherry, sentenced to hang on 16 February 1951 for the shooting of a bank clerk during a robbery.
22. Officially, proceedings were the responsibility of the sheriff, but all of the work was done by prison staff.
23. *App.J.* H-20 (1951) 8.
24. In Dallard's era only a part-time teacher (Dr Dale) was employed at Mt. Eden prison.
25. Cavanagh (pers. comm.). Although only a young officer at the time, Cavanagh felt that the emphasis on a treatment role for staff was unrealistic. He pointed out that a prison officer's main duty is custodial and that this correctly takes priority over anything else.
26. Dallard appears to have distrusted clergymen and at times disparaged them openly. In 1942 the chaplain of Mt. Eden prison was banned from the institution after strong disagreements with Dallard (*NZH* 24 June 1942).
27. *App.J.* H-20 (1951) 12.
28. There were eighty seats in the House in 1951.
29. Perhaps the most famous example of inmate self-government is that installed by Thomas Mott-Osbourne at New York, Auburn, Sing-Sing, and Portsmouth prisons between 1913 and 1920 (see Tannenbaum 1938; ch. 17). From 1910, however, similar systems have been tried at Hofwyle, Switzerland, the Rauhes Haus, near Hamburg, at the Boston house of Reformation, and at the George Junior Republic, Freeville, New York. The subject is dealt with by Grünhut (1948) 287-95, and it is this text,

in fact, to which Barnett referred. Other examples can be found in Fox (1952) 372-3; Glaser (1964) 217-23; Johnson (1968) 554-6; Jones (1965) 245-7; McCleery (1961b); Sutherland & Cressey (1970) 523-5; and Sykes (1958).

30 Visits at the time were permitted for half an hour, once a week.

31 Although Tannenbaum (ibid.) omits to mention it, this was one of the very problems that in the end beset the Mott-Osbourne programme (Korn & McCorkle 1965, 590; Glaser 1964, 218). It is for this reason that inmate self-government systems achieve only precarious stability and their success is usually short-lived (Glaser 1964, 218-223); Johnson (1968, 555); Korn & McCorkle (1965, 590); McCleery (1961b, 171); Sutherland & Cressey (1970, 524-526); Sykes (1958, 119-20).

32 In 1952 Mt. Eden prison also held a DAP (daily average population) of twenty-four women. Throughout this period, as at other times in New Zealand penal history, conditions of confinement for women seem to have been forgotten.

33 The following year it was reported that an inmate — presumably the same inmate — had won his BA in prison (*NZH* 4 September 1953).

34 *NZH* 7 July 1952.

35 Dallard (1976) reports that denim jeans were being introduced in some prisons by the late 1940s.

36 The 1952 *Rules Governing Prisoners for Mt. Eden* stated that men serving one month or less would not qualify for canteen. Full privileges were not awarded until a prisoner had served three months of his/her sentence.

37 *App.J.* H-20 (1953) 3, 6-8; NZPD vol. 300 (1953), 1883.

38 NZPD vol 300 (1953) 1887.

39 *App.J.* H-20 (1954) 6.

40 Ibid. The percentage of Māori in prisons may be rising, with current figures indicating a record increase from about 40 per cent in the 1970s to 50 per cent. In 1952 any person claiming to have half Māori blood or more was considered to be Māori. This unreliable identification method has been replaced by an equally unreliable one, wherein anybody who looks Māori and claims to be Māori is listed as being Māori. Identification techniques may partially account for recorded rises in the Māori prison population.

41 *App.J.* H-20 (1954) 5. The document referred to is *A Penal Policy for New Zealand*, released by the Department of Justice in 1954.

4: PROGRESS AND CONTROVERSY 1954–1957

The election year of 1954 was also Webb's final year as Justice Minister, and the one in which he made his most significant, long-term contribution to penal policy. Since first coming to office five years before, Barnett had been complaining about existing criminal justice provisions and urging new legislation. Revision would be laborious and time-consuming, but work was already advanced by 1953. The following year, bills came before the House, and two important statutes were enacted before the end of the final session. These were the Criminal Justice Act and the Penal Institutions Act.

THE 1954 LEGISLATION

The two Acts of 1954 were the first major changes to penal justice for several decades and their innovative content was great. Unsure about what reception their measures might receive, and hoping to soften any opposition, the Justice Department took the novel step of producing a preparatory pamphlet. This paper, the first comprehensive statement on penal policy in over thirty years,[1] outlined in simple and concise terms the broad philosophy of the Department.

Entitled *A Penal Policy for New Zealand*,[2] the new document outlined general areas of concern and pointed to specific problems which the coming law would address. From then onward, the Department of Justice would be guided by five overriding objectives:

1. To divert people from a life of crime by treating them at a young and malleable age — for example, by expanding the probation service.
2. To use imprisonment only as a last resort for young and inexperienced offenders.
3. To eliminate short sentences as far as possible.
4. Where imprisonment was considered necessary, to bring every possible reformative influence to bear upon a prisoner.
5. To provide lengthy terms of imprisonment for those who persistently defied the authority of law.[3]

It was these principles which underlay the Criminal Justice Act of 1954. In repealing and consolidating a wide range of earlier laws, the statute extended the powers of probation officers and dealt specifically with the custody and release of trainees, prisoners, and the mentally disordered. It tightened restrictions on youthful incarceration, revised and redefined the sentence of borstal training,[4] and introduced a new provision, that of detention in a detention centre. This latter concept — a departure from the short-sentence embargo — was initially to consist of a short, sharp, disciplinary sentence of four months,[5] aimed at the deterrence of young first offenders.

The old sentence of reformative detention was struck out and another one, called 'corrective training', replaced it. Like the borstal term, corrective training was of an indeterminate time of nought to three years. Unlike borstal, however, which only accommodated youths of seventeen to twenty-one years, corrective training was targetted at men of the twenty-one to thirty age group. Both sentences were designed to provide a medium wherein reformative influences could be maximized.

Although corrective training never really worked and was eventually repealed,[6] like borstal and detention centre training (but unlike reformative detention), corrective training was to take place in special institutions — or parts of institutions set aside for their purposes. The proposed National Penal Centre was to have separate accommodation for borstal trainees, and originally the detention centre was planned to operate from either Waikune prison or New Plymouth.[7] Paparua was envisaged as a corrective training centre, and Tongariro[8] was to be used for first termers. This new emphasis on young and naive offenders arose from concern at the increasing flow of younger persons before the courts. It was believed that recidivist rates could be reduced by (a) providing training and rehabilitation programmes and (b) keeping young people away from the influence of established criminals.[9]

Since it was not to be involved in any of the above schemes, the prison at Mt. Eden was only obliquely affected by the new legislation. An effective borstal and detention centre programme, however, would remove most young males from the maximum security environment, and corrective training — which was also designed for young men with weak criminal backgrounds — would relieve maximum security of criminally inexperienced young adults. As such, the residential component of the gaol would be altered. The remaining population of Mt. Eden would be affected, however, by section 40 of the Act, which abolished the distinction between imprisonment with and without hard labour. In operation they had never been satisfactory[10] and it had long been recognized that their differences were academic.[11] In this respect, therefore, the legislation really only gave legal expression to something which was already established in practice.

Another statute which had displeased Barnett was the Habitual Criminals and Offenders Act of 1906, the removal of which he had recommended in his memo to Webb in December 1949. The Act provided that a person repeatedly convicted of various classes of offence could be imprisoned indefinitely by the Supreme Court. No maximums or minimums had been set, and release was determined by the Prisons Board.

Apart from the fact that it was unjust, Barnett felt that the Act failed to work as a deterrent. Of the 113 who had been eligible for the status between 1940 and 1946, 80 per cent had re-offended. Moreover, 66.2 per cent of released habituals were found to have recommitted 'serious' or 'semi-serious' crimes. In 1953 there were 515 men and five women held in prison[12] as habitual criminals and in his report of that year, Barnett had again pressed for repeal.[13]

Barnett's wish was given effect by the legislation of 1954, and the old penalty was replaced by a similar, but narrower form of indeterminate sentence known as preventive detention. Preventive detention was aimed at persistent offenders aged twenty-five years and over. In most cases the term was limited to between three and fourteen years, although those convicted of sex offending could be held for life. In both instances, power of release was given to a newly established authority, the Prisons Parole Board. Although it was popular to begin with, the new penalty was not long favoured by the judiciary. Twenty-eight persons received it in 1956, but use declined steadily from that point. By 1965-1966 only three, and then two persons received preventive detention, and few preventive detainees served anything like the statutory maximum. The majority were discharged from prison less than six years after their date of commitment.[14]

The other significant piece of justice legislation effected in 1954 was the Penal Institutions Act. This Act had less impact than that of Criminal Justice, partly because it applied only to inmates and partly because it added little to established arrangements. The Penal Institutions Act was a collative rather than innovative document, designed principally to reorganize and update earlier amendments. Like Criminal Justice, it was strongly influenced by British example. There were three main areas where Penal Institutions opened new ground.

The first of these in section 21 of the Act allowed the temporary parole of inmates. Initially drafted for compassionate purposes and for prisoners in the last months of their sentences, the main significance of the section is that it laid the legal foundation for what was later to become known as home leave. The latter did not commence until 1965, but it continues as a prominent aspect of the modern penal programme.

The second new area addressed by the Act concerned the administration of the Justice Department itself. As will be recalled, Justice and Prisons had been split in 1919 when Matthews had altered his title from Inspector to Controller-General. Operations of the Departments remained close, however, and in 1933 Dallard had been given the leadership of both. From this point on Justice and Prisons shared the same cloak, and in 1954 the relationship gained legal recognition. Section 3 of the Penal Institutions Act incorporated Prisons again within the Department of Justice. Hereafter, the head of Prisons was called the Secretary for Justice.

In relation to these new measures there appeared a final area where the 1954 enactments made major adjustments to existing law. This concerned the discharge of prisoners before termination of their full sentences. Section 31 of the Criminal Justice Act dissolved the Prisons Board of 1910 and replaced it with a Prisons Parole Board and several Borstal Parole Boards. The scope of the new quorums was restricted, the functions of their antecedent having been extended in 1917 and 1920 to cover nearly all classes of inmate. From 1954, parole boards' influence was generally confined to the cases of lifers, preventive detainees, and corrective and borstal trainees. Most other inmates were now eligible only for remission, which through section 31 of the Penal Institutions Act amounted to up to one quarter of their full terms. As the years progressed, we shall see, provisions for early release were gradually extended once more.

THE PENAL GROUP

The year 1954 was important for Barnett and it saw the fruition of numerous schemes. Not all of them were legislative. Early in the year one of the more effective innovations of the decade took place, and it occurred within the Department. On 16 March an advisory council, initially comprising seven members, was established by Barnett at head office.[15] The committee was chaired by Barnett and its function, as outlined at its first meeting, was to give members the opportunity of keeping abreast with, and taking part in, the formation of future policy. Known as the 'penal group', the body would convene every Monday morning and proceed according to an agenda distributed the preceding Friday.

The role of the group in the subsequent history of the penal system is an eminent one. For sixteen years it was at these Monday meetings that the shape of future policy was wrought. Members were expected to contribute extensively to proceedings and were often required to make lengthy reports on developing concepts. The group's weekly discussions provided a casting block for evolving strategy and as its worth became proven, the scope of its initiatives grew. Membership was increased and insight was supplemented by input from overseas. By the beginning of the 1960s the meetings of the penal group had become the spearhead of what was to prove this country's greatest era of justice reform.[16]

J. R. MARSHALL, MINISTER OF JUSTICE

At the end of 1954 Clif Webb retired from Parliament and went to London to take up the post of New Zealand High Commissioner. In the election at the end of that year, support for the National party dropped significantly. Its majority in the House fell from twenty to ten. National's share of the vote went from 54 to 44 per cent. This loss Marshall attributes principally to the rise of Social Credit,[17] which had made its first appearance as a political party that same year.

The 1954 election brought new blood into Cabinet, and one of the new Ministers was Clifton Webb's understudy and frequent acting Minister, John Marshall. Like Webb, Marshall was a man of conservative sympathies, who had strongly supported the return of capital punishment in 1949–1950. In spite of growing opposition he continued to defend it to the end of his parliamentary life, and his consistency over the issue is indicative of the type of politician he was. Steadfast, honest, and possessed of ideals that he could capably defend, Marshall's career, even before he entered politics, was indicative of a person destined for distinction.

John Ross Marshall was born in Wellington in March 1912, son of the late Allan Marshall, a former district public trustee. He was educated at Whangarei and at Otago Boys' High Schools, before travelling to Victoria

University to study law. In 1934 Marshall gained a Bachelor of Laws degree, becoming Master of Laws in 1935. In 1936 he was admitted to the bar and he completed a Bachelor of Arts in political science the following year. Marshall entered the Army in 1941, where he rose to the rank of Major, serving with the infantry in the Pacific, the Middle East, and in Italy. It was while in the army that Marshall first met a future Minister of Justice, D. J. Riddiford, at that time a Major. He also met a young Intelligence Corporal, then a member of Marshall's division, Robert D. Muldoon. In 1974 Muldoon was to oust Marshall as leader of the National party, during the period of the third Labour Government. In 1946 Marshall entered Parliament as Member for Mt. Victoria (Wellington) and in 1949 was given the State Advances portfolio. He served as Prime Minister, briefly, from February to November 1972, and by the time he retired from politics in February 1975, Marshall had held more Government portfolios than any other person in New Zealand history. He was awarded the Knight Grand Cross in 1974 and died in 1988.

An ardent church-goer, John Marshall, or 'Gentleman Jack' as his colleagues knew him, was a man of strong Christian principles. His sense of right and wrong was clearly defined and rooted in the soil of conservative orthodoxy. Intelligent and quiet-spoken, Marshall's refinement and correctness distinguished him from the treacherous cut-and-thrust of New Zealand politics in the past decade.[18] His intrinsic sense of honesty, decency, and moral responsibility was reflected in his policies, and in the way he behaved in office. By the time Marshall took over Justice at the end of 1954 the new penal method was already maturing. As Minister of Justice and Attorney-General he was concerned with its progress and toured every penal institution in the country to acquaint himself with developments. Barnett — whose relations with Marshall were a lot better than they had been with Webb — was now deeply involved in managing his very complex plan of penal reconstruction. Marshall inherited this novel, but still unproven programme, and while yet unaware of exactly what it involved[19] he accepted the approach and strove to contribute to its success. Of the reforms which took place, both in his own term and while working with his predecessor, Marshall told me:

> I was part of them, I supported them, and accepted collective responsibility for them . . . [Webb] did consult me on most of the major policy decisions that he was working on, and he didn't always take my advice, but certainly we worked very closely together. Consequently I was *in on* the formation of policy at that level.

Although Marshall saw himself as a liberal, that quality had a deeply traditionalist streak to it. His justification of the capital penalty revealed this, as did his general philosophy on life. A present Deputy Secretary who served under Marshall described the former Justice Minister as an 'intelligent conservative'. 'I suppose you could call him a "liberal, 1950s vintage" ', he said. 'But essentially I would regard him as a conservative, although a conservative who was prepared to look at things with a reasonably open mind.'

1955 Developments

So, conformist though he may have been, Marshall's succession to office had no immediate effect on Barnett's ambitions. In the Justice Minister's first year, inmate-staff discussion groups started at Mt. Eden and the first prisoners were released on temporary parole. At the site of the proposed National Penal Centre, the first batch of corrective trainees was received. The first full-time chaplain[20] was appointed to head office to co-ordinate the work of full-time Protestant chaplains elsewhere. The difference between this, and the previous 'one prison, one chaplain' scheme of 1951 was that now the prison chaplaincy would be directly affiliated with, and organized from, the central bureau. The concern with ecclesiastical matters was a direct result of Marshall's own religious beliefs. He felt that spiritual guidance was integral to the reclamation of 'evil minds', and while not particularly religious himself, Barnett accepted the idea, Marshall said, with 'unsentimental sincerity'.[21]

Staffing had always been a problem in the Prisons Department, and even now, manning levels were worse than even Barnett was prepared to admit.[22] A prison psychologist had joined other professionals at Mt. Eden in 1954, but there was still a serious deficiency in custodial complement. Although some felt that the new specialist staff had reduced the ordinary officer's role to that of a turnkey, most personnel, custodial as well as professional, were now expected to assist with inmate activities. Where such activities occurred outside routine hours, staff worked voluntarily and without pay. They were also expected to assist in searching for escapees. For any extra hours involved in this duty they were neither paid, nor were they given time off in lieu.

In 1955, outings for Mt. Eden's inmates were frequent, but membership of the best of inmate clubs and groups was restricted to long-termers. Participation in the prison band, for example, was confined to those doing five years or more. The band gave three concerts, and was allowed entertainment from six outside parties, every year. Trips outside the prison for sport and recreational purposes had become a regular and established part of procedure. Monthly outside visits of the prison bowls team, for example, had been going on for several years.

In the early months of 1955, it may have seemed that this 'Reformative Recreation' scheme would remain a permanent feature of the maximum security scene. Haywood must have felt secure in it because, as events were to show, he paid scant attention to its operation. Barnett, apparently, was confident enough to give the superintendent a fairly loose rein. Marshall, still in his first year of office, seems for his part to have been sure that his permanent head was in control. Thus, he either left Barnett to his own devices, or did not check to see whether the instructions he may have given about the programme were carried out. Whatever the case, an incident occurred at the end of 1955 which was to place the competence of all three men in question. It also spelt the end of outside trips for maximum security prisoners for almost two decades.

The Horton Affair

On 6 December 1955, seventeen inmates escorted by four officers left Mt. Eden prison to play their monthly game of bowls, at the hall of the Hibernian Society in Mt. Albert. Seventeen maximum custody prisoners in the hands of only four officers might today seem a dangerously low security ratio, but at the time there was nothing unusual about it. Reliant as it was on the benevolence of volunteers, the administration had become accustomed to making use of whatever staff it could get. The composition of the team was a little strange. Apparently no care had been taken in its selection, except that longer-sentence men had been given preference. Mt. Eden then held 190 prisoners serving two years and above, and thirty-one of these were lifers. Yet twelve of the seventeen players in this bowls team were serving life sentences, and among them were several well-known figures. One was George Horry, a former habitual criminal, convicted in 1952 of murdering his girlfriend. Also present was Douglas Cartman, whose gallows remission in 1941 had been vigorously opposed by Prime Minister Fraser, and Gordon McSherry. McSherry had been the first man sentenced to hang (and be reprieved) after capital punishment was restored in 1950. The most surprising inclusion in the Hibernian bowls team, however, was the man whose crime had contributed so sensationally to the decision to bring back the gallows. This was the killer of Kitty Cranston: Edward Raymond Horton.

On 6 December the bowls team left the institution in the prison bus at 6.30 p.m., under the eye of the superintendent.[23] Evidently he saw nothing unusual in the composition of the group, or if he did, he said nothing about it. Horton had made eight previous excursions without mishap[24] and there seemed no cause for alarm on this occasion. The escort bus arrived at the hall at 6.50 p.m. and bowling commenced. At about 8.30 p.m. the chief officer, Mr A. Burgess, who was sitting with Horton, went out to the toilet. Unknown to him, Horton followed and once clear of the building, he vanished into the night. According to official reports the disappearance was noticed some minutes later, and, having frantically searched the environs, an ashen-faced chief whispered to his three colleagues that Horton had gone.

When news broke of the escape there was panic in many quarters. Haywood, whose immediate responsibility it was, rang a stunned Barnett and gave him the news. Barnett immediately informed Marshall.

The way Haywood must have felt at this point can only be imagined, but certainly he was in a state of despair. Dan Cavanagh, who was off duty at the time, described Haywood's reaction.

> . . . the phone went at one o'clock in the bloody morning and I staggered over. It was Haywood on the phone and he said, 'That you, Mr Cavanagh?'
> And I says, 'Right'.
> And he said, 'Ah — Horton's escaped.'
> 'Oh yeah? What do you want me to do?'
> He said, 'Nothing.'

So I said, 'All right.' Put the phone down and I said, 'What in the bloody hell did he ring me for?'

The sixth of December 1955, and the days which followed it, were an anxious time for Haywood. When news of the absconding reached the press, Haywood's mistake made headlines all over the country. Although not known for understatement, the comment in *Truth* the next week was probably accurate enough: 'It is no exaggeration to say that a wave of terror gripped many parts of Auckland when it was known Horton had escaped.'[25] Recollection of the Cranston murder only six years earlier, filled many with horror and fear.

In Auckland all police leave was cancelled and reinforcements were flown from as far off as Christchurch and Wellington. Every available prison officer was mobilized immediately to begin one of the biggest man-hunts the country has known.

The following day Sam Barnett, who had recently become Controller-General of Police, took a special flight with Marshall and the head of the Criminal Investigation Branch to supervise search procedures in Auckland. Haywood, of course, was 'on the mat' over the affair and he was, in Marshall's words, 'firmly' spoken to.[26] Haywood was required to tell Barnett and Marshall immediately what had happened and why it had happened. Marshall had known nothing of lifers being included in the outside teams and neither, apparently, had Barnett. Marshall believes Barnett was ignorant of what was going on[27] and MacKenzie, the prison psychologist, who was one of the supervising officers, is sure of it. I asked him whether Barnett knew about the composition of the bowls teams.

> Oh, I'm *sure* he did not. In fact I *know* he did not . . . I know that when Barnett realized that there were twelve lifers on that trip, he nearly blew his top! He really had no idea. I must say, I didn't know either. I thought one or two lifers maybe, but I saw them all trooping out and I looked twice!

How it was that Horton got on to the team, and whether so many lifers had been allowed out before, is contentious. Marshall thought that MacKenzie had something to do with the team's composition,[28] although this MacKenzie denies, saying that like the others, he was merely a volunteer supervisor.[29] To Barnett, Haywood claimed that Horton had only been included after careful consultation with Mr Burgess and the prison psychologist. Of this, MacKenzie told me:

> A question of personal loyalties within the prison arose out of this statement. The superintendent, the chief officer, and I knew that we would be asked by the Minister why Horton had been allowed out, and it was clear to Burgess and myself that we were expected to agree that we had been consulted on the matter beforehand. In fact we had not been consulted, but we felt that by going along with the bowling party without protest we did in fact tacitly agree to Horton's presence.

Some clue as to how the team was picked comes from Percy Anstiss, at that time junior officer at Mt. Eden. Anstiss says that in 1955 the lists of inmates for Reformative Recreation were normally made up by another lifer, Frank O'Rourke, who was also a member of the team. The group was then checked and approved by one of the staff.

> ... the situation developed simply because O'Rourke put the list through to Jim O'Connor (a principal officer), who was supposed to vet the list. O'Rourke always made the lists out, who was to go out.... He wasn't in charge of the band, but he used to handle the welfare lists and all that sort of thing and the activities and who was going out to handle the bowls. He would write the names down and he would take it through to the front. So he just had all these names down. Now I don't know whether it was done purposely or what, but the name was there and Jim O'Connor approved it. He shouldn't have approved it. And he just sent it back with O'Rourke, saying 'approved' and that was it.... I don't think he even looked at the list.

Whether Haywood really did watch the team get on to the bus as he claimed, or not, it seems that the system of determining who was to go out on recreation was flexible, and left pretty much in the prisoners' hands. This theory fits in with the general impression of loose management at this time, as will be expanded in chapter eight.

The reason for Horton's escape and whether or not it was planned has been another matter of debate, and one which likewise has never been settled. MacKenzie, who spoke to Horton after the recapture, said Horton had told him, 'The night was dark, I went down to the pisshouse, there was nobody there, so I just kept walking.' Banks, who knew Horton, said that Horton had told him he just wanted to buy an ice-cream, then got scared and decided not to come back.

> I talked to him afterwards, he said he had no idea of running away completely. He just had a few bob in his pocket — bookmaking it was from — and he just wanted to see how the outside world was doing.

In a confidential memo to Barnett, Haywood said Horton had told him, 'I was fed up with the place, I've served long enough.' Haywood, too, was convinced that the escape was unplanned.

These answers are not entirely satisfactory in that they leave two points unanswered. Firstly, knowing that his escape would jeopardize the recreational privileges he and his fellows were enjoying, would Horton have walked away without any plan of staying out for good? He must have known that recapture would earn him a good spell in solitary, followed by segregation, and he must have known too that outside privileges, for him at least, would then be over. It is quite arguable that he might not have considered these things: he had not shown much foresight in the past. However, when interviewed after the incident, Horton complained to the superintendent of heart attacks.[30] Perhaps, therefore, the reason for the escape was that he felt he might not survive his sentence. If so, he probably had intended staying out and thought about it beforehand.

The other question unanswered concerns his money. Horton said he had about nine pounds in his possession when he left the institution. He had eight pounds on him when he was picked up. In 1955 this was a substantial amount, and one wonders whether he would have taken such a sum had he not intended to run off. Perhaps he would have. Perhaps he considered his 'roll' was safer on his person than hidden somewhere in the gaol. Recreation parolees were not searched and as in most prisons, there was a fair amount of illicit cash in Mt. Eden at the time.

The answers to these questions may lie in the words of one inmate I interviewed. Horton had always insisted that no one had helped him with the escape. This inmate — another lifer — told me he not only knew of Horton's plans, but had helped him formulate them and provided him with some of the cash he had in his possession when apprehended:

> I knew he was going to go a couple of nights before. He told me he was going to go and came to me to ask if I could lend him some money. He knew I had money because I ran the book. I gave him ten or fifteen quid. Later on after he was caught they asked if I'd given him any money. Of course I denied it. Horton denied it too. He was pretty staunch, Horton was.
>
> When he told me he was going I told him to make for the waterfront and swim out to a launch and stay on the launch. He could have stayed on the launch for quite a while, they've always got food and cooking facilities on them and he could have lived well there, in complete safety, as long as he didn't show any light at night. He was going to make for Wellington and I told him to lay low on the launch for a week till the heat died down, then go to Wellington. But I told him not to go direct, he'd have to catch a bus and go via Palmerston North or something. Stay off the main routes, see.
>
> I don't know what he was going to do once he hit Wellington, but he reckoned that once he got to Wellington he'd be set. . . .
>
> Afterwards, when he came back I asked him what happened to the plan. He said he didn't know Auckland very well and couldn't figure out which way the waterfront was from Avondale where he'd escaped. So he hid in the Whau Creek and eventually got caught when he went into [Mayo's] grocery shop in Avondale. They recognized him and rang the coppers.

The ex-prisoner who told me this asked not to be named, but seemed knowledgeable and sincere about the matter. I considered him a reliable informant. When interviewed he was over seventy years old, had served only one sentence of imprisonment (life), and had not offended in the past twenty-five years. He had held a responsible position in the gaol and, most importantly, he had good recall of detail. The information he gave me was free from obvious exaggeration, and as far as I could determine, everything else he told me was accurate. I am inclined, therefore, to believe that his version of the Horton incident is correct.

In any event, in spite of the grave fears which were held in some quarters about what Horton might do while he was at large, the escape turned out to be an anticlimax.

According to Haywood's confidential memo on the incident, Horton, dressed in civilian clothing (as was customary for parolees), lay low in Avondale. On his second night of freedom he emerged and bought some food. Two mornings later he bought more food. Unshaven and bedraggled he was recognized by a store owner, who called the police. Horton was arrested after two and a half days of liberty, and without any fuss was returned to custody.

Recriminations over the incident were swift and the onus of blame was heavy. Haywood, who had to take a major share of it, was devastated. A man he had trusted had let him down in one of the worst ways imaginable. Of Haywood's reaction the prison psychologist said:

> He was bitterly disappointed. Horton had risen up through the band to be a pretty normal sort of person. And Haywood thought, with some justification, that it was due to Haywood's own enthusiasm for the band and having people like Horton in it. Then to be let down like that — and he was *continuously* being let down by people. But to be let down in this dramatic way, with so much publicity in Parliament, and so on, was terrible for him. But I'll say this for Haywood: he did not bear a grudge for very long. Horton was back to his normal status some months after he'd done his solitary confinement.

Haywood, at the centre of the controversy, saw fit to depart from protocol and defend his actions publicly. Questioned about why Horton had been included in the programme by *The New Zealand Herald*, Haywood replied:

> Would you deny him privileges, just because of his crime? If we did that, we would have to do the same with all prisoners, and then the whole reform principle of the New Zealand prison system would collapse. . . . Is a man to be left to rot — mentally and physically and spiritually? Our aim is to prevent that happening and to provide for a constructive future for a prisoner. The privileges which Horton earned are part of that policy.

Haywood then went on to explain that Horton had improved considerably since his inclusion in the team. He had developed interests and had become socially acceptable. Prior to his admittance to the scheme he had been heavily screened and was considered suitable.[31] To the Minister, however, Haywood was contrite and apologetic, accepting responsibility and recognizing his mistake. 'He wasn't in any way trying to justify what had been done', Marshall told me.

Although it was remiss of Haywood to introduce such an extravagant programme without Barnett's approval, by Marshall's own admission regular trips — forty-one in all — had been made by the club over the thirty-one months before the escape[32] and Horton, as noted, had already been out on eight previous occasions. These outings were well covered in the newspapers, so Barnett and Marshall could not, nor did they attempt to, deny knowledge of them. For if Barnett did not know that lifers were included in the teams, he certainly should have.

Marshall, to me, and in his *Memoirs*, insisted that clear instructions about the composition of teams had been given:

> It was ironic that before approving this policy in 1952, Clif Webb had sent the report and recommendation for my comments. I noted on the report that I thought it was a good idea, provided the prisoners to be allowed this recreational privilege were carefully selected, were nearing release, were not classified as dangerous, and were strictly supervised.[33]

When Marshall asked the Department to produce this report, it could not be found.

Whether Webb took any notice of Marshall's reservation and passed the instruction on to the Department is unclear. Irrespective of this, in the end, responsibility for the scandal was the Minister's. For a Minister is responsible for everything that goes on in his Department. As political head everything that takes place does so in his name.

Marshall was aware of this and it was fortunate for him that the incident occurred late in the year, after Parliament had risen. He was thus spared, for a while, the ordeal of incisive questioning and criticism that he would have faced had the House been in session. As it was, dealing with the press was bad enough, and on 22 December he stemmed rising debate by issuing a detailed press statement.

> ... the responsibility, of course, is mine as Minister of Justice. I accept that without any reservations. Horton's escape has brought the whole of the reformative prisons policy under examination. ... I would emphasize that Horton's escape is not a failure of [that] policy, but a wrong application of it; I am confirmed in the view ... that a prisoner of Horton's character and history should not have been included in the outside visits scheme, whatever his behaviour in prison.

After commending the work of Haywood, Marshall said that a serious mistake had been made (by Haywood), although Haywood should have been given explicit instructions from head office.

> The inclusion of a prisoner of Horton's character and record in the privileges scheme was not contemplated, nor was it known by myself or senior officers of the Justice Department.
> I cannot possibly defend sending out [a party of twelve lifers] under any circumstances.

In short, although he accepted full responsibility for the escape, Marshall did not agree that he was to blame. He criticized Barnett, for failing to deliver specific instructions to prison officials, and Haywood, in the absence of superior direction, for using his fullest discretion (and poor judgement) in composing excursion parties.[34]

The House did not convene again that year and when it next sat in May 1956 the opposition capitalized on the issue. Pilloried by Labour party President Arnold Nordmeyer, Marshall stood up and accepted unqualified responsibility for the affair. He agreed that he should have known about the outside trips and conceded every point Mr Nordmeyer made.[35]

The way Marshall handled the Horton matter was a reflection of his moral character, although this too had not been without its critics. At least one newspaper called for his resignation[36] and there were rumours that he might do so. But he did not resign, and calls for a public inquiry went unheeded. Instead, an internal investigation was held, and its results were those published in the December press release. Thus, argues Polaschek, several questions were left unanswered: should Marshall have known who was included in the team? Did his senior officials fail in their duty of knowing what went on in their

Department and of advising their Minister of the possible consequences of policy? Could the Minister reasonably have been expected to have prevented the incident by controlling his Department more closely?[37]

Polaschek takes Marshall to task for not resigning and for identifying and blaming his subordinates. Constitutionally, Polaschek considers the practice incorrect and basically unsound.[38] However valid Polaschek's arguments may be, Marshall certainly showed greater respect for constitutional decorum than has been observed more recently.[39] As far as the prison system — especially the maximum security prison — was concerned, the Horton affair had numerous repercussions. The first, of course, was that Reformative Recreation was immediately stopped. Some months later the ban was revised and a new policy in regard to lifers was issued. In summary this provided that:

1. Lifers would only be granted recreational privileges on the recommendation of the Minister, and generally would not be considered until they had served at least ten years and received a favourable recommendation from the Parole Board.
2. Lifers would only be transferred to other institutions with the approval of the Minister of Justice, on recommendation of the Parole Board and the Secretary.
3. Lifers could now only be employed in quarries and other outside-the-wall projects with the approval of the Secretary.
4. Persons imprisoned for crimes of violence would be held in the maximum security prison and would only be employed outside the walls with the written approval of the prison superintendent.
5. No prisoners were to leave the bounds of Mt. Eden, Mt. Crawford, or Paparua prisons without the approval of the Secretary for Justice. No outside activity was to be initiated without his prior consent.
6. The chief aims of penal policy, as outlined in 1954, would still be (a) protection of the community; (b) the punishment of crime; (c) the reformation of criminals.[40]

So although the total ban on outside visits was relaxed with the statement of 5 May, fear of a repetition was great. The result was that for nearly twenty years, no teams ventured again beyond the confines of a maximum security institution.

The legacy of the Horton escape was heavy, but it was not entirely ruinous. The incident had embarrassed Marshall and perhaps in an effort to redeem himself he now set about reviewing penal policy as it had been operating since 1954. On 20 August 1957, Marshall presented *The New Penal Policy (The Second Phase)*[41] in a speech to the House of Representatives. The object of the 1957 adjustment was to temper the broader principles which had been forged in 1954. Marshall announced that the probation service would be expanded. Alcoholics and borderline mental defectives would be diverted from prison where practicable. Perhaps most importantly of all, Marshall presented his intention to establish within the Department a research unit, to investigate the causes and treatment of crime. This was effected the following year.

Maximum security was not forgotten either, and changes were to occur there also. In September 1957 Marshall announced that the Government was intending to replace Mt. Eden as soon as possible. However, because of the fear that another scandal — any scandal — would reopen wounds only freshly healed, there would be an emphasis on a tightening-down of management. I. J. D. MacKay, at that time an administrative officer in the Prisons division at head office and a member of the penal group, was transferred to Mt. Eden as deputy superintendent. While it is not unusual today for exchange to occur between head office and prison personnel, in 1956 it was less common. There was a suspicion among some, inmates included, that the appointment of MacKay to Mt. Eden, for a number of months in 1956, may have been more than simply to enhance his career prospects. Evidently this feeling had some substance, for when I spoke to him about MacKay's transfer, Marshall replied:

> Well, the Horton escape occurred just prior to that, didn't it? And I was not pleased with the administration of the prison. . . . That was the background to MacKay's being sent there . . . to make recommendations for improving the administration.

Haywood, no doubt, understood the reason behind the appointment, as well as its threat to his sovereignty. Whether MacKay's presence had much impact on his administration, though, is moot. It was during this time that officers began getting paid for the work they did outside routine duties, although this had already been recommended by Haywood in December 1955. Shifts were also rostered, so that officers working late at night were assured of proper rest before coming back to work.

Otherwise, the effect of MacKay, except in so far as he was able to report on matters to head office, was apparently not very great. MacKay was relatively junior and Haywood was too strong and shrewd a man to be perturbed by an underling from Wellington. So in spite of MacKay's presence, things seem to have gone on fairly much as Haywood wanted them. As superintendent, Haywood's word was still law as far as day-to-day managerial decisions were concerned. Anstiss, who worked on good terms with Haywood, said of the relationship:

> Haywood was too powerful for him. There were certainly divisions of opinion. Haywood let MacKay do 'his thing' as long as MacKay didn't interfere in the way Haywood ran his gaol. I mean, that's obvious.

Anstiss gives an example:

> [MacKay] wanted to rip out all the cords and flexes and — they had them everywhere, you know. That was part of the privilege. You could have a light over your bed. String of flex and a light over your bed. Now that was a privilege [that] was when there was nothing else. And he wanted to rip them all out and, you know, he called me down and said, 'Get rid of all these things,' and I said, 'I think you'd better talk to Mr Haywood first.'

I knew bloody well what would happen! Haywood called him in and said
— [laugh] — and nothing happened.

There were other disagreements as well. But MacKay, a young, professional public servant with only three years in Justice,[42] was far out of his depth among the intrigues of maximum security. Haywood, the seasoned gaoler, resented interference and did not hesitate to make his feelings felt. An example of Haywood's treatment of MacKay, which is also revealing of the former's cunning, was related to me by Cavanagh:

> ... there was lots of little wroughts pulled up there, there was a fellow called MacKay sent up there to try and sort things out. Haywood must have known what it was about, and I used to feel sorry for MacKay because although he'd probably been told what the hell was going on, he was too much of a gentleman; he didn't know how to handle Haywood.... One time a fellow called Egan, who was an old crook in the old sense of the word, was taken down the Magistrates Court about something — he ran away I think — so they took him down to the Court and said, 'Oh, what did you want to do this for?'
> And he said, 'On account of the new deputy superintendent we've got. I don't like him.'
> So MacKay said, 'Did you read the bloody newspaper?'
> I said, 'I was at the Court. I heard it.'
> He said, 'You heard what this bloke said?'
> I said, 'Yeah.'
> 'I didn't even know him. I've never spoken to him.'
> So I thought, how do I go here? I've got to be loyal to old Haywood. So I said, 'Aren't you awake?'
> He said, 'No.'
> I said, 'It's a put-up job.'
> He said, 'Who would bloody well put up a thing like that?' — Because he wasn't used to that sort of thing, you see.
> And I said, 'I don't know. You work it out.'

When I asked if he reckoned Haywood put this guy up to it, Cavanagh shrugged and grimaced and replied: 'Well. You know. Sometimes your mind boggles at it, eh?'

MacKay did not stay long at Mt. Eden and some months later he was returned to head office. He became Inspector of Prisons in 1957.[43]

ELECTION YEAR

Excitement over the Horton affair was soon over, and it was not long before the prison was back on its normal path — minus, of course, the outside excursions. Haywood proceeded under constraint, while Barnett, having also taken control of the police in 1955, was now dividing his time between two Departments. The Horton scare must have preoccupied him for some of this period, and it was not until the following year that some of the proposals outlined by Marshall in his speech of August 1957 were put into effect.

In line with the policy of trying to keep mental patients out of prisons, the Mental Health Act was amended in 1957 to allow prisoners to be remanded to psychiatric institutions for examination. In March that year the first conference of prison psychologists took place. As a result, it was resolved that medical facilities for gaols would be centralized, and in April 1957 the first Director of Prison Medical Services was appointed.

However, 1957 was an election year and much of Marshall's energy, apart from working on the new policy, was taken up in the campaign. Sidney Holland, who had been Prime Minister since 1949, was becoming ill, but it was not until 20 September, amid a great wave of speculation, that he finally retired from office. Keith Holyoake became Prime Minister and John Marshall, on 2 October, was appointed his deputy.

The popularity of the National party, due to some extent to the natural attrition of eight years in office, was receding. The emergence of new leadership at the eleventh hour did not improve the party's chances. Nash and Nordmeyer, the opposition leaders, were two of the best-known politicians in the country. They offered financial relief and an immediate hundred pounds rebate to every taxpayer if they became the Government.

The ballot was held on 30 November and it was evident, even from early returns, that there was a swing away from National. At the final count, Labour had taken six more seats, giving it an overall majority of two. With the Speaker appointed, in December 1957, Labour edged into power with the narrowest of possible leads: a working majority of one vote in the House. The National party vacated office before the end of the month.

NOTES

[1] The last before it was Matthews (1923).
[2] Dept.J. (1954).
[3] Ibid. 6.
[4] Prior to this legislation, the borstal term was 2 to 5 years, if imposed by the Supreme Court; 1 to 3 if imposed by a lower court. Under the new law, all terms would be 0 to 3 years, with date of release to be determined by the Borstals Parole Board.
[5] In 1959, before the sentence was even put into effect, it was reduced to three months.
[6] Corrective training was repealed by section 2(1) of the Criminal Justice Amendment Act, 1963.
[7] *PGM* 19 April 1955. New Plymouth had ceased to be used solely for sex offenders in 1952.
[8] The Tongariro Prison farm complex consisted of two institutions: Hautu and Rangipo.
[9] The pernicious influence of 'hardened' criminals upon the criminally naive had been commented on in Sutherland's text of 1947. In New Zealand the problem had been recognized even by Hume and Dallard, although neither did a great deal about it. The need for a workable rehabilitation scheme, with specialized training for certain offenders, was a persistent theme in annual reports from 1951 onwards, and featured prominently in the Department's publication of 1954.
[10] Ministerial dissatisfaction with the distinction between hard labour and other forms of imprisonment had been expressed as early as 1924, a point noted by Barnett in his annual report of 1953. Indeed, imprisonment without hard labour seems to

11 have been largely forgotten after the introduction of reformative detention in 1910 (Webb 1982, 14-15). In 1953 Barnett began to campaign for the replacement of both reformative detention and hard labour (*App.J.* H-20 (1953) 18).
11 For an excellent discussion of the history and distinctions of the sentence, see Webb (1982) 11-16.
12 In 1954 the national DAP of prisons was 1154.38.
13 *App.J.* H-20 (1953) 8.
14 Webb (1982) 72.
15 Although membership fluctuated over the years, founding members were: S. T. Barnett, H. R. Sleeman, P. K. Mayhew, A. F. McMurtrie, C. J. Caughley, K. Menzies, and I. J. D. MacKay.
16 The minutes are held at National Archives, and cease on 17 November 1969.
17 Marshall (1983) 211-12.
18 Marshall conceded that he was pleased to be out of the politics being played in New Zealand since the mid-1970s (Hooper 1985, 44).
19 In his *Memoirs*, for example, Marshall (1983, 220) said he did not even know, at that time, that the term 'warder' had been discouraged in official usage several years before.
20 This was the Revd L. C. Clements, from the National Council of Churches.
21 Marshall (1983) 228-9.
22 The annual report had commented that only 50 per cent of applicants were successful (*App.J.* H-20, 1955).
23 Confidential memo from Haywood to Barnett 9 December 1955. There is doubt over whether Haywood really was present when the team left.
24 Official statement of J. R. Marshall 7 December 1955.
25 *Truth* 14 December 1955.
26 J. Marshall (pers.comm.).
27 Ibid.
28 Ibid.
29 D. MacKenzie (pers. comm.).
30 Confidential memo from Haywood to Barnett 9 December 1955.
31 *NZH* 8 December 1955.
32 Official statement by J. R. Marshall 22 December 1955.
33 Marshall (1983) 227.
34 Official statement by J. R. Marshall 22 December 1955.
35 NZPD vol. 309 (1956) 723-6.
36 Marshall (1983) 227.
37 Polaschek (1958) 222.
38 Polaschek (1958) 223.
39 I refer here to the Central Otago irrigation project where, in 1984, a Minister accepted no responsibility for inaccurate Ministry of Works costings. Instead, all blame was placed on public servants. Marshall's acceptance of responsibility in the Horton matter receives commendation from Scott (1962) 126.
40 Public statement by J. R. Marshall 4 May 1956.
41 Marshall (1957).
42 MacKay had moved to Justice from the Department of Internal Affairs in 1953. He was thirty-six years old in 1956.
43 MacKay rose to the position of Deputy Secretary for Justice, before being appointed a member of the State Services Commission in 1971. In 1972 he became Director-General of Social Welfare, and died in office in 1976.

5: THE TROUBLED YEARS 1957-1960

In spite of a solid majority of votes (Labour headed National by 4.1 per cent), Labour's hold on the electorate in 1957 was weak. A loss of one seat would have resulted in a hung Parliament. A loss of two, a National Government. Labour was keenly aware of its precarious tenure, and it was this basic insecurity which gave a good measure of conservatism to its administration. In the one area where Labour did attempt extravagance — taxation — it stumbled badly.

We shall see that the Labour party's intention in this, its second innings of Government, was to avoid controversy as far as possible. Labour's aim was to cling to its office and try to strengthen its hold for a second term. The sweeping reforms the party introduced during the 1935-1949 period would be repeated in 1957-1960.

H. G. R. Mason, Minister of Justice

The new Minister of Justice once again was H. G. R. Mason, the man who had headed the Justice Ministry in Labour's first Government. Now nearing the end of a long political career, Mason had first entered Parliament thirty-one years beforehand. A vegetarian and a theosophist, at seventy-one he was one of the oldest Members of Cabinet.

From an early age, Henry Greathead Rex Mason had enjoyed success and distinction. Born in Wellington in 1885, H. G. R. Mason was the son of Harry Mason of the Government Printing Department. Educated initially at Clyde Quay School, in 1898 Rex Mason won Queen's and Education Board Scholarships. From there he attended Wellington College, where he won Turnbull and Rhodes Scholarships, as well as the Barnicott Memorial Prize. He matriculated in 1900 but remained at school, becoming Head Prefect and Dux of Wellington College in 1902. Furnished with a Junior University Scholarship he attended Victoria University, where he took a BA in 1905 and an MA with Honours in mathematical physics the following year. In 1909 Mason graduated with an LL B.

At this early age Mason's distinction was already noteworthy. In 1910 a journal called *New Zealand Freelance* published an article on Mason, listing his impressive achievements.[1] By 1910 he was chief clerk at a Taranaki law firm, then he moved to private practice in Wellington. He shifted to Auckland and became Mayor of Pukekohe from 1915-1919, before entering Parliament in 1926 as Labour Member for Eden.

Before 1935 Mason was President of the Labour party, but relinquished the title when his party came to power that same year. During the early Savage-Fraser period, apart from those of Justice and Attorney-General, Mason also controlled the Departments of Education, Native Affairs, and Public Trusts and Patents at various times. At the age of eighty-one he was compulsorily retired from politics and was awarded the CMG in 1967. At the time of his death eight years later, Mason was the last surviving member of the first Labour Cabinets.

As an individual, Rex Mason was generally liked and respected even by his political foes. Dallard's very close affection for him has already been noted above. Mason's party opposite, Sir John Marshall, remembered him similarly as:

> ... a man of peace and sweet reason; a tall, gaunt, ungainly figure, striding through the corridors of power in uncongenial company. ... It is perhaps going too far to say that he kept the Labour Party on a straight and narrow legal path, but their path was straighter and narrower than it might have been if he had not been there.[2]

Gwen Watts, wife of former National party Finance Minister, Jack Watts, remembered Mason with similar affection:

> I had listened to his speeches in the House and never once had I heard him make a spiteful remark about an Opposition Member. ... His sincerity was so obvious that I could not work up any animosity, political or personal, against him. ... He was likeable, intelligent, and persevering.[3]

Patricia Webb described Mason as a person with 'very strong ideas' who was 'easy to work with in that you knew exactly where you stood with him'.[4]

During Mason's first period very little was achieved in the area of penal reform. Apart from Mason's basic lack of interest in the subject, this was attributed largely to spartan economic conditions and to the imposing conservatism of his permanent head. Now conditions were somewhat more favourable. Mason had inherited a system in the throes of change, energized by a head whose vision and enthusiasm far surpassed those of his predecessor. Would the reformative interest which Mason had failed to display in his first term now be kindled by the energy of his new Secretary?

The short answer is that they would not. The reasons, though various, are not hard to find. To begin with, it will be remembered that Mason had opposed the appointment of Barnett in 1949, but that Barnett had been given the post, none the less. On reaching power, he had done little work on the penal system until after the change of Government. Then, with the acquiescence of his Minister, Barnett had set about tearing down and rebuilding a programme which Mason had so badly neglected.

The Department had been soundly rebuked after the war for the poor state of the prisons. Together with Dallard, Mason had denied anything was amiss. When Barnett, amidst a blaze of publicity, had exposed conditions as they really were, criticized the head in whom Mason had held so much faith, and commenced restructuring the Department, Dallard's neglect, and Mason's lack of concern had been laid bare for all to see. These facts did not go unnoticed in the House, nor were they ignored by the media. Now the ageing Mason became foster father to an infant conceived in his absence, and his lack of affection for it is perhaps understandable.

When Mason returned in 1957, Patricia Webb, although not particularly well acquainted with him, said that Mason was no longer the person he had been. Mason was not only old, but a man upon whom the years of political combat had taken their toll. Finlay remembers him as one who became increasingly illiberal with age, leaning more to the right as he neared retirement. At seventy-two, Mason's energy and capacity for innovative thinking were becoming limited. Cameron suggests the same:

> I think at that time it was unfortunate although inevitable I suppose that he should have been Minister of Justice again because he was really, I think, as far as reform went, he was getting pretty old and I think he was largely limited to two or three of his hobby horses that he hadn't been able to get done in his previous term.[5] . . . on the whole, I suppose, the range of his interests, the span of his concentration must have been very much diminished from the early period and he wasn't really very interested in the general revision of the Crimes Act that he was landed with. He tended to leave an awful lot of the play to the Department and really confined his concentration to two or three matters that he was really interested in. . . . he really didn't seem to have very much ambition. That was my impression during the 1958-59-60 period.
>
> You know, most of the things that he was keen on had either been done or weren't going to be done. He still retained his very, very good analytical mind, but he really seemed to me rather 'past it'. . . . he really didn't want to do a great deal.

However, apart from Mason's personal lack of interest in penal reform, the climate was unsuited to a progressive activity. Overcrowding, as we shall see, was becoming a preoccupying problem. Within twelve months of coming to power, Labour's cotton-thread majority was already beginning to fray under the strain of the 1958 'black budget'. It could hardly stand the weight of more controversy. During the second Labour Government, Mason says, the policy of Nash was to 'do nothing, and offend no-one'.[6] This contributed to Mason's record of ineffectiveness in office.

MASON AND THE PARTY LEADERSHIP

For the latter (and greater) part of the first Labour Government Peter Fraser was Prime Minister,[7] and for the whole of the second the chief was Walter Nash. Mason did not get on with Fraser — whom he once described as a sadist[8] — and he detested Nash. However, he believed that the personalities of his leaders were the key to understanding both of the Governments he served in.[9]

Fraser had been the only member of the first Labour caucus strongly to oppose the repeal of capital punishment. The Government's reluctance to abolish is largely attributable to this fact. Nash did desire abolition — albeit somewhat equivocally[10] — but penal reform was not one of his great concerns. Ministers interviewed in the early 1970s by Sinclair unanimously agreed that Nash did not interfere with the running of their portfolios.[11] Yet the Prime

Minister did have a big say in major issues and was reluctant to delegate authority.[12] Nash, although lacking any sense of leadership, had a great love for power, said Mason.

> It was impossible to rationally discuss anything else with Walter. And at any point where any Member of Cabinet tried to discuss his decision, this was used by Walter as an attack on a divine decision. And to disagree with Nash was an illustration of your immorality because you were trying to buck a divine decision.[13]

While Mason's opposition to Nash was considerable, it was well known that Nash himself was not over-endeared of Mason either. In 1953, while Nash was away attending the coronation of Queen Elizabeth, Mason had led an unsuccessful movement to depose him.[14]

The reason that Mason, in spite of his opposition to Nash, was given the important portfolio of Justice was probably that he was uniquely qualified to do it. Mason was the only lawyer in the Labour caucus. Martyn Finlay was out of Parliament and Mason, with fourteen year's experience behind him, was really the only choice. After complaining to Nash that the portfolios of Attorney-General and Justice were not enough for him, he was also given Health. This, he later believed, was only to prevent Mabel Howard from getting it.[15]

The antipathy of party leadership towards Mason was well known, even in opposition circles. Marshall referred to the 'uncongenial company' Mason shared while he was in politics,[16] and to me he agreed that Mason was opposed by Nash. Although they depended on him — as the only lawyer in Cabinet — to keep within the rule of law, he got little support from his colleagues, Marshall said. In an interview with L. B. Hill in 1968, Mason lamented the powerlessness of MPs in the face of opposition from caucus, and as far as this went, he said, Nash was caucus.

Mason considered both Nash and Fraser to be obstructionists, who resisted almost everything he wanted to achieve. Before anything could be done, he later claimed, no matter how trivial, Nash wanted a Treasury report made on it. Apart from delaying the proposal, this procedure almost always precluded it from being approved. The inertia of Fraser and Nash, and, at the end of the first Labour Government, falling prison populations, were the major reasons Mason gave for so little penal progress having been made in the period of his office. Subsequently, this same inactivity was blamed on rising musters.

Generally Mason's years in politics were unhappy ones, and those of 1957 to 1960 were perhaps more so than previously. These were troubled times for Justice, and the huge problems that presented themselves were not easily overcome. Most of the developments which took place after 1957 had already been in the pipeline when National retired. The opening of the research unit, establishment of a prison officers' training school at Pt. Halswell, conversion of the Rolleston Army Detention Barracks for use as a prison, and the opening of a new dormitory at Waikeria all occurred in 1958, but had been planned and initiated by the previous Government.

The new approach towards medical services in December 1958 had been started with the appointment of a Director of Prison Medical Services in 1957, and been alluded to by Marshall in August. There was little done that was new in 1958, but to be fair to the Labour party, it must be said that had National remained in power, events would probably have been much the same. In National's final term a crisis had been developing that left little room for manœuvre. Over a period of three years it began to dominate almost every resource the Department had.

OVERCROWDING

The problem which was of most immediate concern to the Justice Department from the mid-1950s onward was overcrowding. Prison populations, since reaching a baseline in 1948-1949,[17] had been rising steadily and Barnett had been alarmed by the fact as early as March 1954.[18] It was not until 1955-1956, however, that the gradual annual rise became a 20 per cent leap, and this was followed by another leap, then another, and another, and another. In five years the daily average population of prisons grew by half, from a little over 1200 in 1955 to 1850 in 1960.

In an attempt to find a solution, in July 1956 P. K. Mayhew, newly appointed Director of the Penal Division, had been asked by Barnett to assess the situation and to try to explain it. The muster at that stage lay around 1330 — about 300 more than the previous year. Mayhew noted that regardless of transfers the population of Mt. Eden, which hitherto had seldom gone above 300, was now 350. At other institutions there were only a total of thirteen vacancies. Mayhew was unable to explain the situation, and he could envisage no quick solution.[19] The following month the Director drafted a confidential letter to all superintendents, telling them that overcrowding was a nationwide problem and asking for comments on what might be done about it.

Although in July 1956 the accommodation maximum of 1320 had been surpassed, by December it had dropped to a high, but manageable, 1302. However, the daily average of 1304 was still well up on the previous year. The jumps were disproportionately weighed in favour of young offenders, and Invercargill borstal had become so packed that a wing at Wi Tako prison — a semi-open institution for adult first offenders — had to be converted to take the overflow.

By February 1957 the position was unimproved, and Barnett wrote a long letter of caveat to Marshall, explaining the urgency of shortages. Referring to the inaction of Government, he acknowledged that other needs had necessarily taken precedence. He listed the 'alarming' recent increases in the numbers of men and youths confined in penal institutions. In the year up to November 1956 these had leapt by nearly 15 per cent, from 1110 to 1270. Although he could not positively account for them, Barnett believed the rises were due to:

1. A slight increase in police efficiency.

2. An increase in population, especially among the adolescent and young adult age groups.
3. Longer sentences being handed out by the courts.
4. An increase in the total volume of crime.

Each of these factors was valid and reflected similar developments in the United States and Europe. In the light of them, Barnett argued, there was no reason for complacency about the future. On the contrary, he said, the picture showed 'that our prison and borstal population is going to increase, and that it is likely to increase very rapidly.'[20]

Prophetic though these words would prove, the 'rapid rise' came even quicker than Barnett could anticipate. After the letter had been drafted, but before it could be posted, Barnett was prompted to append a desperate appeal:

> Since the attached letter was drafted, the position has worsened. The population today has increased to a figure that gives one spare bed in New Zealand, and in at least three institutions, men are sleeping on the floors of the wings or assembly halls.
> I understand that there will be resistance from the Ministry of Works and the Cabinet Works Committee to any attempt to get expedition. I hope that you may be impressed with the real need for urgency and that you will lend your full support to the endeavours we are making through the ordinary channels.

There really was no alternative to an extensive building programme and this is what Barnett wanted. John Robson, Barnett's Assistant Secretary, realized too that money for prisons would not be got easily. There was to be an election in 1957 and the best way of raising finance would be to create a perceived need for it among the public. Barnett agreed to Robson's suggestion of a news campaign to set the plan in motion, and it was decided to launch the scheme at an address to the Manawatu Justices' Association ten days later. The speech, duly delivered, was well covered in the press, and was reinforced in Barnett's annual report. Having noted that he had been predicting increases since 1953, Barnett wrote that there was no alternative to starting a large-scale building programme. An ambitious construction proposal, designed to provide 700 more beds, was then suggested.[21] A number of the ideas had been put into effect by the end of 1958.

Response to the campaign was positive. Letters of support were received from the General Secretary of the Public Service Association, and the problem was generously treated in a series of articles in *The Evening Post*[22] and *The New Zealand Herald*.[23] In Parliament a question about what National was doing about the building programme[24] drew a reply from Marshall that he would pull Mt. Eden down 'tomorrow' if he could and that the institution would be demolished 'as soon as possible'.[25] Meanwhile, plans for its replacement at the National Penal Centre would be going ahead.[26] In fact, at a meeting of the Penal Group five months before, amended plans for the centre had already been detailed. The National Penal Centre, which in April 1955 had been designated for 1200 inmates, would now be built for a maximum of only 1000. Under the new scheme, a medium security facility was no longer contemplated, but there was to be a first offenders' prison for 200, a maximum security unit, and a large borstal which would be constructed down the road, away from the main complex. Accordingly, the name of the institution would be changed from National Penal Centre to 'National Prison Centre'. Along with the new penal programme, this acute and apparently irreversible muster situation is what Mason faced when he took office at the end of 1957. As noted, the following twelve months saw the fruition of various rescue schemes that had been initiated by the previous Government. By March 1958, thirty beds at Rolleston had been made available and the new dormitory for Waikeria was underway, but one wing at Wi Tako was still being used as a borstal.

However, in spite of these remedies, overcrowding had meant that some of the provisions of the 1954 legislation were failing or had still not been given effect. The detention centre had not yet been established. Initially New Plymouth had been envisaged as a detention centre, but, on 2 October 1956, Marshall had been forced to announce that the prison's conversion had been postponed due to its need for the use of adults. New Plymouth was to remain a possibility, but by February 1958 Barnett had three more: Wi Tako, Rolleston, or Waikune, with Waikune now the most favoured. A new wing was to be built at that institution, but conversion to a detention centre would be impossible unless the population crisis abated. Unfortunately this was not to occur.

Corrective training was another scheme seriously affected by the crisis. Waikeria, which in January 1955 had begun to take corrective trainees, was by 1957 too full of borstal boys to accommodate everyone received. Places for them at other institutions were arranged, and this effectively meant that the corrective sentence, for these inmates, would be abandoned. They would serve their terms no differently from others given imprisonment.

Reflecting these frustrations, Barnett's report of March 1958 gave little cause for comfort. Prison populations were continuing to rise and the situation was, he said, 'exceedingly grave'.[27] Practically every possible improvisation had been made yet there was still insufficient room for the ever-increasing numbers. Some cells had three men in them, while classrooms, common rooms, and dining halls were being converted to dormitories. Because of the time it takes to build new institutions, no relief was expected in the near future.

In a memo dated March 1958, Barnett informed his new Minister of the considerable difficulties being faced. In 1957 the population had risen as high as 1441. During the winter of 1958 it was expected to reach 1550. Mt. Eden prison, ideally suited for no more than 300-320 inmates, had recently held as many as 396. Men at Mt. Eden were sleeping in dormitories and three to a cell. The basement, which normally took fifty beds, now housed 103. The maximum accommodations the system could sustain at any one time would be 1421, and with a projected peak of 1561 in 1958, the situation was considered serious.

In May, another memo was sent to controlling officers announcing that although virtually all beds were occupied, prisons must prepare for another 5 per cent increase in the next few months. Superintendents, asked to furnish details of possible extensions to their institutions, reported back in July that by utilizing areas like common rooms, laundries, bakehouses, chapels, store rooms, and so forth, accommodation could be increased to 1783. Mt. Eden reported that the schoolroom and the chapel could be converted to take another 30, giving a capacity of 440.

By this time, Mt. Eden's population was well over 400, and as the largest receiving institution in the country its dispersal problem was the most serious of all. A well-published escape in June 1958[28] increased pressure for a solution, and Mason responded by blaming his own party for its lack of activity. The escape was due to overcrowding and understaffing, he said,[29] and he was relying on the Government for a remedy.

Pledges for relief were not long coming and in September a plan was announced for the erection of an auxiliary prison at Auckland. Mt. Eden was to be stripped before summer, and overcrowding there would cease. Surplus inmates would be transferred out, possibly to a new gaol at Whangaparaoa.[30] Two weeks later the Director of Prison Medical Services, Dr W. H. B. Bull, and Mr H. H. Buchanan, a consulting psychiatrist were invited to report on the prison with a view to pruning its population. After inspecting the institution for five days, interviewing the inmates and examining their files, the two concluded that only nineteen of the 450 or so inmates at Mt. Eden could be transferred safely. They therefore recommended that the female division in the north wing extension be vacated and men moved in their place.

The suggestion was not wholly new. Early in 1958 Barnett had proposed Dunedin as an alternative for women, and, following the June escape, Barnett had spent five days in Auckland studying this same idea. The institution, at that stage holding 450, only had facilities for 275. On 24 July it had been announced in *The New Zealand Herald* that the women would be shifted out, so the Bull-Buchanan report really only confirmed what Barnett already knew.

Two weeks later, action was taken. While the small wooden women's division inside the walls remained for remands and short-termers the rest were moved, initially to Arohata, until accommodation could be arranged elsewhere. The old Dunedin gaol, closed forty years before and used since by the Police and Defence Departments, was then reopened for the detention of females. Although — in Barnett's own words — it had 'nothing to commend it . . . and much to condemn it',[31] Dunedin remained the main custodial facility for women for the next fifteen years.

These measures were really nothing more than a stopgap and did little to curb what was developing into an almost uncontrollable problem. Barnett's report of 1959, fifty-six pages long, was the most disconsolate of all he was to write. The daily average prison population had increased from 1139 to 1537 in the four years from 1954, Barnett revealed, indicating a rise of 35 per cent. At the time of writing, the muster was 1700; close on 25 per cent up on two years earlier. There had been a 400 per cent jump in borstal populations since 1950, and a 123 per cent hike in the four years since 1954. On a per caput basis, in the decade since 1948, the ratio of youths aged between fifteen and twenty who had been sentenced to borstal training had grown by almost 300 per cent. Māori offending had risen dramatically, especially among the young. In 1958, 39 per cent of all those received into borstals were Māori.[32] By the end of the 1950s, therefore, New Zealand had developed a fully fledged youth and race crime problem.

The figures went on, and all of them were bad. At the present rate of expansion, Barnett wrote, the prison and borstal population of 1955 could be expected to have doubled by 1975.[33] Barnett certainly had no cause for optimism in 1959, and his report betrayed his despondency. He commented how in the past he had deferred his needs for money to the more important ones of hospitals, schools, and roads. It was no longer the case, however, that the necessity for penal spending lacked immediacy. Unless a great deal of capital was outlaid, he predicted, there would be a 'complete breakdown in the prisons, borstal, and probation services'. '. . . that we must build — and do so quickly — I think is undeniable', he wrote.[34]

This was the message of a man who, only ten years earlier, had set out to revolutionize penal policy. Now, frustrated and embittered by a retinue of failure, Barnett attacked the system which had once been his cynosure.

> In truth, we have no national penal policy. We are imitators. We have to be convinced by experience in other countries before we will venture. New Zealand has a reputation for independent thought and courageous initiative in the social field. That is certainly not true in the penal field, although we have singular opportunities to develop our own practices and policy.
>
> We do not command international attention in the penal field. Few nations would come to learn from us. True, we have made some advances in recent years, but few could be said to be characteristic of a young country exercising its own national attitude toward crime and criminal offenders.[35]

Corrective training was failing. For some reason the courts did not favour it,[36] which was just as well, Barnett reflected, because there was no room for corrective trainees anyhow. The detention centre scheme had not even started, again due to population pressure. But the final casualty of Barnett's regime was also his largest and most ambitious project: the National Prison Centre.

THE NATIONAL PRISON CENTRE

The National Penal Centre was first announced in 1951. Because of the size of the plan it had inevitably been a long-term one, but it conformed to Barnett's idea of concentrating large numbers of inmates and specialist services into a fairly small area. This was to maximize the efficiency and economy of its operation. By 1957, population increases had made it obvious that the centre would never be large enough to cater for all types of inmates. The scope of the project had then been reduced, and its name changed to 'National Prison Centre'.

Barnett was convinced that the concept was a good one, but as early as April 1955 Mayhew had expressed doubt about it.[37] This reserve apparently spread, and two years later it had been claimed in Parliament that the National Penal Centre was no longer favoured.[38] Assistant Secretary Robson, for one, had never been particularly keen on it, and in November 1958 when Dr Bull attacked the idea of large institutions,[39] Robson supported him. The increasingly vocal Robson noted that the committee was 'generally agreed' that smaller prisons were by far preferable to a large, single complex.[40]

Although unhappy about it, by 1959 Barnett had succumbed to the inevitable. Population increases had become so overwhelming that the notion of a prison centre at Waikeria, he conceded, would have to be abandoned in any case. The centre, which would now have a maximum population of 935, would be devoted solely to the treatment of corrective and borstal trainees. Accordingly its name would be changed again, this time to 'National Youth Penal Centre'.[41] By early 1960, plans for two new cell blocks and two villas for fifty boys each had been drawn up, and work on them was due to begin shortly.[42]

BARNETT RETIRES

Samuel Barnett retired from office in July 1960 and his report of that year was the last he ever wrote. He noted then that several main problems faced the penal system of New Zealand. The substance of these were:

1. That one third of our prison population was under twenty-one.
2. That just under one third of it was Māori, and in the under twenty-one category, more than a third was Māori.
3. Overcrowding in penal institutions.
4. Insufficient staff.
5. That New Zealand did not have a single modern penal institution. All of them were old and archaic.

Populations had continued to rise. There had been a 47.1 per cent rise in DAPs over the previous five years, 20.9 per cent in the last two, and an 8.6 per cent rise during the latest twelve-month period. The future could not be foretold.

An extensive building programme was planned, but, apart from a new women's prison at Paparua to be begun in 1961,[43] nearly all work involved additions to existing structures rather than the erection of new ones. (Moreover, an examination of the 1960 proposals today reveals that most of them (a) had already been effected at the time of the report; (b) became delayed for five years or more; or (c) were never carried out at all.)

The 1960 report was another long one and despite the building plan, Barnett's departing message was discouraging. There looked to be no escape from inflated prison musters. Schemes for dealing with the young were failing in one case and had not started in the other. Even the borstals were not working and youthful offending was now a greater problem than ever. Barnett's answer to crime in general — the National Penal Centre — had been abandoned. And in the face of opposition from his probable successor, it looked as if even the youth penal centre might never be built.

For six years after taking office, Barnett had lived on the crest of a reformist's wave. Then for five years following, he had seen his hopes drown in a tide of prisoner influx. That Barnett had made great progress since 1949 is without doubt. And the fact that he achieved so much less than he might have was due to matters largely out of his control. Notwithstanding these truths, Barnett left the Department of Justice a disillusioned and unhappy man. A decade before he had entered, flushed with enthusiasm and vigour, determined to create a penal system to stand among the best in the world. Now, with eleven year's work behind him, a sadder, wiser man no longer believed in the hope of rehabilitating people behind bars. Those who did become law abiding, he felt, did so more often in spite of their imprisonment than because of it: 'It is a fallacy to expect prison of itself to improve men; it is just as likely that resistance to crime will be weakend as a result of prison experience.'[44] For Barnett, the dream of the therapeutic prison had vanished. The zeal of the reformist had succumbed.

NOTES

[1] *New Zealand Freelance*: 19 March 1910.
[2] Marshall (1983) 261.
[3] Watts (1969) 80.
[4] P. Webb (pers. comm.).
[5] This is probably fair comment. Some of Mason's particular interests were decimal currency, the Mareo murder case, homosexual law reform, abolition of capital punishment, and licencing reform.
[6] Mason (1968). Lack of commitment had long been something of a problem for Nash. Speaking about Labour's stand on the May 1951 waterfront strike, Nash had won lasting derision for claiming that his party was 'not for the watersiders, nor . . . against them' (*NZH* 15 May 1951. Through this action he became labelled by the National party as a person without opinions.
[7] Fraser took over from Labour's first Prime Minister, Michael Joseph Savage, in March 1940.
[8] Mason (1968).
[9] Ibid.
[10] Sinclair (1976) 204.
[11] Ibid., 341.
[12] Ibid., 339–41.
[13] Mason (1968).
[14] Sinclair (1976) 292–4.
[15] Mason (1968).
[16] Marshall (1983) 261.
[17] DAPs for 1948 and 1949 were 1005 and 1024.
[18] *App.J.* H-20 (1954) 6.
[19] Minute Sheet, P. K. Mayhew 25 July 1956.
[20] Urgent memo from Barnett to Marshall, 11 February 1957.
[21] *App.J.* H-20 (1957) 10–11.
[22] *EP* 14 March 1957; *NZH* 17 May 1957.

23 *NZH* 17 May 1957; *NZH* 17 August 1957; *NZH* 29 August 1957.
24 NZPD vol. 313 (1957) 2404.
25 Ibid., 2407.
26 Ibid., 2406.
27 *App.J.* H-20 (1958) 11-12.
28 I refer to the escape/s of R. Maketu. These will be covered in chapter 7 ('Maketu' is a pseudonym).
29 *NZH* 16 August 1958.
30 *PGM* 1 September 1958. The institution at Whangaparaoa was never built.
31 *App.J.* H-20 (1959) 11.
32 *App.J.* H-20 (1959) 13. Māori are usually said to constitute of about 9 per cent of the total population. However, as indicated earlier, there are problems with definition. (A good discussion of the difficulties encountered in defining 'Māori' appears in *NZH* 5 May 1986, 6.)
33 Barnett's pessimism was not unwarranted: the populations, in fact, more than doubled, rising from 1210 in 1955 to 2601 twenty years later.
34 *App.J.* H-20 (1959) 5.
35 Ibid.
36 For example, only sixty-three people had received the penalty in 1958 (ibid.).
37 *PGM* 19 April 1955.
38 NZPD vol. 313 (1957) 2405.
39 *PGM* 17 November 1958.
40 *PGM* 15 December 1958.
41 *App.J.* H-20 (1959) 25, 26.
42 *App.J.* H-20 (1960) 15.
43 This institution was in fact delayed for many years and did not eventually open until 1974.
44 *App.J.* H-20 (1960) 26.

6: POWER AND POLITICS AT MT. EDEN 1950-1960

Population problems were not all that Barnett had to contend with during his last few years in office. Within maximum security itself, cracks were also appearing, and although these were worsened by crowding, many were caused by administration. Imaginative though the liberalization programme was, little thought had been spared for the practicalities of government. Staff resources were already stretched to capacity and logistics demanded that greater freedoms could not be effected without an adjustment in management.

ORTHODOX GOVERNMENT

An old adage among prison officials is that there are two ways of running a gaol: with the inmates, or without them. That is to say, inmates may be included in the control of a prison, or they may be excluded. This was Haywood's choice after 1951 and it points to a basic dilemma of all totalitarian systems: that is, how best may a small group maintain control over a much larger one whose essential interests it opposes? At Mt. Eden this problem was sharpened as musters grew and as the reformative recreation programme forced extra duties upon staff. Prison officers were no longer expected merely to preserve custody and discipline, they now had to be therapists as well. The difficulty of pursuing these twin goals is well recognized. In fact, considering the amount of conflict they produce, Cressey has commented that one of the most amazing things about modern prisons is that they 'work' at all.[1]

In the classic penal situation, control is maintained by the wielding of absolute authority. Power of security personnel is total; inmates have no say in policy or management. Degradation ceremonies, described so well by Goffman[2] and Garfinkel,[3] together with ritualized social difference and deference patterns, are established to create a gulf between the rulers and the ruled. Familiarity between staff and inmates is condemned; displays of emotion or human weakness are avoided. At Illinois State Penitentiary in the 1930s, for example, an officers' rule book specified that:

> Officers, guards and keepers shall not, under any circumstances, allow prisoners to speak to them on any subject not immediately connected with their duties, employment, or wants.[4]

New Zealand's penal tradition is no different. At the old Lyttelton Gaol (1861-1920), rule 20 stated, 'wardens or guards, unless when acting as overseers, are not to hold any communication with prisoners except on matters of discipline'.[5] Under Hume, one of the early Penal Regulations read:

Every officer shall . . . bear in mind that discipline is the purpose of the prison and his employment therein, and he must understand that by failing to enforce submission to the authority with which he is entrusted he will prove himself to be incompetent in the performance of his duties.[6]

Such methods are typical of traditional systems, and we have seen that in New Zealand a rigid control profile was maintained right up to the time of Haywood. However, because personnel resources are often stressed in prisons, staff-inmate distinctions cannot always be maintained, and sometimes convicts have to be co-opted to assist the maintenance of order. This process, which takes place in the interests of effective administration, may occur officially (that is 'formally') or unofficially (i.e., 'informally').

FORMAL CO-OPTATION

Formal co-optation, also seen in school prefectorial systems and industrial worker management committees, takes a special form in the autocratic context of detention. Examples of the latter include the 'honour rolls' of some American prisons, the 'red bands'of the English prison,[7] and the well-known inmate hierarchies of Nazi concentration camps.[8] Inmates in these instances are systematically enlisted as part of the official control mechanism.

As noted, the co-optation of prisoners is often necessary when authority structures are weak or overtaxed. In the United States in the 1960s, there were a number of State systems which even employed armed inmates to guard other convicts. In one example cited by MacCormick, 2000 inmates were administered by only ten custodial personnel.[9] Such practices are generally only a cheap way of solving a serious organizational problem. However, not all forms of formal co-optation are of this type. The councils at Sing Sing and Mt. Eden were established to assist rehabilitation goals, not just pragmatic ones. Barnett's hope was to encourage respect for law and order by giving prisoners some control over their own affairs. This is a principle which Donald Cressey called the 'retroflexive reformation' principle. Cressey believed that after a period of alienation from society, eventual reinvestment with some authority nourishes sympathy for community objectives.[10] A similar reaction is seen in the concentration camp[11] and is one I observed also at Waikeria detention centre in the early 1970s.

Whether inmate participation ever affects the long-term behaviour of criminals is doubtful, but we have seen that Barnett was keen enough on the idea to press for a prisoners' council at Mt. Eden in 1951-1952. Even after the experiment had failed, he continued to defend it. Only opposition from other members of the penal group held his enthusiasm in check. The prefect system has never been established in New Zealand and as we know, even the councils were short-lived.

Informal Co-optation

Systems of formal co-optation are seldom used in top security and are regarded today as unsound. None the less, inmate power is often used unofficially in a number of ways. In free society, the trade union movement is one of the clearest examples of co-optation which is informal. Although employers have no hand in the setting-up of workers' associations, the will of unions must be considered because of the threat it poses to production.

For large prisons the considerations are similar. Although seldom organized into true associations, inmate hierarchies may be potentially disruptive; an ever-present threat to routine. If the desires of the powerful are not accommodated, levels of tension can rise. Leaders may catalyse rebellion and in the interests of stability some of their requirements need to be recognized. Cloward sums the situation up this way:

> Limitations on power in the one system compel adaptive or reciprocal adjustments between the two systems . . . In effect, concessions must be made by the officials to the inmates . . .
> In the final analysis, stability depends on reciprocal adjustments between the formal and inmate systems.[12]

A prisoner élite develops which is tacitly recognized by administration. In return for its co-operation, support is given by the authorities. Practices such as granting leaders direct access to senior staff, permitting their possession of scarce or prohibited articles, and overlooking the way they enforce control, are examples of concessions which endorse élitist power. However, these privileges are secure only if leaders can maintain peace and good order, and to this extent the relationship is symbiotic: in return for its indulgence the administration secures the allegiance of a conservative, but independent, power base. Stability is reliant on harmony between the official and the inmate systems. Each needs the other and has an interest in the status quo.[13]

So whether formally instituted or not, inmate élites are created to serve, or become seduced into serving, the purposes of officialdom. Because they arise out of managerial weakness, however, their balance can be uncertain. Power relations may fluctuate and a lack of organized control generates its own tensions. Organizational goals are lost as the system struggles to overcome internal conflicts. At Mt. Eden, these stresses increased gradually, to the point that they could not be resolved without resort to violence.

Stratification and Privilege at Mt. Eden

In the early years of the fifties, the accomplishments of Haywood and Barnett received generous treatment in the media. However, published reports, although extensive, were also superficial and failed to notify the very important changes taking place at the prison's base. Advanced and imaginative — even by today's standards — though Barnett's ideology may have been, it was certainly not without its problems. An inexperienced superintendent, an untested method of management, and a 'treatment' programme which nobody really understood were all ingredients of a system that became unwieldly and uncoordinated. Crucial to the new style of management was Haywood's 'open door' policy. The superintendent mingled with, and spoke freely to, inmates in his institution. This was something that had never been done before and the rapport he developed with the men allowed an insight upon the day-to-day events of the prison.

Convicts were open to the new boss because he came to be seen as far more approachable than any of his predecessors had been. One of the keys to Haywood's method was his interview procedure, which previously had been a formal ritual involving a written request followed by an officer escort, saluting on arrival at the governor's desk, and a brief statement of business. Haywood dispensed with ceremony, and would interview prisoners on the spot, without any formal application and without other officers present. His door was always open to certain inmates too, and, although their identities were known, information about to whom this leave applied was never recorded.

For the most part, direct access to Haywood was limited to selected long-termers, who in classic fashion became a privileged élite. Many of these men regularly supplied information about goings-on in the gaol, and got powerful and superior status in return. MacKenzie told me:

> There are several ways of running a prison as you probably know, and one thing about Haywood was that he ran Mt. Eden through a hierarchy of prisoners. He regarded his staff with ill-disguised contempt most of the time.

I replied that I was very interested in this aspect and asked if he definitely co-opted the power of inmates in the running of the gaol.

> Oh yes, he would have his 'barons' in his office one at a time. Close the door, nobody allowed in. And he and the bloke would talk for an hour together, and he'd get all the gossip. All the stuff about the gaol. And about the screws from these blokes. And they were cunning fellows, they weren't suckers by any means. They knew what they were at. They made their own alley good.

The staff themselves, knowing who the favoured men were, felt compelled to acknowledge their status. Speaking of a prisoner who I shall call 'Maketu', Anstiss said:

> Well he was a tough cookie. He'd been around a long time, and this was the problem, you'd run up against prisoners who'd been in gaol a long time. . . . And the staff — how could I possibly stop [Maketu] from going to the boss? He'd say 'I'm going to see Haywood,' you know, and I'd say, 'Well!' — and I knew bloody well that — 'OK!'

The powerful men really had the run of the gaol, MacKenzie told me, 'And no officer would dare stop them going through a grille or something like that. An officer would be likely to get told off if he refused to let O'Rourke or [Maketu] go through from one wing to another.' During official interviews, Cavanagh confirms, it was common for Haywood to see some inmates alone and shut his door when they arrived.

In this way Haywood gained an exclusive understanding of the gaol's internal business, allowing him to make decisions without fear of contradiction. Haywood had favourites among the staff too, and they were privy to some of this information. Like the favoured inmates, these adjutants enjoyed Haywood's indulgence only as long as he could be sure of their loyalty. Percy Anstiss was such a man. Anstiss claims that Haywood used him frequently for duties which, because of their sensitive nature, he would not normally request of an ordinary officer, Haywood, for example, was always careful about who dealt with his prisoner informants. Only certain staff could be trusted with them. Anstiss illustrated how he and a few others were used.

> There would be several times when he would stop everybody from going into the prison. The prison would be locked up.
> He would call me in and he'd say, 'Go into the east wing, cell so-and-so', and you'd be told something, and you'd go back and tell him. So all the staff would wait outside and I'd go into the wing alone. Take the keys in.

I asked, 'And go into the inmate's cell?'

> Yeah. . . . It was done quietly. You might even have to do it at night, something like that. That's the sort of situation I found myself in, many times.

The result was twofold. Firstly, it alienated officers to some extent, who felt that privileged prisoners had greater access to the boss than they did themselves. Several commented on this and one, when asked about it, replied:

> Yeah. Yes, that's so. They certainly had a lot more say in things than I would have liked personally. They would have access to him that we didn't have. We didn't have unlimited access to go in and see him and talk to him. But inmates did and I'd say that caused a fair bit of resentment.

Another, who had worked at Mt. Eden since the late thirties, had 'no doubt at all' that staff felt their authority was being undermined by the system of privileged access.

The second effect of the Haywood regime was that it strengthened the power of the inmate hierarchy. As shown by McCleery[14] and Michels,[15] the control of communication is an important aspect of power. Informed groups may affect decisions at the top and monitor the flow of information. Thus, at Mt. Eden, a man in the superintendent's grace had leverage within the system and a degree of understanding of its workings. This influence and this knowledge were beyond the reach of the average prisoner as well as that of low ranking staff.

So along with the other privileges, life was made easier for the inmates Haywood favoured. Not all of them were 'shelves';[16] some were just trusties or persons essential to the running of the gaol. Cooks and bakers, for example, had to be reliable, responsible people. Trusted men might be given jobs outside the walls, or assigned useless tasks so they could be locked up later at night. Men with responsibility or freedom of movement were issued with special clothing to distinguish them from the rest. Most inmates still wore brown or white moleskins but trusties,[17] known as 'Blues', wore blue denim trousers. Frank O'Rourke, the lifer who made up the list for the fated Hibernian Hall bowls trip, was one of this category. He was one of the most powerful inmate personalities, and he ran the Māori concert group. Another was Pat Bennett, a trustie who later embarrassed Haywood by burgling John A. Lee's[18] bookshop in Mt. Eden, while working outside the walls.

Job changes, cell changes, and a myriad of other favours available at administrative whim were easier for Haywood's trusties. MacKenzie gave me an instance:

> One baker, to my certain knowledge, who had homosexual tendencies, asked the superintendent if he would allow him to share a cell with one of the bakers' assistants, which he was. A young fellow was moved in with him. And nobody mentioned it at all — you couldn't mention these things.[19]

At Mt. Eden, privilege bought greater privilege. The elect had a better chance of membership in the prison band. Bandsmen, as we shall see, enjoyed considerably more freedom than the rank and file. Favoured men might also be allowed a radio, which most others were denied. Former inmate Banks told me that Haywood used to keep a box of tobacco under his bed, which prisoners called the 'narks' rations'. The tobacco, Banks said, was for rewarding informants, although established convicts like himself could also ask for a 'fig'[20] from the narks' rations if they ran short during the week.[21]

Advantaged inmates were sometimes exempt from standard procedures. There was an unstated rule that certain cells should not be searched. I asked Lashmar about it:

> I think it was understood. It was understood that some cells you didn't muck about with.

I asked, 'what was the reason for that?'

> I suppose they were people who were useful to the administration [laugh]. You know, it *couldn't* have been any other reason!

The differential privilege system at Mt. Eden was varied and pervasive, and staff either participated in it or tried to ignore it. Inmates assisted it and got out of it what they could. One of the more extraordinary examples of inmate opportunism took place in the prison quarry.

In the Mt. Eden quarry, stone was obtained by blasting it off a rock face in the evenings. Loose rock would be barred-down off the face the following day, and large boulders broken up into manageable sizes by inmates using spawling hammers. The rock was stacked, and finally carted to a stone crusher where it was reduced to gravel. This work, apart from the blasting, which was usually done by staff, was all carried out by inmates. The prison gravel was stored at the insitution and sold to private contractors, whose trucks were loaded from a large hopper attached to the crusher. These activities were controlled by staff, but due to the large numbers of inmates in the quarry and the small staff ratios, supervision was not close. An inmate loaded the trucks with scoria and an inmate sat in the quarry's tally office and filled in the books. Because of the lack of control over output and because trusties were seldom searched when they came through the gates at night, cash settlements between them and the contractors were relatively simple.

The result was a system wherein, by clandestine agreement, trucks received large and regular overloads of metal from the man operating the hopper. In payment he would gain a fraction of the true value of the extra metal from the contractor, with money paid either in cash or into a private account. There were at least three inmates involved in the graft from the quarry — one operating the hopper, the other two in the tally office. Several inmates and staff whom I spoke to had direct knowledge of these happenings, and one prisoner, who was close to the centre of it, described in detail the way the system worked. Only a small number of men was involved in the quarry operation, and the affair is thus more important as a consequence of administrative style than for any effect it had on the organization as a whole. Of greater significance is the matter of discipline. Like the privileges network, discipline at Mt. Eden under Haywood was irregular, and equally a product of his command. It too was flexible and informal, and geared to a variety of requirements.

INFORMAL DISCIPLINE

The disciplinary powers of prison administrators have a firm statutory basis and boundary. Offences by inmates and measures for dealing with them are set out in the Penal Institutions Act 1954, sections 32–6.

In the 1950s, punitive options available to staff ranged from loss of privileges to cellular confinement with restricted diet.[22] The superintendent had, and still does have, only a modest disciplinary capacity compared to that of the Visiting Justice.[23] Since 1941 there has been no legal provision for corporal punishment. Where assaults are concerned, officers in prisons possess the authority of police constables,[24] and are bound as such by the provisions of the law. However, these limitations notwithstanding, the unofficial method of reward at Mt. Eden also had its converse. A tradition of *de facto* punishment became established there, the operation of which often overrode formal procedures.

Coercion of prisoners in adult gaols, especially maximum security, is today comparatively rare. Impromptu discipline is frowned upon. It is in this respect that the penal institutions of the 1950s differ greatly from today, for at Mt. Eden, where Haywood's administration encouraged *ad hoc* measures, violence between inmates and staff became increasingly common. Here the maintenance of order by force took three major forms: co-optation of inmate leaders, fighting between staff and inmates, and the beating-up of prisoners by staff.

Co-optation of inmate leaders to assist with rule enforcement, as discussed earlier, often occurs in prisons. Such enlistment takes various forms and the pattern employed at Mt. Eden was integral to its operation overall. Then, as now, *direct* employment of inmates for disciplinary purposes was unusual. Typically, use of inmate justice was tacit, and followed a pattern of casual reciprocity. Tom Hooper, carpentry instructor under Haywood, explained how the process operated in his shop in the late 1950s:

> I might get a new bloke come in. Smart arse. Wouldn't work anywhere else in the prison and just a shit stirrer. He'd come into my shop. He'd be working there, refuse to work, or whatever. I never had to do anything about it. I may have done once or twice. But some of the inmates'd come up to me (nod at the prisoner), and,
> 'Alright, Boss?'
> 'Oh, he's giving me a bit of a pain in the arse. Not pulling his weight. I'll have to get rid of him.'
> 'She's right, Boss.'
> Next thing you'd see him, he'd have two black eyes and a fat lip, you know. So they used to protect me.

I asked, 'Did that happen much in those days?'

> Yeah, there was quite a bit of it.

On one occasion, inmate Maketu ingratiated himself to this same instructor by warning him of a planned escape from the workshop. 'If the escape had gone off I was in the shit and have to answer a lot of questions,' the officer said. '[Maketu] sort of protected me in that respect, you know. Yeah, he was all right.'

Most staff I spoke to held similar views, seeing men like Maketu as 'rough diamonds', useful to administration if handled correctly. Handling such men 'correctly' meant maintaining good relations with them, overlooking some of their infractions and permitting them liberties which other prisoners were denied. Gaol staff did not view this as an immoral system. Prisoners who worked with staff were not seen as traitors or scabs, but merely to be acting in the interests of everyone. 'The old lags just didn't want anybody upsetting routine,' they said. 'They wanted a smooth-running gaol.' This may have been so, but, at Mt. Eden, defending routine meant safeguarding élitist privileges. In using their power to discipline others, co-opted leaders served the interests of administration. They protected the status quo and in classic style, they earned the favour of their bosses.

Fighting between inmates and staff was also common at this time. Unsurprisingly, some officers refute any suggestion of staff assaults on inmates. Former officer Moyle, for example, described tales of staff-inmate violence as a 'pack of lies', created by people with axes to grind. Others were candid that violence did occur, and all prisoners I knew agreed.

Haywood had a number of staff who were physically well equipped, sometimes known as the 'heavy brigade'. These men were also those most often chosen for hanging duty.[25] Physical prowess or fighting ability in staff was an asset to Mt. Eden's administration. Percy Anstiss, who claims to have had many fights with inmates during his service at the prison,[26] tells about the strange way in which he was recruited:

> ... my recruitment was based on the fact that an inmate burst into the room when I was being interviewed by the superintendent and I just automatically stood up and it was a reflex and I put a headlock on him, because it all seemed to be happening in front of me and I dropped him to the floor.... [When it was over] the old superintendent just leaned over and he said, 'Well, when can you start, lad?' That was it. That's really all that was involved.

Fighting between inmates and staff was by most accounts accepted, and the more physical officers seem to have resorted to it freely. Anstiss continues:

> I wouldn't hesitate. If a guy wanted a fight, it was on. If he threw one I threw one back at him! ... But the spin-off was, of course, you got these young thugs who say, 'You wanna have a go?' and you'd think, 'Christ!' and there were some incidents which everybody watched with apprehension.

Lawrie Hargrave, an employee who at the time held a senior dan grade in Judo[27] was well known for solving disputes with violence, a fact to which he admits. Hargrave worked in the quarry and, because of his reputation, was often challenged by inmates. When this happened he accepted. 'We had a couple of incidents where a guy would, say, step me out, you know, say he was going to knock me over and I'd say, "Oh, well, if that's what you want — if I get hurt, I fell off the rocks, if you get hurt, you fell off them." '

At Mt. Eden in the 1950s this too was quite usual. Convention dictated that injuries sustained in fighting would be explained as having happened accidentally. Hargrave says he had twenty to thirty fights with prisoners during his twelve years at Mt. Eden and only once was he charged with assault. Inmates confirm that Hargrave often fought with prisoners.

Apparently fighting did not always go the officers' way. In a legendary incident, one of Mt. Eden's best-known, toughest, and most respected officers, Case Tuyt, went into the detention block to 'sort out' a powerful man called Ray Twist. Tuyt came back, the story goes, with his cheek and jawbone broken. The tale, related to me many times, holds that according to protocol Tuyt took the matter no further, and Twist was only charged with his original offence. Haywood apparently tolerated this behaviour and probably condoned it as well. A number of incidents are known where fights broke out in front of the superintendent. When the mess had been cleaned up the rule remained: nobody had seen anything.

Beating up of prisoners by staff of superior numbers or size was given less support by Haywood. In these cases, inmates were unwilling participants in conflict. It was inevitable, in a context where combat between staff and inmates was accepted, that a degree of bullying would occur. The real frequency of this type of activity is difficult to judge, though, because it usually took place out of the sight of witnesses and because injuries may have been inflicted, or said to have been inflicted, in self-defence.

There are a number of clues to support the view that the bullying of inmates in Haywood's time was greater than it was later on. In a fictionalized account of the institution in the 1950s, former prison officer Michael Burgess[28] depicts a gaol where intimidation of inmates is usual. In one example a prisoner, terrorized for information by the novel's hero, is advised to explain his swollen face by saying that he fell down the stairs.[29] The image which Burgess creates is of an institution where the impromptu 'disciplining' of inmates is routine.

Although Burgess' account is probably exaggerated, the style of interaction he describes is familiar. Some of Mt. Eden's officers were recognized 'bash-artists', and their unusually high levels of aggression were recognized by both inmates and staff. One officer, who also became a British Empire middleweight boxing champion, was well known for his penchant for assaulting the disobedient.[30] Another, by the name of Price, frequently fought with prisoners and enjoyed amusing himself at their expense. As an example Anstiss told me of watching one day while Price forced a weak inmate to eat all his canteen purchases at once.

Bashings of inmates by groups of officers also occurred, but probably not so often. Arthur Burgess, Haywood's chief, hated violence and censured Anstiss for fighting so much with inmates. After a prisoner, who had attacked a staff member with a milk bottle, was found beaten unconscious in his cell, Burgess is reported to have been 'shocked and very angry' about it.[31] Even Haywood was selective about the violence he condoned. Having once discovered that staff had entered the isolation unit and injured inmates in the cells, he too became agitated.[32] Although nothing was apparently done in these cases, it is plain that the manifest abuse of power was not tolerated.

In most cases assaults on inmates were associated with gross insubordination, attacks on staff, or escaping. Such incidents were, and I imagine still are, more common where youths are concerned than adults. Older criminals often talk of the hidings they received at the hands of welfare and penal authorities when they were youngsters. Staff violence was frequent at Waikeria detention centre when I was there and is still visible at the corrective training centres which have replaced them.[33] According to one former borstal screw, borstal escapees in the fifties and sixties were always bashed up when they were caught.

> If they'd take off they'd take 'em in and give them a hiding. You know, a couple of big screws holding him while another punched him. You know all that. . . . I used to find that bullying of inmates was common. They hid behind their keys. They hid behind their authority.

'Who was getting done up?' I asked.

> *Always* the guys that escaped. The guys that were getting done up were the guys who might have embarrassed a screw in one way or another, to show them, 'next time you do it, this is what's coming'.

Others agreed that the beating of young escapees was fairly general, and it was probably fear of retaliation, as much as anything, that made violence towards Mt. Eden's adults less frequent.

THE INMATE SUBGROUP: AN ANALYSIS

Violence at Mt. Eden, although common, was thus less oppressive than it could have been. The level and type of violence in maximum security was largely a result of Haywood's administration. Haywood was a stocky, tough man, whose quarry background had accustomed him to latitude in the handling of inmates. In the light of his personal history, prisoner–staff relationships at Mt. Eden are quite easily understood.

The inmate social system itself is more complex and begs closer examination. What is particularly puzzling about it is the lack of reaction to some of the manifest abuses of power which are reported to have occurred at this time. For in the decade after Dallard's retirement, although escapes and other crises were not uncommon, there is not a single case of major collective rebellion at Mt. Eden.

Price's force-feeding of a prisoner is one clear instance of an action which might easily have sparked a response. On another occasion this same officer tried, unsuccessfully, to dupe a new inmate into helping assemble the scaffold which was to be used to hang a prisoner called Fiori.[34] Inmates objected, but took no action. Around the prison and on the parade ground a colleague of Price told me that Price used to address Māori prisoners as, 'you black pricks' and 'you fucking black bastards'. In the 1950s, 20 to 30 per cent of Mt. Eden's inmates were Māori, yet it seems that nothing was done about Price's behaviour — by the administration or the inmates.

The impunity of staff was mirrored in the character of inmate leaders. Just as certain staff habitually abused their authority, so did leaders exploit their positions, often to the detriment of others. Some of the gaol's most prominent figures were men whose commitment to the inmate body was low. Capable and shrewd, they managed to use personal influence to their own advantage, without visible regard for the rest.

With a hubris born of unrivalled power, Maketu, for example, adopted an air of haughty contempt over the tiny society at Mt. Eden. A meal queue was a nuisance — to march to the front of — a newspaper a luxury to be guarded jealously — before being torn up. Maketu's presence, generally, was predatory, alienative, and disruptive, and his status, founded on physical strength, lacked qualities of leadership or integrity. Yet his behaviour was for many years accepted by the bulk of his community.[35]

Although the time factor makes detailed analysis difficult, there are two main reasons why power relations at Mt. Eden were able to develop the way they did. Firstly, the political consciousness of the general public, and so the criminal community itself, was low in the 1950s compared with later years. The fifties were a time of national conservatism, when fears of Soviet expansion

— among other things — led to the emolument of traditional values. In this era of chauvinism the perceived alternative to orthodoxy was Communism or worse. Detractors of establishment precepts bore a heavy social stigma. Deviants — political, sexual, criminal, or otherwise — were badly stereotyped and a record of waywardness was best left concealed. Out of this mood of conformity grew a mentality, from which not even prisoners were exempt, which accepted the rule of authority. Within the gaols as well as outside them, defiance of State power was less frequent than it became in a later period of activism and change. Prison indiscipline followed the national pattern and in the fifties tended to be brief, individualistic, and lacking in principled objectives.

The basic respect which administration enjoyed was capitalized upon in the establishment of convict leaders. Enchanted by a recognition they had never before encountered, leaders at first found it easy to identify with their guardians. However, the concerns of these men — as witnessed by the inmates' council — soon transferred to the enhancement of their own interests. Arising out of this, then, the second great factor which fragmented the community was the gaol's organization. The emergence of a pro-administrational élite at Mt. Eden was obstructive of organized dissension because, not only was budding restiveness deprived of catalysing leadership, but existing (co-opted) bosses were able to monitor inmate activity from a point inaccessible to officialdom. By controlling resources, privileges, and the channels of communication, anomic disunity among the rank and file was maintained. In securing its own privileges, the élite served a conservative function. Although the system was stable for a time, it was not long before it began to break down.

NOTES

[1] Cressey (1961) 2.
[2] Goffman (1968).
[3] Garfinkel (1968).
[4] In Tannenbaum (1938) 324.
[5] Reprinted in Gee (1975) endpapers.
[6] Quoted in Dept. J. (1969a) 29.
[7] Morris & Morris (1963) 249-50; Klare (1960) 31-2.
[8] Kogon (1958) 259-60.
[9] MacCormick (1973) 136-7.
[10] Cressey (1955-1956).
[11] Abel (1951) 154; Bettelheim (1943) 447-8; Kogon (1958) ch.10.
[12] Cloward (1968) 94.
[13] The conservatism of inmate élites is covered well in Cressey & Krassowski (1957-1958); McCleery (1961b); (1975); and McCorkle & Korn (1970) 412.
[14] McCleery (1975).
[15] Michels (1962) 16.
[16] A 'shelf' is an informant.
[17] Anstiss tells me there were between six and eight trusties at Mt. Eden at this time.
[18] John A. Lee was a noted Labour party politician and author, who championed the cause of the underprivileged.

19 Dan Cavanagh confirmed that Haywood was blasé about homosexuality. 'Haywood couldn't care less', Cavanagh said. 'If they were that way inclined, he just let them get on with it.'
20 A 'fig' is a one- or two-ounce ration of tobacco. A two-ounce ration is also called a 'double'.
21 The existence of 'narks' rations' was confirmed by Tom Hooper, an officer at the prison at the time.
22 Restricted diet was abolished in 1981 by Penal Institutions Regulations 1961, Amendment no. 3, s. 25.
23 Visiting Justices are District Court Judges or Justices of the Peace who visit institutions to adjudicate over appeals against superintendents' judgements or over charges laid under s. 32(2) of the Penal Institutions Act.
24 Penal Institutions Act 1954 s. 8.
25 Hanging duty involved escorting condemned men to the scaffold. It also involved pinioning the prisoner's arms and legs, fitting the white hood and adjusting the noose around his neck. A separate duty was 'special duty' or 'death watch'. This entailed sitting with the prisoner in his cell. All condemned prisoners were attended twenty-four hours a day in four shifts of six hours. Officers who had done 'special duty' with a prisoner were never chosen as his gallows escort.
26 Several prisoners I spoke to remembered that Anstiss often fought with inmates — even though sometimes he lost.
27 At the time Hargrave held a third dan black belt. He currently ranks sixth dan and is vice-president of the International Federation of Judo.
28 Michael Burgess is a *nom de plume*.
29 Burgess (1964) 69–72. Although written as fiction, the book contains many recognizable characters and anecdotes.
30 This was Tuna Scanlan, New Zealand professional middleweight champion 1957–1964; British Empire middleweight champion 1964.
31 D. Mackenzie (pers. comm.).
32 Ibid.
33 Corrective training, abolished in 1963, recommenced in entirely new form in April 1981.
34 Fiori was the first man hanged (in 1952) after the restoration of capital punishment in 1950.
35 A similar style of domination is described among the 'barons' of English prisons. See, for example, Klare (1960) 33–4, and Morris & Morris (1963) 233–37, 244–47.

7: THE PRISON BAND 1952–1960

In December 1955, Haywood had seen the freedom given to sports groups at Mt. Eden backfire, when Horton escaped from the bowls team. This incident resulted in the cancellation of outside excursions by maximum security prisoners. The recreational programme within the prison survived, however, and various groups continued to visit the institution to meet with inmates. The most prominent of the Mt. Eden clubs was the prison band.

THE BAND AND ITS FUNCTIONS

The prison band, as mentioned above, was conceived of in February 1952 at the request of the prisoners' council, and formed the following month. Mrs Haywood, wife of the superintendent, had agreed to assist with concerts, and twelve men had initially been selected to form a concert party. From this point on the prison was treated to fairly regular recitals and plays by outside groups, and the band itself also performed regularly to visiting audiences. By April 1955, it was reported that nine concerts per year were permitted at Mt. Eden. Six of these were from outside teams; three were concerts put on by the inmates themselves. In addition, there were performances from local drama societies.

In the 1950s the inmates' band, known to some as 'Haywood's Brass Band', became increasingly important in the routine of the institution. Its significance grew even greater after the outside sports programme was withdrawn. The prison band was introduced as part of a planned programme of reform, and in this respect it performed a number of functions.

First of all, as a 'reformative' agent, the band gave prisoners an appreciation of music. It encouraged the dozen or so fortunate to be selected for it to become proficient in the playing of a musical instrument. Horton, for example, was one who benefited greatly from his association with the band and before long became an accomplished trombone player. The prisoners were proud of their group, revelling in the sense of achievement their newly found proficiency gave them.

Secondly, the prison band and the visiting concert parties improved the morale of prisoners, thus serving the gaol's custodial interests. Moreover, bandsmen themselves spent many hours learning to play their instruments and practising for their performances. As long as men were thus actively employed, they were less likely to cause difficulty elsewhere. More importantly, after the Horton affair, acceptance into the band became the greatest and most restricted privilege available. It was largely because of this that it became so easily dominated by long-termers and criminal leaders.

Thirdly, the band was a good public relations agent. Its concert performances were generally meritorious and, especially in its earlier years, were frequently reported in the press.¹ The media were unanimously supportive and the Mt. Eden prison band became the liberal penal policy's new showpiece. Such healthy publicity was of no mean value at a time when the credibility of liberalism was still suffering from the Horton incident. Gaols everywhere were straining under the pressure of population increases. The publicity generated by the band helped divert public attention from the accommodation crisis, which by this time was beginning to foil progress in so many other areas.

Apart from these services, the prison band had another role which was not directly related to penal policy at all. Yet its significance is crucial because without it, Haywood's job would have been more difficult and events would have turned out differently: the establishment of the prison band provided an interest for Haywood's wife, and helped to stabilize their marriage.

THE SOLITUDE OF THE SUPERINTENDENT

It is rare for those who have not worked in a prison to understand fully the handicaps which such a job imposes. In a small country like New Zealand,² it is difficult to go anywhere without meeting past acquaintances. This is especially true of the cities. In 1952, 90 per cent of all prisoners came from urban areas. One third of them were from the Auckland province. Auckland's prison, which in the later fifties held between 400 and 450 inmates — that is, one quarter of the total prison population — was by far the largest penal institution in the dominion.³ Auckland at the time was relatively small. Throughout that decade, the population of the city and its boroughs remained below 500,000 and the country's high rate of incarceration⁴ meant that, in public places, contact with ex-prisoners was likely. Prison officials often found such encounters embarrassing. If an ex-inmate was aggressive or drunk, the incident could be painful as well.

In consequence of this, prison employees — then as today — normally mixed in a close social circle. They were seldom seen in public. As today, superintendents were in a worse position. Because of their authority and duty, they could not fit easily into the world of ordinary staff, so their social milieu was even more restricted. Consider the case of Haywood. For almost the whole of his career, he had worked at Paparua prison in the south. After twenty years' service he had transferred to the North Island and become superintendent of the country's only maximum security gaol. When Haywood came to Auckland in 1950, therefore, it is safe to assume that he would have had few, if any, close acquaintances in that big city. His responsibility was greater than that of any other superintendent and he could hardly have associated freely with his subordinates. There were no other penal institutions nearby, the closest being Waikeria, some 150 kilometres distant. Furthermore, it was Haywood's onerous task at that time, to organize the custody and execution of the condemned. His was the country's only hanging gaol.

Reviewing these circumstances, the invidious position of Haywood can partially be appreciated. Whenever he ventured off Justice property, Haywood exposed himself to the possibility of encounters with men and women he had locked up, and the relatives of those he had hanged. In his task, the executioner had the blessing of anonymity. The role of Mt. Eden's superintendent, on the other hand, was publicly obvious.

As far as Haywood was concerned these very great social disadvantages must have been offset by the distinction, the authority, and the kudos of his office. Haywood, at least, had a successful career. His wife did not. Because of the Haywoods' position they did not often go out at night; while during the day, Mrs Haywood was left on her own. She is rumoured to have had contact with women's associations, but she was childless, and therefore was denied the preoccupations of parenthood. Under these conditions, it is easy to speculate that the marriage would have come under strain.

Haywood loved and needed his wife — she was the closest companion he had — and it is comprehensible that, when the opportunity arose to provide her with an interest inside the prison, Haywood happily accepted it.

THE SUPERINTENDENT'S WIFE

Haywood saw his wife as being well suited to the task of organizing the prison band, and it was a task which she pursued with enthusiasm. The Mt. Eden concert party became one of the most important aspects of her life, and its members, the family she could not have. For Ettie Haywood, the band became more than just a means of relieving a monotonous lifestyle. It filled an important emotional need in her. She developed a strong maternal affection for some of the men. She spoiled them, she put her arms around them, she brought them little gifts.

The prisoners reciprocated and easily slipped into the role she had created. In some cases, Mrs Haywood became the indulgent mother they never knew; a sympathetic voice in a world of men calloused by deprivation and hardship.

So the emotional rewards the men took through membership in the band were considerable. The power they gained over Haywood, through their friendship with his wife, was significant also. As far as the band was concerned, Mrs Haywood wielded extraordinary power; a fact which was widely known but seldom spoken about. She was a strong and demanding woman, whose will often conflicted with her husband's. This was noticed by Anstiss:

> She would want to do something and he would say, 'Oh, no, that's bloody silly.' And she'd get on stage and she'd bawl him out. Here's the guy in charge of the toughest gaol in New Zealand, and he's saying, 'Oh [grunt], whatever you say.' And he'd just sit there and he'd glower. And everybody'd be quiet, you know.

It was Haywood's acquiescence towards his wife, more than anything, that allowed the freedom of the band to extend as far as it did. Although she was a civilian with no official say in institutional matters, junior staff felt compelled to concede to her wishes. Ettie Haywood was personally involved in the band's selection and through her influence a dozen or so men acquired a peculiar status in the prison. Mrs Haywood, staff and prisoners told me, had almost complete control over the concert party, and officers were reluctant to interfere with its activities. As far as the band was concerned, her word was law. Hargrave commented:

> Haywood had gone on with [the band] against a lot of flak because it was said at the time that if you were in the concert party, you couldn't be touched. And I think there was a grain of truth in it, you know. . . .
> [Mrs Haywood] was very much the producer, the director, of the whole thing. 'I'm running this show', and it was very true that the inmates — La Mattina and a couple of others — played up to her. They could get a status which was supposed to be untouchable if they did anything wrong.

Dan Cavanagh, an experienced officer with a down-to-earth manner and solid practical principles, was incredulous at the situation he saw developing.

> I couldn't believe it. Looking back, at the time, it sort of developed gradually. Everything stopped for this bloody band . . . the place just became a shambles, with the whole place centered round this bloody band of his. And then she got into the act and you know, 'he's such a nice boy', and they'd only have to ask for something. You know, some fellow would say, 'Oh, the laundry doesn't wash my singlet and underpants properly.' And before long would come a little parcel: new underwear. All this sort of shit, you know.
> I'm no psycho, but I understand they were a childless couple, and I think she used to get involved with prisoners there in a sort of maternal sense. And the fellahs would just — take advantage of it. Bloody awful, eh. Embarrassing. And she couldn't see it eh. *He* could see it, and I believe there was some fairly big arguments between them over it — but she always got her own way.

One officer who had been a professional musician assisted Mrs Haywood with the concerts and his opinion was the same: 'The concert party was almost a sacred thing,' he said. 'It really was the number one feature.'

The existence of the band and the degree of influence Mrs Haywood and her protégés gained through their association drew considerable resentment from staff. One eloquent and poetic officer satirized the new system in verse. Written to the tempo of *The Song of Hiawatha, The Song of Hiahori* tells in fourteen humorous stanzas the story of a young Māori offender and how he becomes 'reformed' by involvement in the prison concert group.

Most prison personnel I contacted felt that a superintendent's wife should never be involved in prison affairs, and wondered why it had been allowed at all. Mrs Haywood was not a bona fide official and really had no rights there at all. Haywood himself was not an artistic type and many saw his (often reluctant) acceptance of what went on as merely catering to her whims. The prison psychologist told me that he had been thoroughly against Mrs Haywood's involvement because of the controversy it was producing among the staff. Unlike uniformed personnel, he was able to express his disapproval, and this, he said, made him unpopular with Haywood. Anstiss remembered too that officers knew they could ill-afford to criticize openly what Mrs

Haywood was doing. So did the inmates. Haywood was resolutely defensive of his wife's activities.

> That was the ultimate. It you *really* wanted to fall foul of Haywood [the best way] was to say something derogatory about his wife. And that was the end of you. You were *down*. He wouldn't stomach that at all. He wouldn't have anything derogatory said about her.

Officers I spoke to described Mrs Haywood as a pleasant, kind-hearted person, but lacking in practical knowledge or common sense. The matron of the women's division likened her to a fantail, and said she felt that Mrs Haywood was erratic and often lived in a fantasy world. Others doubted her motives. Hargrave told me:

> I got the impression that she was *socially* into that scene. She'd married a senior prison officer . . . and it was in those days — it had started to be the 'in' thing to get interested in inmates and institutions. And she was very active in that sort of a group. She used to bring her knitting circle or whatever they were and I think, like most of the officers there, I didn't really go along with it. . . .

'Did you think she was being a bit 'trendy', or something like that?' I asked.

> Yes, I did, yes. . . . I thought it was a 'trendy', 'in' thing at the time.

Underlying discontent was of course invisible to a public, enchanted by the novelty of entertainment by 'dangerous criminals'. As time progressed, the antics of the band became an increasing source of vexation to many of those working inside the walls. As her excitement about the performances grew, Mrs Haywood began taking part in them. She dressed up, and even danced, 'with no great merit at all', MacKenzie recalls. 'It became impossible, and really divided the prison to a great degree.'

There was discontent among the inmates too. Jealousy festered among the many who were excluded. While the bandsmen played, to the applause of public and press, the rest of the prison sat listening in silence, locked inside their draughty, stone-walled cells. It seemed hardly justifiable to them that a small, hand-picked élite should have so much, while they the majority, got so little.

From early in the band's history, security began to suffer. Lax administration allowed concerts to become the cover for other activities. Prisoners cut a doorway at the base of the stairs which led to the stage. As visitors filed unsupervised into the hall, the girlfriends of some of the men diverted from their proper course and disappeared through the doorway. They rejoined the main visiting party again as it filed out.

The band and the concerts also became a major avenue for liquor. Drink could be smuggled in by visitors, or brewed by prisoners in the stage area, which was seldom searched by staff. One band member told me that use of alcohol before and during band functions was common. This sometimes had embarrassing consequences for administration. One particularly bad instance was given by Cavanagh:

> At the opening of the bloody concert they all assembled in front of the audience and they struck up a tune.
> 'God Almightly!' Dinny said to me, 'Look at those men, they're all pissed!'
> And they were all discords, they all looked a bit silly, eh . . . What happened was, prior to the concert starting, they'd prevailed on whoever was running it to let them go up and practice in the hall. Well the stuff was already there in tea-cans or some bloody things. And they all got bloody *loaded*. All home brew, it was. It was an absolute fiasco. Some bloke even fell out of the roof! He was the prop-man and he fell down right in the middle of a recital. Pissed as a chook.

Worrying though such incidents must have been, they were not reported in the papers and were not, in themselves, a great threat to security. The very loose management of the prison itself, however, did contain serious weaknesses, and nowhere was this looseness and this weakness more pronounced than in the special cell reserved for members of the band.

THE MAKETU AFFAIR

As discussed in chapter seven in the second half of the 1950s the penal system of New Zealand began facing a problem of chronic overcrowding. By mid-1958, Mt. Eden prison, with facilities for 275, was accomodating between 400 and 450 inmates. Due to a shortage of space, prisoners were sleeping in common areas and crowded into cells. Up to three men inhabited single cells and there were up to a dozen in association quarters. While head office was aware of the hazards that such living presented, there was no other means of dealing with the crisis.

Because they were not as strictly bound by hours of lock-up, because of the faith and trust vested in them, and because of the special privileges and interests which bandsmen had in common, Haywood permitted them to live communally. They were allocated to association cell 'A', with their instruments and personal effects. Situated in the basement of the east wing, the unit was known as the 'band cell'.

In the previous chapter I spoke of cells at Mt. Eden which officers were reluctant to search, for fear of upsetting Haywood. One such cell, which officers never searched, was the band cell. No specific direction was ever given that the band cell was not to be interfered with, but prison staff knew that to search it would be to invite reproach. If anything was disturbed, or the band's equipment damaged, inmates would complain to Mrs Haywood. She in turn would complain to her husband, who would chide the officer concerned.

Because of Mrs Haywood's influence, therefore, personnel felt it best to leave the band cell alone. Hargrave expressed his reluctance to search the bandsmen's cell in these terms:

There were things there, and unless they were *exactly* in the right place the inmate would be upset. . . .
They had cells with a lot of personal possessions which generally speaking weren't turned over thoroughly. Yet there was nothing in the rules which said that they were to be treated any different from anyone else.

Officers in general agreed. It was particularly true of association 'A'. Cavanagh comments:

We weren't allowed in the cell. Officers were not allowed in that bloody cell. Because it was the band cell. They weren't allowed in otherwise the inmates would complain. You know their *gear* had been *damaged*. 'Oh, shit!'
But there wasn't anything in writing. He was shrewd enough not to put it in writing, but, 'no need to go in *there*, they're all right'.

One wonders how, with the experience of the Horton affair behind him, Haywood could have allowed this to happen. Escapes had always drawn publicity and were of prominent reader-interest in the newspapers.[5] Prison breaks between 1950 and 1954 had in fact dropped and were 23 per cent lower than they had been between 1945 and 1947. In 1954-1955, however, escapes had begun attracting sizeable comment and, in spite of the fact that *prison* breaks were low,[6] overall abscondings were high. Borstal figures boosted the total several fold,[7] so that by 1958 the number of escapes was higher than it had been for years.[8] Attributed largely to overcrowding,[9] the high escape profile was an embarrassment to the Department of Justice. This feeling was added to in February 1958 when two prisoners broke out of a Mt. Eden association cell, by sawing through their bars and climbing over the perimeter wall.[10]

The cell the bandsmen were in contained twelve, all of them long-termers. A number were serving indeterminate sentences of preventive detention or life, and they were all security risks. In the light of these facts, and of publicity over the February escapes, Haywood's blasé attitude is perplexing. Perhaps Haywood had enough confidence in his informants to feel sure that any devious activity would be reported. This is not really acceptable. Haywood must certainly have known by then that his intelligence network was fallible.

Perhaps his trustfulness was due to a confidence that his men would do nothing to jeopardize the rare status they enjoyed. This too, is hardly satisfactory. Two years before, Horton had thought little about the consequences of his escape, and now he was a member of the band. The men who got drunk before concerts showed no great reluctance to gamble with their privileges either. Haywood must have seen this. Perhaps the power of Haywood's wife was even greater than we have indicated. Conceivably, it could have been that her influence over him was so effective that the risk of escape seemed preferable to the certainty of recrimination from her, if he ordered the cell to be searched.

The answer to the question we can never know. What is clear in hindsight, though, is that whatever his reasons, Haywood made a serious blunder, probably the greatest of his career, in not ordering that the bandsmen's cell be kept under strict surveillance. For on 14 June 1958, the institution once more was shaken by headline news: prisoners at Mt. Eden had been discovered to have been escaping from the gaol at night, committing crimes in the city, and returning to their cells before daybreak.[11]

The leader of the escapees was a man serving a sentence of preventive detention. A persistent criminal and a known absconder, he was one of the prison's most dominant personalities. He was also one of Haywood's principal confidants: the man we have referred to as Maketu. Maketu was not one of the original bandsmen, in fact it was because of his troublesome reputation that he had been deliberately excluded from it. Although he had always desired to be a member, Mrs Haywood had expressly forbidden it. According to Maketu's version, he threatened to create a scandal about Mrs Haywood's involvement in the concert party and have the scheme closed down, unless he was allowed to take part in it. He joined the band soon afterward.

Once established it did not take Maketu long to become a recognized leader in the party. He was moved out of the separates division where he had been housed and into association cell 'A'. It was here that he conspired with others to formulate an audacious plan to add even greater comforts to his lifestyle. The plan was to break out of the cell under cover of darkness, scale the walls, spend the night in town, and return to the prison before morning.

The official record of the escapes, as presented to the Minister after their discovery, holds that Maketu only absconded on a single occasion. Later press reports confirm this, and state that the bar which had been removed to facilitate the escape had only been cut for three to five days before it was found. The bars, Haywood said, had been 'tested' as recently as June 6, and found to be secure. They were also 'shaken' every day, but the shaking had not been enough to dislodge the one that had been cut.[12] The difference between the 'testing' and 'shaking' of cell bars was not explained.

In fact, Maketu escaped from the cell several times. The loose bar had been cut for a number of weeks before it was discovered, and at least four of the dozen men in the cell went out on various occasions. Maketu claims that one of the cell's bars had been cut through in a night using a hacksaw blade, which Archie Banks told me he had supplied.[13] The bar was sawn at an angle so it could be replaced more easily, and the cut was hidden with a mixture of plasticine and quarry dust, forced into the fissures in the steel. This mixture was exactly the colour of the bar and effectively disguised the cut during daylight hours.

After the bar had been freed on the first evening, Maketu and a lifer called B.M. ventured out of the unit and clambered up a drain-pipe, on to the prison roof. Having surveyed the environs, they then returned to their cell. The exercise was repeated a few nights later. It was after this point that plans for the later escapes were made in the cell, beneath a blanket with a funnel protruding out of the top for ventilation, and a candle providing light.

It was not until three or four weeks after the bar had been cut that B.M. and Maketu first mustered the courage to try scaling the wall. Some grappling hooks had been made in the plumbers' shop, along with some lengths of plaited rope. These were hidden under a loose floorboard in the bandsmen's cell. The first night the men ventured over the wall these grappling hooks were used, but the glass embedded in the top of the wall cut the cord through, causing B.M. to fall heavily and injure his arm. Another rope was then fetched and used to assail the wall a second time. This attempt succeeded, and the men descended the opposite side via steps which led to the sentry tower, before running down the road dressed in basketball uniforms. The two returned the same way as they had exited and were safely back in their beds before daybreak.

Following the initial mishap it was decided that grappling hooks were too dangerous, so for subsequent excursions an alternative scheme was hatched. Four lengths of three-quarter inch galvanized pipe, used for weight-lifting, now had bushes fitted into their open ends. The pipes could then be pieced together, with pins holding them in place. This gave the rods an overall length of five metres. With the top pipe bent over to form a hook, the ladder could be rested over the brim of the wall for the men to climb.

As they were made, these pipes were transported from the plumbing shop hidden under blankets on a clothes trolley, and in the meantime movements of the night-watch were being studied. It was discovered that some staff patrolled erratically, others regularly; but once the habits of the night shift were known, it was easy to work out a safe time to move. While exiting from the gaol was now relatively simple, getting back in was still a problem, because the escapers could not see into the yard from outside. A different method had therefore to be devised, and this was facilitated by a torch which the prisoners carried to signal when they were ready to come in. Seeing the torch flashing from a high point outside the gaol, inmates inside were able to indicate when the coast was clear by using a lighted candle. The escapees could then ascend the wall and traverse the yard without fear of detection.

This plan worked well and, according to Maketu, was used four more times before the scheme was eventually discovered. Maketu told me that a total of five trips were made out of the institution, in addition to the two reconnaissance forays along the roof. MacKenzie and Banks both believed that only three escapes had been made. Questioned about the discrepancy, Maketu insisted that there had been five complete escapes, and that a total of four prisoners had gone out: himself, B.M., I.K., and S.R. It was decided by Maketu to guard against discovery by strictly controlling who went out. Edward Horton, although a member of the band, was not in association 'A'. Banks says he could have arranged for Horton to be moved into the band cell, had he wanted to take part in the escapes. Haywood, for one, may therefore have been lucky that Horton chose not to participate, and remained in his own cell for the whole of this period.

Although the escape scheme went undetected for several weeks, the superintendent's first hint of something amiss came not long after the men began absconding. He received a phone call from someone outside the prison, informing him that Maketu had been seen the previous evening, eating in a city restaurant. Apparently Haywood dismissed the report as a case of mistaken identity. Another clue came when the grate of the prison boiler became clogged with melted glass. This was from empty liquor bottles, which the escapers had brought back full from the city. Having consumed their contents, they had then tried to destroy them by throwing them into the furnace. It was not until 11 June, four days before news of it appeared in *The New Zealand Herald*, that specific information about the escapes reached Haywood. Haywood was told that prisoners were leaving association 'A' and committing crimes overnight.

MacKenzie reports,[14] and Maketu believes, that the disclosure came from an inmate. Cavanagh thinks it came from Archie Banks' wife. Mrs Banks was friendly with Mrs Haywood, and Banks himself was bandmaster and conductor of the concert party. It was Banks who had secured the blade to cut the bar, and he told me he had some involvement in the planning and covering-up of the escapes. Mrs Banks knew about them because Archie Banks used to give Maketu letters to deliver to her. He would have a reply back from her by 4.00 the next morning, Banks told me.

On Wednesday 11 June, after receiving his tip-off, an incredulous Haywood called Cavanagh and Byrne into his office. He ordered them to search association 'A'. A few minutes later, in stunned silence, he received news of their discovery. They had tested the bars, one of which had fallen out when pushed. They then searched the cell and found the pipes, stored in pieces. Looking further beneath the floor-boards, they found contraband: civilian clothing, a milkshake machine, a radio, and hundreds of cigarettes and bars of chocolate, hidden inside a mattress. These had been stolen during burglaries, committed while the men were at large.

Haywood summoned Banks, and asked him what he knew about it. He told Banks that he, Haywood, would probably lose his job. Banks was transferred almost immediately to Mt. Crawford. Maketu was also interviewed. According to MacKenzie, when Maketu was first spoken to he admitted leaving the institution with one other prisoner, to visit his own wife and the other prisoner's girlfriend. However, he told Haywood he would take full, and personal responsibility for it.[15]

In Wellington, Barnett was notified, but by the time he arrived at the prison, crucial evidence was already being destroyed. Hooper told me that most, if not all, of the contraband was taken outside and burnt. Cavanagh says he saw an officer called Randolf, in the prison stoneyard, burning civilian clothes found during the search.[16] No mention of civilian attire or contraband was made in the official report to the Minister, although reference to the radio, and to unspecified amounts of chocolate, cigarettes, and tobacco appeared in the press.[17] The official report to the Minister said that (trusties') blue denim trousers had been of assistance to the escapers.

When Barnett arrived at Mt. Eden, Haywood knew his own job was in jeopardy. So, Barnett knew, was his. The Horton affair had left stains on the dossiers of both men. The present incident was much worse. A month before, in the early hours of 9 May, three men fitting the descriptions of certain individuals in the prison band had raped a woman in Cornwall Park, after brutally beating her boyfriend. The assailants, who were carrying torches and masquerading as policemen, had then escaped without trace.

At the time, the Cornwall Park event had stirred great publicity and comment in newspaper correspondence. As soon as the escapes were discovered, speculation about the identity of the rapists began afresh. The possibility that the assaults may have been the work of maximum security prisoners was immediately raised, and unconditionally rejected by Barnett.[18] Barnett, as Controller-General of Police as well as Secretary for Justice, had considerable say in both Departments. It was essential to his career that these serious allegations about the prisoners be discounted. Timing of the escapes and the numbers and identity of the escapers thus became critical factors.

When Barnett arrived in Auckland he went immediately to Mt. Eden prison. On arrival at the office of the superintendent he called for Maketu to be brought from his cell. Barnett and Haywood then questioned Maketu about the escapes. There was nobody else in the room. I spoke to Maketu at length about this matter. At first he told me nothing about the meeting with Haywood and Barnett. Later, however, he requested I turn my tape off, but allowed me to make notes while he related in detail what had taken place.

Barnett spoke, Maketu said, and he told Maketu that he could not let anybody out of prison and that whatever happened he, Maketu, would have to do his time. He told him that both his own, and Haywood's jobs were 'on the line' and he asked the prisoner who he would rather have as superintendent — Haywood or somebody else. He then suggested to Maketu that he had gone out alone. Maketu agreed that this was the case. After he had spoken to Haywood, Maketu said he was seen by some high-ranking policemen (whom he named). The policemen told him he had to tell them the truth, the whole truth, especially in respect of the rape. Maketu insisted that the rape had taken place before the men made their first trip out, and he told them his wife would confirm this fact. The police then threatened to arrest Maketu's wife for conspiring in the escapes.

The police talked about the bad publicity which would be caused if it became known that more than one man had escaped. They said that they did not think the justice system would be able to stand such a disclosure. The interrogators then asked Maketu how many men had gone out of the prison. Maketu told them only he had. They asked him again, and again he replied that only he had escaped. They asked him where he had gone. He said he had gone home. Under further questioning Maketu admitted that he had also gone to Carlaw Park, where he had burgled the canteen.

At this point the police took Maketu to see his wife, to confirm the date of his first escape. They told Maketu that they were going to charge him but he was assured he would only receive a short sentence, running concurrently with the preventive detention he was already serving. The sentence would be given to appease the public, they said.

When he was returning from court after sentencing, the police told Maketu that they had known all along that S.R., I.K., and B.M had gone out, and that Maketu had been lying to them. Evidently the other three had broken under questioning and confessed details of where they had gone and what they had done. However, the police said, they did not want these facts revealed because of the bad publicity it would create. 'They knew every move I made and who I went out with,' said Maketu.[19]

Results of the internal inquiry were submitted to the Minister six days later. This report, as related earlier, made no mention of any contraband

and spoke of a single escape by only one inmate. Escape apparatus had not been discovered during daily cell searches, the report claimed, because an inmate from the cell had reported sick each day for the previous two weeks. The pipes had apparently been concealed under his mattress.

When he appeared in court a few days after the discovery, Maketu pleaded guilty to escaping and to two charges of breaking and entering. Mr McCarthy SM said he would sentence Maketu to a concurrent term of two years' imprisonment, 'for what it's worth'.[20] The Magistrate knew his action could be nothing more than nominal, since the Act under which Maketu had been originally sentenced prohibited a court from imposing a cumulative penalty on one of preventive detention.[21]

Maketu had received his term of three to fourteen years on 23 November 1956. The extra two years was passed on 14 June 1958. He was released from prison on 1 June 1961, after serving four and a half years of his original sentence and only three years after the second had been given. Maketu's early release, in spite of his conduct record, raised speculation that he had been rewarded for his co-operation in helping to soften what could otherwise have been a ruinous scandal.

Repercussions, however, could not be avoided, since it was widely known within the service that a cover-up had occurred and that both Haywood and Barnett had been involved. From the time Barnett had taken over the Police Force in 1955,[22] there had been controversy over the appointment. The police had had a tradition of promotion from within the service, and the choice of a civilian leader was unpopular. From the outset, the position had been 'temporary' although it had, by 1958, lasted some three years. By June that year, Barnett's police administration was already under question, and his relationship with Mr P. G. Connolly, the new Minister of Police, was strained. Embarrassment over the Maketu incident and dissatisfaction with Barnett's handling of it, now added weight to the antipathy upon him.

On 18 June, barely a week after the escapes had been discovered, it was announced by the Minister of Police that Mr S. T. Barnett was being replaced as Controller-General of Police by Superintendent William Spencer Brown. The replacement would be effective immediately. The reasons for the new appointment were given that the previous station had only been provisional and that the Labour Government had always favoured in-service promotion in the police force anyhow.[23] When asked about it, Barnett refused to confirm that he had been forced to resign; however, his removal was rumoured to have followed a refusal by the Minister to respond to an ultimatum from Barnett regarding the conditions of his employment.[24] Whatever the specifics of the event were, it is plain that there was a direct relationship between the scandal at Mt. Eden, and Barnett's sudden departure from the police force. Had the Justice leadership been political, or had the full facts of the matter been known, Barnett would likely have been sacked from there as well. However, the playing-down of the incident was effective enough to allow both him and Haywood to retain their Justice positions.

As soon as the report of 17 June was ready, Mason flew to Auckland to inspect the institution personally. A number of new security measures were immediately installed. A recruitment campaign was announced, and the night guard strengthened by four, to reach a total of fourteen. A full-time 'bar-tester' was appointed. It was recommended that bars themselves be reinforced with hardened steel rods. Floodlighting was to be improved and urgency was to be given to preparing a new plan for Paparua prison, near Christchurch. Part of this institution was to be converted to a maximum security unit to ease the pressure on Mt. Eden. Two new medium security blocks for seventy-five men each were to be constructed at the National Prison Centre at Waikeria, and on September 1 a plan was announced to build an auxiliary gaol at Mt. Eden before the summer.[25] Steps would be taken to strip Mt. Eden of surplus prisoners, and overcrowding at the institution would cease. In late September the women's division was vacated and men were moved in their place. By 17 June, all faults were reported to have been remedied, but, none the less, a 'security expert' was assigned to make an independent study of the prison and its procedures.

The 1958 report, discussed by the penal group on 28 July, stressed the need for even more efficient use of floodlighting, and suggested that a police dog might be of use in assisting night watchmen in their work. It was commented by Barnett that overall morale had improved since the elimination of 'privileged groups'. Doubtless this referred to the dispersal of the bandsmen, although to what extent the élitist system as a whole had changed is unclear. The investigation had obviously awakened Barnett to the fact that much was amiss at Mt. Eden, so one would imagine that adjustments were a priority. Inmates and staff report that following the Maketu inquiry a much firmer grip was taken on aspects of management, particularly in so far as security was concerned.

The recreation programme went on in spite of all this, and bowling, education, and sport with outside teams continued. Even the band kept functioning. Still under the tutelage of Mrs Haywood, the programme of entertaining outside visitors was maintained as well. In September 1959, for example, the prison band gave a successful series of performances at the gaol, which ran over a period of five nights. The popularity of the band, and the quality of its performances had not been visibly affected by the escapes. If anything they had improved, and the band's activities earned renewed interest. When a series of recitals was given in December, up to 450 people attended each session, and many disappointed patrons had to be turned away from a packed house.

Conclusion

So while, after a number of months, routine at Mt. Eden had calmed, it was doubtless too, that as far as head office was concerned, Haywood's credibility had suffered. His personal confidence also must have fallen after having, once again, seen his professional judgement fail.

When the situation as a whole is considered it is difficult to attribute the bandsmen scandal to any single factor. Overcrowding, staffing shortages, ministerial indifference, and Haywood's own injudiciousness, all had a part to play. It must be remembered, none the less, that Haywood was not an experienced administrator when he took over Mt. Eden, and that he had no knowledge of, or training in, the running of rehabilitative programmes. Even after the new approach had begun, prison officers were scarcely able to operate it. They were not informed about what the official policy was. All they did was what they were told.[26]

When he began to apply the new penal policy at Mt. Eden to an impossibly expanding population, Haywood was able only to follow the vague directions of head office. There was no model to guide him and he was forced, very much, to rely on his own intuition. Spurred by favourable reaction in the early years, and then steered by the needs of overcrowding, Haywood's response was a combination of experimental humanitarianism and practical resourcefulness. Lacking sufficient personnel, yet eager to implement the esoteric philosophies drafted in Wellington, Haywood made use of the limited facilities he had. The result was a system which functioned well for a time, but which became increasingly difficult to handle.

So, the Maketu escapes, like that of Horton, were a direct result of administrative folly. Once more, the Minister was uninformed about what was happening in his Department. Once more, Barnett had given Haywood a free rein in the running of his institution. As before, the calamity which befell the organization was caused by a style of management which gave inordinate discretion to a convict élite.

Looking back from a distance of over two and a half decades, that Barnett and Haywood saw fit to conceal the more odious aspects of the Maketu affair is both understandable and fortunate: understandable because, had the full facts of the escapes or of the cover-up been made public at the time, the careers of both men would have been destroyed; fortunate because had this occurred, the penal system of New Zealand, which had advanced so far, so quickly, would doubtless have succumbed and miscarried. No positive purpose would have been served.

As it happened, the regime faltered but ran on, and although it was not to encounter another ditch as deep as that dug by Maketu, the road into the sixties was pitted. For the repeated difficulties which confronted Haywood were by this time beginning to show. It is apparent by the end of the fifties that the superintendent's health was failing, and his ability to cope was becoming strained. Events and circumstances at Mt. Eden in the final five years of Haywood's reign were indicative of an empire in decline.

NOTES

[1] The prison band became a particularly newsworthy topic in the years of 1954-1955. In *The New Zealand Herald*, for example, articles about the band appeared on 6 July 1954, 23 July 1954, 9 April 1955, 22 June 1955, and 20 December 1955.

[2] Throughout the fifties the population of New Zealand was about two million. It now has about three and a half million inhabitants.

3 In 1958, for example, the DAP of Mt. Eden prison was 415. Invercargill borstal was next, with 217, followed by Waikeria, 208.
4 In the 1950s the ratio of prisoners per 100,000 mean population rose from 54 to 73. New Zealand soon achieved one of the highest prisoner ratios in the Western world, behind the United States, with a ratio in 1988 of 100 per 100,000.
5 This was particularly true of Wellington's *The Evening Post*.
6 The Annual Report for 1957 indicated prison escapes were at their lowest in twenty years (*App.J.* H-20 (1957) 11).
7 In 1956-1957, for example, 80 per cent of all escapes were from borstals. Borstal escapes were four times higher than prison figures in 1957. (*App.J.* H-20 (1957) 11-12).
8 There were sixty-three escapes in this year compared with the normal annual mean of about fifty.
9 *App.J.* H-20 (1959) 12, 14.
10 The reference is to the escapes of J. G. A. Shanks and F. H. Johnson, two recognized security risks, who absconded from their cell on the night of 11 February, 1958. The incident created great publicity and a major police alert in Auckland (*NZH* 12 February 1958). After recapture Johnson received a four and a half year extension to his sentence, while Shanks was given a term of preventive detention.
11 *NZH* 14 June 1958.
12 *NZH* 16 June 1958; *EP* 16 June 1958.
13 The official report describes the bar as having been cut in a lunch hour. After the escape, an inmate demonstrated how mild the security steel was by cutting through a bar with a serrated table knife in about twenty minutes (MacKenzie 1980, 59).
14 MacKenzie (1980) 59.
15 Ibid.
16 This was widely known also, among the inmates. See for example MacKenzie (1980) 36.
17 *NZH* 16 June 1958.
18 *NZH* 14 June 1958; *EP* 16 June 1958.
19 The true identity of the rapists was never established and the officer in charge of records at Auckland Central Police Station told me that the file on the incident would by now have been destroyed, along with those of other unsolved crimes. Although Maketu denied to me that any of his men had complicity in the rape, by his own admission, the cell bar was already cut on the night the assault took place. It was also widely believed at Mt. Eden that the band cell escapers had been responsible for it. The names of two of those identified to me by Maketu as having been his accomplices were given to me independently by former prisoners as having been involved in the incident. Maketu's name was not mentioned. It must be remembered that a total of four men escaped from the cell on various occasions, but only three were reported to have been at the scene of the rape.
20 *NZH* 16 June 1958.
21 Criminal Justice Act 1954 s. 30.
22 Barnett had taken over the police following a scandal over the previous head. See *Report of the Commission of Inquiry into the Police Force* (1954-1955).
23 *NZH* 19 June 1958.
24 *NZH* 23 June 1958.
25 The auxiliary prison was not built until 1966.
26 My informant, Lawrie Hargrave, was particularly emphatic about this point, and he made it several times.

8: THE DECLINE OF HAYWOOD
1958–1964

The Effects of the Hangings

Of several factors which, in the end, led to the fall of Horace Haywood, one of the strongest, and certainly the catalyst of many of his later troubles, was capital punishment. The law reinstating hanging for murder came into effect on 1 December 1950. In February 1951, only eleven weeks later, Gordon McSherry was sentenced to death. His speedy reprieve met a public outcry and it was evident that future vacillation would not be tolerated. There had been a record eight convictions for murder in 1950. If this trend continued the death sentence would soon be confirmed.

The carrying out of the death sentence and the case for and against it have been argued elsewhere, and it is not my intention to reproduce them here.[1] Between 1951 and 1957 the sentence of death was passed twenty-two times for murder in New Zealand, and was effected on eight occasions. All hangings took place at Mt. Eden. Of interest is that the existence of the gallows had a certain, albeit minor, effect on Mt. Eden's routine, and a number of larger consequences for prisoners and staff.

There was wide variation in the responses of those associated with the hangings. Some were clearly tormented by events, others accepted them as inevitable and normal. Haywood had only been superintendent for five months when the first execution took place and it is ironic that it was during a period of penal enlightenment in New Zealand that this Draconian measure was revived. We recall that it was in the belief that he could competently address himself to both requirements which had favoured Haywood as superintendent in the first place.

In the 1950s, as previously, the decision of whether or not a death sentence would be carried out was normally made by Cabinet, sitting in its capacity as the Executive Council. The case for consideration was prepared by the Minister of Justice,[2] and legal issues were explained to sitting members by the Attorney-General. When the Executive Council had made its decision, its duty was then to advise the Governor-General and to recommend that the penalty either stand or be commuted to life imprisonment.

After capital punishment was restored the first person hanged was William Fiori, convicted in February 1952 of slaying his employer and his employer's wife, for money. Fiori, described as a 'borderline feeble-minded person',[3] had lodged no appeal against conviction; however, a recommendation for clemency had been made by the trial jury. Having received the prisoner after his sentencing, therefore, the authorities anxiously awaited the Executive Council's verdict. From the time of the verdict, Haywood had had close contact with Fiori. The trial had been conducted in Hamilton and Haywood himself was a member of the escort party which had transferred Fiori to Auckland.

The prisoner had been condemned on 14 February and seventeen days after, on Monday 3 March, the Minister of Justice, Clif Webb, announced that the case would not be considered that week, as important documents had yet to be received. Two days later, it was revealed that Cabinet had decided there were no grounds for commutation.[4] The law required that execution take place within seven days of confirmation by the Executive Council. Prisoners, of course, knew that Fiori was due to hang, and trouble had already threatened over erection of the scaffold. The gallows mechanism and the rope, with a sandbag attached to it, had been tested and the prisoner was weighed every twenty-four hours from the time the sentence was confirmed, so that nobody would know on which day the execution would take place.[5] Every morning a resounding crash echoed through the prison as the steep trapdoor was opened. Moreover, the condemned man was held in the west wing, whereas the scaffold was situated at the opposite end of the gaol, in the basement of the east wing. This meant a long walk down hollow, stone corridors for the hanging party. The imminence of a hanging and the instant of the drop would thus be advertised clearly.

The authorities were unsure of how prisoners would react to an execution, and the actual date, when it was fixed, was kept a close secret. Although no hanging had taken place in New Zealand for some eighteen years, old lags remembered that they happened in the mornings. They were unaware that the schedule had changed: executions would now be carried out at night.

On Thursday 13 March 1952 a notice was posted on the walls inside Mt. Eden. It said that by favour of Kerridge Odeon, a movie was to be shown that evening. Most of the men attended this rare event. Archie Banks, however, remained in the basement bathroom where he was employed as a painter, and he remembers talking to Fiori when he was brought down for exercise. Fiori seemed quite normal, Banks said. Banks was then asked to vacate the area so the condemned man could shower.

Banks was never to see Fiori again. For at 8.03 that night the trap in the east wing slammed open and Fiori was dead. While prisoners had sat enjoying their evening's entertainment, a man had taken his final walk up the seventeen steps to the scaffold.

It was not until the following day that prisoners heard the radio news that the hangman's work had been done, and a feeling of revulsion swept through them. Although to do anything was now too late, inmates resolved never again to be gulled. The movie ploy could work but once, and the administration realized that in future, routine would have to be different.

It was exactly eighteen months before the next execution took place, by which time there had been some refinement. It had been decided that hangings would now take place an hour earlier — at 7.00 p.m. — and that all inmates would be locked up before the event, with their judas holes[6] closed. A long seagrass mat would be rolled out between the west and east wings, to muffle the footsteps of the entourage as it arrived. The hangman — a retired policeman, who wore a buttoned-up raincoat with a felt hat pulled over his eyes to avoid recognition — always came an hour earlier, and waited by the scaffold for his victim. Although the gallows doors had now been padded in an attempt to muffle their opening, there was still no way of preventing inmates from knowing whenever an execution was imminent.

After Fiori, the second man hanged was Eruera Te Rongopatahi, in September 1953, and a little over three months later, two weeks before Christmas, Harry 'Darkie' Whiteland was executed. Whiteland's last words were to strike a chord of disquiet amongst official observers: before the white hood was fitted, he wished everyone a 'Merry Christmas.'[7]

It was another eighteen months before any more hangings took place, then in the five months between July and December 1955, Frederick Foster (7 July), Edward Te Whiu (13 August), Eric Allwood (13 October), and Albert Black (5 December) were all executed in close succession. Black, whose final moments were reported in detail by a *Truth* reporter, also wished observers a 'Merry Christmas.'[8]

Most of those who were hanged went quietly and this made the procedure easier for staff. They were spared what they dreaded — the prospect of a prisoner demonstrating violently against his fate. As an administrative expedient, therefore, prisoners were offered a dose of morphine and phenobarbitone four hours before their final walk, and always arrived at the scaffold pinioned at the elbows and knees, with hands strapped in front of them. In more than one case the sedative given was so heavy that the victim had to be supported on his feet. However, the mistake was not often made, and most were conscious and lucid when their final moment came.[9]

The eighth and last execution was that of James Bolton, fifteen months after Black. Bolton's case was disturbing; he differed markedly from the young nondescripts who had gone before him. Bolton was sixty-eight years old, a successful farm manager convicted of poisoning his wife. He vigorously protested his innocence to the end, and many believed he was, in fact, blameless. On 18 February 1957, the day he was scheduled to die (although he was unaware of it at the time) Bolton requested an interview with the superintendent and the prison psychologist. He made an impassioned plea for his life.[10] However, the hour of execution had been set, and in order to avoid embarrassing scenes, Bolton was brought to the gallows so heavily sedated (yet still emotionally distraught) that he had virtually to be dragged through the gaol by staff. Scenes like this affected everyone.

The Impact on Inmates. The impact of these ghoulish proceedings on inmates is difficult to assess. Certainly the majority were against capital punishment, but they were remarkably acquiescent about their feelings. Recalling the discussion in chapter 6, a collective sense of political, social, even moral commitment among maximum security prisoners was lacking in the 1950s. There had been resistance to assisting with the scaffold, and to attending films when hangings were scheduled. On such evenings, an expectant silence fell over the whole institution as the hour of seven approached, and there was some banging on cell doors afterwards. However there was no effective, organized action, and as far as inmates were concerned, routine very quickly returned to normal.

Some prisoners were troubled by executions more than others, but feelings were more personal than public. When Eddie Te Whiu was condemned at

the age of twenty his two elder brothers, Cecil and Henry, were inmates of
the institution where he was held. In 1977, I spoke to Cecil Te Whiu about
it, and he told me that his sadness and bitterness about Eddie's death would
be with him for the rest of his life.

On the whole, however, inmates took events fairly lightly. Maketu, for
example, described prison atmosphere at the time of a hanging in this way:

> Oh, on the day of a hanging, everybody spoke about it, eh. And next day the
> screws would say, 'Oh, he went like this,' or 'he didn't have nothing to say',
> or 'he just hung his head'. That's all it was, just general conversation around
> the place. That's all. Nobody felt ill at ease. Nobody I knew felt ill at ease.
> I suppose some of them might have, but kept it to themselves.

Cavanagh confirmed too, that apart from the prison being a bit quieter, an
execution did not seem to affect the prisoners. Some even took a macabre
interest in them. Although everyone was supposed to be locked up at the
time of hangings, it was easy for some to stay out of their cells, especially
those who worked in the kitchen. By climbing on top of the elevator which
brought meals up from the kitchen to the wings and lifting a flap at the
base of the door, the arrival of the hangman, the escorting of the prisoner,
and the removal of his body on a stretcher could be watched. Procedure,
identity, dress, and demeanour of the hanging party were accurately described
to me by two prisoners who had watched through the gap in the door themselves,
and by one who had been friendly with others who had. The prisoners saw
Bolton 'crying . . . dragging his feet . . . blabbering and frothing at the mouth'
as he was shepherded to his appointment in the east wing basement.[11]

Two of those who watched in this way were lifers; one was Les Shortcliffe,
convicted of killing his young son in 1950, immediately before the passing
of the Capital Punishment Act. Shortcliffe, described by MacKenzie as 'violent,
unpredictable, and insensitive',[12] always said he had hopes of being able to
recognize the hangman so he could kill him after he was released.[13]

The Impact on Non-Institutional Personnel. In spite of their supposed deterrent
function, executions were veiled in secrecy and the Justice Department took
pains to allow as little publicity about them as possible. However, they were
inevitably fairly public affairs. Apart from prison personnel, the hangman,
the sherriff, and the doctor — all of whom played active parts — a limited
number of spectators was allowed. Any Justice of the Peace was permitted
to attend, as were various representatives of the police. News delegates too
were invited to witness proceedings, provided they exercised 'restraint' in their
published accounts. Condemned men were allowed to receive religious
instruction from a minister of a denomination of their choice, and to request
that he attend their hanging as well.

Like the inmates, the effects of executions on witnesses were varied. One
Justice of the Peace is reported to have attended every one.[14] He was studying
their 'psychological effects', he said.[15] The majority were less intrigued. The
Truth representative who attended Black's execution was clearly repelled by
it, and this was reflected in his detailed (and somewhat un-'restrained')
account.[16] Monsignor A. H. Hyde, parish priest of St Benedict's Catholic

Church reports that there was some reluctance among the clergy to attend the condemned, because of the terrific emotional impact it had upon them. Some had to be pressured into 'doing their share' of what they clearly considered an unpleasant duty.

Father Leo Downey, Director of Catholic Social Services and a man of great strength and compassion, attended all of the four men who requested Catholic instruction before their deaths.[17] Most priests were less resilient and due to the strain it had on them, Hyde insisted that no *padre* should ever attend a hanging alone. He, himself, attended two — those of Te Rongopatahi and Black — with Downey, and described the first at the most unnerving experience of his life. The morning after it, Hyde told me, he was so distressed he could not stand up straight when he got out of bed. It took Hyde a week or two to get over the effect which the experience had upon him.

Amongst the worst affected were the sheriffs,[18] the official directors of proceedings. Their job was to command the superintendent to 'surrender the body of the prisoner' and later to signal the hangman to pull the trap lever by raising the right hand, with the warrant of execution held in it.[19] During the 1951–1957 period, three different sheriffs officiated over the hangings, and each of them was seriously affected.[20] One suffered a physical collapse at an execution and had to be held up by two prison officers while he performed his duty. Another so dreaded executions that he began turning up to them drunk.[21] Each of them went sick — for periods of up to four months — after an execution had taken place, suffering symptoms like shock, anxiety, nervous exhaustion, and duodenal haemorraging.[22]

The Impact on Prison Officers. Officers I know reckon that amongst prison staff, opinion was divided about fifty-fifty on capital punishment, but tended more towards abolition as the number of executions mounted.[23] Personnel chosen for hanging duty were replaced if they displayed any signs of nervousness, and only the most stoic, reliable, and accordant were considered. The job was not for the faint of heart. When Foster refused to raise his chin for the noose, one of the party later boasted of having stuck his fingers in the terrified man's eyes to force his head back.[24] D. D. Price, who volunteered for hanging duty, openly defended the penalty. Apart from trying to trick a prisoner into assisting with work on the gallows, he also went to lengths to cajole a distressed Father Hyde into witnessing a practice drop. Where executions were concerned, MacKenzie told me, Price was a 'tough, unfeeling *bastard*'.

The Justice Department *Memorandum to the Joint Committee on the Capital Punishment Bill* (1950) stated that executions in the past had been 'an unpleasant duty which (warders) faced without flinching'.[25] So they tended to be in the 1950s. Dan Cavanagh, who attended all eight executions, was totally unmoved by them. As far as he was concerned, there was no emotional involvement because officers did not feel individually responsible. 'There was nothing *personal* in it,' Cavanagh argued, they were only carrying out the sentence of the court. To Engel, Cavanagh has explained that the older officers who took this duty had been through the depression, the war, and some had seen internment in prisoner of war camps.[26] These experiences had inured them to suffering, but he did concede that an air of despondency would prevail in the prison whenever an execution was scheduled.

The detachment of Cavanagh and others was not shared by all, for prisoners under sentence of death had to be monitored day and night. Unlike the hanging party, which had no contact with the prisoner before his final escort, those assigned to 'special duty' (or 'death watch') had intimate involvement with him for weeks on end. They sat with him while he ate, while he slept, while he prayed, and while he wept, and they got to know him well before he was taken away. Two of the death-watch volunteers[27] I spoke to expressed considerable concern for, and empathy with, the condemned. However, Antiss, who also volunteered for the duty said that he did not think anybody had any particular concern for it. 'It was just another job that was there to do, and you did it,' he said. 'Didn't worry me.'

Owing to emotional pressure the death watch never walked with the hanging party, but those who counselled the condemned to the grave found a dispassionate stance impossible. Hyde got to know Te Rongopatahi well before he escorted him to the scaffold, a mistake he was not to repeat when he later attended Black. MacKenzie comments too that he got so acquainted with Foster that when the time for execution arrived, he grieved for him as he would a close friend. In future he vowed to play a more impersonal role. When he was asked to witness the execution of Allwood, therefore, MacKenzie made a point of diverging from his previous practice of keeping in contact with the victim.[28]

The Impact on the Superintendent. The person who the hangings affected most deeply and irrevocably was Horace Haywood. Unlike most others, Haywood's involvement with the condemned was both intimate and inevitable. To him fell the task of seeing that the health and welfare of capital cases were attended to while they awaited their fate, and of ensuring, if sentence was confirmed, that it was carried out without a hitch. His was the most stressful job of all, as it was upon him that the entire burden of the execution ritual fell. The superintendent was in close contact with a condemned man and his family from the time he was sentenced to the time his corpse was taken from the institution, the day after his death.

Unlike the sheriffs, who the hangings affected so badly, Haywood could not recover by claiming months of sick leave. However, it is said that he would often be absent from duty for several days after an execution took place. Unlike the sheriffs and the hangman also, Haywood had to live close to other prisoners, many of whom knew, or who were related to the condemned men.[29] Haywood had no escape from the anguish, the fear, and the suffering of a condemned man or his family. These matters were directly his concern. When a reprieve was granted, Haywood quickly relayed the news to the death cell. When one was denied, it was Haywood's job to inform the prisoner that his sentence had been confirmed and that he would face the gallows within seven days. On the evening that the execution was scheduled, it was his grisly duty to have the prisoner prepared for death. The transfer of the condemned from his slot to a holding cell in the east wing basement, the pinioning of the limbs, the fitting of the hood on arrival at the scaffold floor, and finally the positioning of the noose, tightly around the victim's neck; all of these crucial tasks were the superintendent's ultimate concern. If anything went wrong, it was he who would have to answer for it, and it was his conscience that would be plagued.

Such solemn duty, such horrific responsibility, weighed cruelly on the mind of the ruthful Haywood. Yet, over a period of almost seven years this remained his task. On twenty-two occasions, Haywood had to receive and care for persons condemned to death by the courts, without knowing whether the sentence would be carried out or not. Eight times over that seven-year period, Haywood accompanied men on their final journey. The effect on the superintendent was marked. Each execution took its toll, although the impact of some was greater than others. Prisoners like Black and Allwood, who faced their destinies bravely, made the job easier. Those who crumpled made it difficult to bear.

Foster is reported to have feigned madness in a pathetic and transparent attempt to avoid the inevitable. As noted, he attempted to resist the rope at the last moment. There was Bolton, physically and emotionally devastated, dragged to his end by sturdy prison guards. These instances taxed the stoicism of even the most implacable ot Mt. Eden's staff, but it was Eddie Te Whiu, a youth of barely twenty years, who was the most distressing case of all. The Te Whiu family was well known to the staff at Mt. Eden prison, and Eddie's case had evoked considerable sympathetic comment.[30] To make matters worse, the boy's people wailed for him outside the prison buildings as they awaited the Executive Council's decision and it was well known that Haywood was taking it badly. Anstiss recalls:

> That was a very emotive situation, it was a terrible drain on Haywood. The whole family crying and shouting in the courtyard. . . . I can particularly remember a situation with Te Whiu's father with his big hat on. And his mother and all the carry-on, crying.

Even Anstiss, who insisted he was unmoved by hangings, admitted he was unsettled by the case of Te Whiu. Of its effect on Haywood, he said:

> Oh, he just looked sick, you know. I think it was mainly because, well, he could've handled it himself if he'd been left alone. But it was the continual pressure, the constant pressure. Pressure from the media, pressure from people who didn't want people to be hung. And he was a guy who involved himself with people, so he became genuinely involved with the guy and with the visitors, you know, with their families, this sort of thing.
>
> So he threw himself right into the whole deal. And he took the brunt for everybody. You know, sheriffs were going down like flies, and he assumed all the responsibility for everything.

Although the institution as a whole gradually became used to executions, Haywood's burden became more difficult as time went on. A Government publication reproduces a statement by MacKenzie:

> Each execution was attended by tension which mounted throughout the institution as the day and the hour approached. But it was noticeable to me that the tension

was less each time. There seemed to be a growing acceptance amongst the inmate population. The tension and anxiety among those who had to carry it out grew no less, however, and the effect on the Superintendent was cumulative....[31]

The Superintendent always adopted a show of bravado, but the effect on him was all too apparent. He became, and looked, a lonely, ageing man, carrying a burden that grew heavier as the days passed. It usually took him several weeks to become his old robust self.[32]

In a personal letter to Engel, Barnett recalls that he was in the superintendent's office when the Executive Council's decision to hang a young Māori came through. The superintendent is reported to have burst into tears.[33] Haywood made no secret of the fact that he disliked executions and an official, reporting on the matter to head office, wrote, '... adverse effects have been noticed in this superintendent, *so that he is minimising the effect executions have on himself*'.[34] Just what these words refer to is uncertain but a clear implication is that Haywood was taking refuge in alcohol.

For the purpose of executions only, alcohol was permitted in the prison for use by the hanging party. When proceedings were over the party customarily retired to the superintendent's office for a much needed drink. It was release through this means upon which Haywood became increasingly dependent. Although the extent of Haywood's drinking is impossible to assess, MacKenzie confirms it became a problem, especially in the later years.

In his own publication MacKenzie writes that, having observed the hanging and cremation of Allwood, he and the superintendent made their way back to the institution, 'via a pub and two double whiskeys'.[35] This is unremarkable in itself, but to me, MacKenzie said of Haywood:

> He was as rough as guts, tough as they come, and yet in later years, as the hangings went on, he began to deteriorate. You know, he'd get pissed up as every hanging came on, he drank a lot to try and cope with his own feelings about it. He wasn't a tough man at heart ... the pressure of hanging showed on him very markedly and he began to deteriorate from then on.

Knowledge, and speculation about Haywood's problem was rife in the prison and did little for his professional image. Haywood's drink was whisky, a fact which even inmates knew. One convict accused Haywood of drowning his emotions in booze whenever a hanging was due, and was later put on charge for telling Hargrave, 'Yeah, Haywood would be still sucking on his whisky bottle after telling the bloke to pull the lever.'[36] Another told me that he had heard Haywood hated hangings so much, 'he used to get steamed up on a bottle of whisky before he would even attend'. I.S., the lifer who knew Haywood so well, said of him:

> Haywood was a good fellow until he started going funny. He went to a hanging in Fiji and when he came back he was off for three weeks. The screw told us he was lying in bed blind drunk, throwing empty whisky bottles at the wall.

This is the type of rumour that spread quickly at Mt. Eden, and even the youngest prisoner knew about it. Ray Brunell, only fifteen years old at the time, remarked:

> That's what they used to say about Haywood, you know, he was a *shocking* lush. It was all around the gaol, but I was too young to realize what they were talking about.

The end of capital punishment for murder in New Zealand came in 1957. Although it was not formally withdrawn until the 1961 Crimes Act,[37] the Labour party, which defeated National at the end of 1957, automatically commuted all death sentences to life imprisonment. Seven lives were spared in this way. Haywood was able to breathe easier during this period, sensing that the penalty would soon be revoked.[38] When National was returned in 1960, with a Justice Minister ardently opposed to hanging,[39] Haywood waited anxiously for abolition.

Haywood's Last Years

Capital punishment was only one of many stresses in Haywood's later years. In 1955 there had been the population crisis and the Horton escape. Later that year, four more men were hanged. Constant pressure took its toll and the ageing superintendent's health began to break down. Even after the Crimes Act, the deterioration continued, perpetuated by continual administrative complexities. Overcrowding remained one of them, and a rash of escapes was another. After 1960, populations continued to rise. National DAPs increased from 1850 in 1960 to 1895 the following year. Then they stabilized for a while before shooting upwards again after 1966.

In spite of Barnett's 1958 commitment to 'strip' Mt. Eden and cease overcrowding by summer, musters continued to grow as well. Whereas the DAP at Mt. Eden in 1958 was 415, it rose to 421 the following year and to 423 in 1960. The number in maximum security stayed around 400 for the remainder of Haywood's career. The years 1958–1962 were peak periods, and in 1960 there were more inmates in maximum security than at any other time in the history of New Zealand. This coincided with a breakdown in the inmate élite system after the escapes of Maketu in 1958. In 1959, there were a number of serious assaults on staff at Mt. Eden. In April, a nineteen year old prisoner received an extra six months for assaulting an officer, and in May a thirty-four year old man received a six-week extension after attacking the deputy superintendent. Finally, in June, while Barnett was inspecting the prison, an attack by three prisoners in the bootshop left four officers injured. One of the assailants later said that the incident had been caused by poor treatment of inmates.[40]

Overcrowded gaols was a national problem, and trouble was not restricted to Mt. Eden. Soon after Dunedin women's prison was taken over, trouble broke out there, and a month later, in December 1959, there was rebellion at Invercargill borstal. The Justice Department attributed all these disturbances to overcrowding, poor conditions, and to the administrative difficulties arising therefrom.[41] However, it was a number of months before the first really serious incident occurred. It was an event which not only foreshadowed the birth of a new decade, it also heralded a fresh set of political dynamics at the maximum security prison.

We have seen that in the 1950s the system of internal controls at Mt. Eden and the apoliticality of prisoners prevented orchestrated opposition to authority. The group assault of June 1959 was perhaps the first indication that the inmate control system was beginning to break up. In April 1960 an incident took place which set the stage for major developments in maximum security for years to come. On Sunday 10 April 1960, an unexpected and futile attack was made by at least six prisoners on staff at Mt. Eden prison. The apparent and hopeless objective of the attack was to obtain a set of keys with which to escape. Three officers received minor injuries in the bid, and one, a fractured cheek bone.[42] Later that evening, four officers entered the cells in the east wing detention area. Four suspects from the escape attempt were systematically beaten up, suffering injuries like blackened eyes, swollen faces, and weals and bruising to their bodies. These were later explained by staff as having been sustained when the inmates 'fell down the stairs'.[43]

The day after the fracas, it was discovered that the Sicilian, Angelo La Mattina, the first man sentenced to death after Labour took over in 1957, had disappeared. According to routine, all inmates were assembled in the main yard that afternoon while a general search of the prison was made. No sign was found of the prisoner, and it was assumed he had escaped. However, when inmates were ordered to return to the wings, 300 of the 400 or so in the institution refused, insisting on an inquiry into the beatings of the night before. The call was denied, and although about sixty came in for their evening meal, 240 remained in the yard overnight. It was not until the following day that the men, assured of an inquiry by the inspector of prisons, agreed to break their strike.

The promised investigation was conducted by a visiting magistrate and was typical of many of its genre. It found that inmates who had reported attacks on themselves or on others were mistaken or lying, that no assaults on any prisoners had been made, and that the demonstration had been caused by a small number of influential men, bent on misleading the majority into a meaningless action. No recrimination against any officers was called for, but definite discipline was recommended for the ten strike organizers.[44] There was no connection between the strike and the escape of La Mattina, who was found seven and a half days later, hiding in the attic of the west wing. The strike, it was felt, was caused first of all by the assembly of the men in the yard after the disappearance of La Mattina; and secondly by the absence of the superintendent, Mr Haywood, who had been on leave at the time.[45]

Had the superintendent been available, his authority and bearing may have prevented the attacks, or at least the retaliatory assaults. Maketu, who played a leading role in the strike, believed that no detention block beatings would have occurred had Haywood been there, and that, in any case, his accessibility would have removed the need for any strike. Temporarily in charge was I. P. Mathieson from Wi Tako, Wellington. Mathieson was not known to staff or prisoners and could not readily be approached. Moreover, he was implacable. When the inmates had demanded an inquiry on 11 April and asked to see some civilian representatives, Mathieson had categorically refused.

The events of 1960 marked a significant turning point in the history of Mt. Eden. It minted a style of prisoner solidarity and a readiness to rebel which had hitherto been unknown. However, it also delivered a telling blow to the credibility of Haywood. At his first meeting as Justice Secretary and chairman of the penal group, Robson said, in apparent reference to events at Mt. Eden, that 'sloppiness' in prison administration would no longer be tolerated. From then on there would be an emphasis on firm discipline in all institutions.[46]

It was now that the internal controls, which had been deteriorating since the Maketu affair, began to break down in their entirety. Under the watchful gaze of head office the era of inmate élitism was quickly being dragged to an end. The changes it ushered were cardinal. MacKenzie remarks:

> In Mt. Eden a new disciplinary regime turned the previous regime upside-down in the early 1960s, perhaps too quickly and too dramatically. The hierarchy of inmates who had been permitted to run the jail were stripped of their informal authority, which was now vested in custodial staff. Inmate morale fell and staff morale rose.
>
> In the early 1960s new tensions were appearing which led eventually to the riots and fire of 1965. The reformative and remedial attempts of the 1960s had come to naught.[47]

There was a general tightening of discipline. Leaders were stripped of their power. Maketu was again taken off privileges, and musical instruments and radios were confiscated. Activities of the band, although not stopped completely, were arrested for a while before continuing in a much more restrained fashion.[48]

Publicly, the trouble at Mt. Eden between 10 and 12 April 1960 was attributed, once again, to staff shortages and overcrowding. On 15 April, a number of security measures were announced, which included increasing the staff complement at weekends by half and restrictions on prisoner movements. There was also to be (another) full scrutiny of routine. Once again, it was promised that the prison population of 400 would be reduced to something like its proper complement of 275. But eleven months later the gaol still contained 419, about the same as at the time of the strike. Complaints about overcrowding and understaffing continued.

So apart from small adjustments, the problems which had been blamed for the 1960 trouble remained. More importantly, because the privilege system was breaking down, co-operation of the powerful disappeared. These inmates now often became the bitter and frustrated opponents of change. Their strong negative influence led to escalation of serious disciplinary infractions as the 1960s progressed.[49]

The idea of a National Penal Centre at Waikeria was scrapped completely in 1960, but it was now evident that the buildings at Mt. Eden — among other things — were totally unsuited to maximum custody. The internal upheavals and the rising sensationalism of escapes were becoming intolerable. In December of that year, a decision was made to construct a completely new institution for the purpose of high security.[50] When it would be available was uncertain, and it was another eight years before the proposed facility was opened. In the meantime, Mt. Eden would have to suffice.

Despite the new restrictions, Mt. Eden continued to run into trouble. Although Barnett had boasted about the years of 1959-1960 having the lowest percentage of penal escapes in a decade,[51] and although this rate was to fall even further in the next three years,[52] it was the quality rather than the quantity of escapes that really mattered. For as absconding rates were dropping, escapes from security grew more frequent than ever. On 3 February 1961, inmate Trevor Nash disappeared from the Mt. Eden prison engineers' shop. Nash, serving seven years, had been convicted in 1958 of what was then New Zealand's biggest theft: a 19,875 pound payroll. The escape caused great alarm, and Nash was further distinguished by becoming the first wanted criminal to have his face flashed on New Zealand television screens. Nash was at large for longer than any maximum security escapee had ever been. His recapture, in Melbourne some 158 days later, was thought of sufficient importance to warrant interrupting a debate in Parliament with the news. Nash had simply walked out of the institution via the quarry, which, apart from security, was now facing difficulties of its own.

On the evening of 12 April 1961, a shot of gelignite set by an inmate under supervision sent a 16 kilogramme rock flying 200 metres in the air, and through the roof of a building at Auckland Grammar School. The next month, because of danger to the school and to the adjacent Colonial Ammunition Company factory, all blasting at the quarry ceased. Crushing of metal continued for a short while with reduced worker complement. However, manhandling rock off the quarry face proved uneconomical, and the whole operation was closed down in 1962.

The quarry had been central to social organization at Mt. Eden. It had also provided the major source of employment: with three gangs of around thirty men each, it occupied about one quarter of its population. After May 1961 many of these were out of work. Several months later, in August, fire swept through prison workshops, extensively damaging the tailor and bootshops. These industries had to be closed also pending reconstruction, and unemployment worsened. A surplus of idle men spending their days in small, bare, exercise yards or locked in cells was now a permanent feature of the gaol. Many of those who did have jobs could not be given enough work to keep them occupied. Inactivity fostered boredom, frustration, and an increase in unrest and refractoriness.

In January 1962, two inmates attacked two officers, blackening the eyes of one of them. In April, a fight between prisoners at roll call broke into a general mêlée when an official tried to intervene. Members of staff were assaulted and two were injured. Seven inmates were later charged over the incident. Two days later, rumours that twenty prisoners planned a strike resulted in sixteen policemen and two naval officers armed with tear-gas being rushed to the institution. Although nothing eventuated from it, it was clear that dissatisfaction was rising. Asked about the reasons for unrest, Haywood refused to comment.[53]

Just over a week earlier, La Mattina and a prisoner called Eddie Tell had escaped from their cell overnight.[54] An inquiry established that the pair had been able to obtain a hacksaw blade[55] and, having cut their bars, climbed over the perimeter wall. In a public statement, the Minister of Justice blamed the escape on gross overcrowding and the inadequacy of Mt. Eden for maximum security.

Although it was never made public, part of the blame was directed at Haywood. After the Maketu affair, head office had instructed that all prisoners who were escape risks should be placed in the most secure cells. Moreover, in a previous inquiry it had been recommended that the 'tower' cells (from which La Mattina and Tell had broken out) should not be used for security risks at all.[56] Since both of the escapees fell within the 'risk' category the superintendent had clearly disobeyed his instructions. As far as head office was concerned, Haywood's basic error was cell allocation, together with a 'certain looseness in procedures'.[57] From that point of view the only value of the escapes was that they highlighted the need for a new institution. Indeed, in a statement issued directly afterwards, the Minister had again pledged to build a new prison as soon as practically possible. In the meantime there was another security spree at Mt. Eden. Utmost care was to be taken in the placement of inmates. Coils of barbed wire were set along the top of the outside wall. There would be greater attention to the searching and supervision of security risks.

The measures had little effect. In April, La Mattina and Tell had been out for two days. Only five months later, on 19 September 1962, a weary Minister had to report that the Sicilian had absconded once more. Ignoring the new barbed-wire defences, La Mattina had scaled the wall at 8.35 in the morning using a home-made rope.[58] When he was recaptured (with a broken foot) an hour later, La Mattina vowed he would get out again.

Another inquiry followed. Once more MacKay visited and another scrutiny took place. From now on, four armed sentries would mount a twenty-four-hour watch over the perimeter, and a new sentry box on the north wall, where La Mattina had gone over, would be constructed. In order to ensure the vigilance of sentries, they would be changed at hourly intervals. Deficient staff complements would be augmented by transfers from other institutions. A special guard was placed in the tailors' shop, where La Mattina and Nash now worked.

Between 1950 and 1958 there had been a total of only twelve breaks from Mt. Eden. Now, although only thirty of the 400 or so men there were classed as potential risks and in spite of constant attention to security, escape rates were burgeoning. By early 1963, the number of escapes since 1950 had increased to twenty-seven. Whereas in the 1950s, administration could often rely on being tipped off about a planned break before it took place, the destruction of institutionalized informing meant that this was no longer the case. Increased discipline alienated staff from inmates and solidified the opposition of prisoners. Haywood and Barnett began to face each week in trepidation of what might happen next.

On 1 October 1962 an escape attempt by four prisoners was foiled, and less than three weeks after the new security measures were announced another occurred. Frank Matich, a trustie serving four and a half years for violent and property offences, disappeared from an outside work party. Recaptured thirty-six hours later and classified thereafter as a security risk, Matich's escapades were still not over.

On 17 May 1962, George Wilder, serving four years for property crimes, broke out of New Plymouth prison. The event was recorded in an insignificant paragraph on an inside page of *The New Zealand Herald*.[59] Amidst a growing wave of publicity and a huge man-hunt, Wilder managed to elude the police for sixty-five days before eventually being recaptured. It was the guile and daring that Wilder displayed in avoiding his pursuers which appealed to the sympathies of New Zealanders, creating what was to become a legend. However, the search was time-consuming and expensive, and the courts remained resolutely unimpressed. Wilder got an additional three years for the escape and the offences committed while at large. He was then transferred to Mt. Eden to serve the remainder of what was now a seven-year term.

Because of the security threat they presented, Wilder and other risks were held under tight conditions in Mt. Eden's east wing basement. Frank Matich was another of these, having been confined in the block since his bid for freedom ten months previously. Two later arrivals were Patrick Wiwarina and Reuben Awa, who had run from Hautu prison in December 1962. To begin with, Wilder remained quiet at the institution. However, it was not long before he was involved in an audacious plan. In the early hours of 29 January 1963, Wilder chiselled his way out of his cell in the east wing basement. Using a home-made key, he then unlocked Matich, Wiwarina, and Awa. Having done this, the four retired to Wilder's cell, where they waited. Shortly before dawn, officer R. H. Grubb unlocked the door leading to the detention block and entered it for a routine check. Unaware of the break and taken by surprise, Grubb was engaged by the four men and knocked unconscious before being bound and gagged. His keys were used to gain access to the hanging yard, where he was dragged into a toilet recess and left.[60] The prisoners then climbed over the perimeter wall using a knotted sheet. Although they were spotted on the wall and fired at with two loads of buckshot, the escapees made good their departure. Awa, Matich, and Wiwarina, however, were recaptured quickly. Wilder, experienced bushman that he was, managed to elude the police for a further 173 days before finally being arrested in a bush hut near Taupo.

This sensational break, one of New Zealand's longest-lived maximum custody escapes, was again widely publicized and the six-month search for Wilder made big news. *Truth* launched a campaign for tighter security[61] and George Wilder, 'champion jailbreaker',[62] became immortalized in song.[63] He is still this country's most celebrated escaper.

Humiliated and ridiculed, the authorities once more viewed the achievement dimly. Wilder, given a cumulative six-year term, was returned to prison with thirteen years to serve. In Parliament Hanan was asked by a member of his own party how Wilder had got away so soon after his last recapture.[64] Lost for explanation, the Minister could do nothing but repeat lamely that Mt. Eden could no longer be considered effective for maximum custody. Yet again, security procedures would be looked at. Again it was insisted that measures to hasten the building of a new prison were being made.[65] This soon became a main priority. In the meantime a separate, secure unit for a dozen of the top escape risks was intended for inside the prison. Plans for the unit were drafted, and by August the preliminary design was complete.

HAYWOOD'S TRANSFER

By the time Wilder was returned to Mt. Eden in July 1963, the old superintendent had gone. On 2 April there had been another fire, this time in the mailbag shop, and barely two weeks later, Haywood was on his way to Paparua. While it was officially announced that Haywood had sought the move for 'personal reasons', it was widely believed that the transfer was, in fact, a demotion.[66]

The prison had never recovered from the administrative reorganization which had followed the 1960 strike. Since then there had been a procession of unexplained fires, attacks on staff, and other signs of unrest. Mt. Eden's annual escape rate of the 1950s more than doubled in the 1960-1963 period. In the twelve months before Haywood's transfer there had been a total of nine attempted break outs. There is no doubt that head office had been concerned for some time about the way things had been going at Mt. Eden and there is no doubt either that Haywood, at the time of his transfer, was a sick man who was losing his ability to govern. Alcohol, we already know, played a part in Haywood's downfall.

MacKenzie agrees that Haywood's decline began during, but continued after, the hangings. His condition worsened steadily. Haywood began falling asleep on duty and could often be seen dozing in the sun in some quiet corner of the gaol. Molloy remembers occasions of seeing Haywood, having arrived for work, parked in his car outside the wall and fast asleep at the wheel. The superintendent was once unavailable for an urgent call because he had nodded off in his car.

Haywood is said to have suffered from a chronic bronchial complaint and he began losing a lot of weight. MacKenzie and Molloy believed that Haywood was not only physically sick, for there were signs of mental instability as well. The prison psychologist said that Haywood had become paranoid and that by the end, his fears had increased 'beyond reasonable bounds'. He put heavy bars across the windows of his house, and he festooned the doors with locks. He took to carrying a pistol around with him. When he was eventually transferred, a trusted inmate, assigned to assist with the packing of Haywood's luggage, was amazed to find three boxes of brand-new tools amongst the superintendent's personal gear. These he recognized from the engineers' shop. I asked the psychologist whether he had heard any rumours that Haywood may have been involved in dishonesty and he said he had not, but he added, 'I think he deteriorated fairly badly, latterly, and anything could have happened. . . . I wouldn't be surprised at anything.'

Whether or not the story of the tools is true, Haywood undoubtedly left Mt. Eden a broken man. The weight of responsibility he carried, the hangings, the escapes, and finally the loss of control, combined to destroy the spirit of one who had worked hard and diligently in the service of the State. Haywood's lateral transfer and his new vocation as assisting Inspector of Penal Institutions was a sinecure. In the end the superintendent knew his administration had failed. His transfer was tantamount to an advertisement of that fact.

Haywood's condition at the time of his move can only be speculated, but the blow to his morale must have added to his decline. In February 1964, not ten months after his arrival in Christchurch, Haywood suffered a 'brief illness', and died.[67]

At Mt. Eden, Haywood's replacement was an entirely different kind of person; a man with a reputation for disciplinary toughness. In deep contrast to his predecessor, the new superintendent was uncompromising, inflexible, and ruthlessly orthodox in his approach to prison administration. His eye for order had made him a popular contender for this, the acme of prison jobs. He was, it was felt, the kind of man needed to steer the institution off its stormy course, and back into safer seas.

NOTES

[1] These arguments are well documented in the UK's *Report of the Royal Commission on Capital Punishment*, and in a book by the Chairman of that Commission (Gowers, 1956). Useful literature on the New Zealand executions appears in Engel (1977) and in Department of Justice (1974) 65-81. Marshall (1983) 221-4 presents an argument for capital punishment. The political debates that led to its re-introduction and final repeal appear in NZPD vol. 293 (1950) 4282-315, 4331-72, 4377-441, 4510-42, 4624-32; NZPD vol. 328 (1961) 2684-712, 2715-35.

[2] In the pre-1935 period, Dallard (1976) claims he prepared the cases that recommended for or against remission of the penalty. The Cabinet, he says, generally followed his suggestions.

[3] This part of Barnett's description written in a memo to the Minister of Justice (Department of Justice 1974, 65).

[4] *NZH* 6 March 1952.

[5] Dept.J. (1974) 70.

[6] The 'judas hole' is the spy-hole set in the steel cell doors.

[7] MacKenzie (1980) 78.

[8] *Truth* 14 December 1955.

[9] Allwood, for example, is reputed to have refused medication and pledged that he would 'die like a man' (MacKenzie 1980, 81). Allwood was the only one of the eight who showed bitterness. After sentencing, Allwood's cell guard had to be doubled because of a threat he made to 'take a screw' with him before he died.

[10] D. MacKenzie (pers. comm.).
[11] R. Brunell (pers. comm.).
[12] MacKenzie (1980) 72.
[13] Maketu (pers. comm.). Shortcliffe died in his cell at Waikeria in 1965.
[14] MacKenzie (1980) 80.
[15] *Truth* 14 December 1955.
[16] Ibid.
[17] These were Fiori, Te Rongopatahi, Te Whiu, and Black. The humanity and strength of Downey is legendary among prison inmates, and he was also a friend and adviser to Barnett (A. H. Hyde, pers. comm.). He died in October 1982.
[18] Sheriffs were chief registrars of the Supreme Court.
[19] Engel (1977) 65.
[20] MacKenzie (1980) 76-7.
[21] P. Anstiss (pers. comm.).
[22] MacKenzie (1980) 76-7.
[23] Father Hyde also told me he initially supported hanging. His mind changed, however, after he had witnessed one.
[24] MacKenzie (1980) 77, 83.
[25] Quoted in Engel (1977) 60.
[26] Both Tuyt and Cavanagh, for example, had been POWs.
[27] No officer was ever forced to take the death watch.
[28] MacKenzie (1980) 80, 81.
[29] Both Fiori and Te Whiu had previous criminal records. Three of Eddie Te Whiu's brothers, as well as his sister, had been inmates of Mt. Eden at some stage, and at least two of his brothers were imprisoned there at the time of his sentencing. Where Fiori is concerned, a few days after his death his wife was arrested and gaoled for receiving proceeds taken from her husband's murder victim.
[30] Te Whiu became the subject of an anti-capital punishment ballad by poet James K. Baxter (Baxter 1960). Engel (1977, 58) believes the case is partially responsible for the rallying of organized opinion against capital punishment. This pressure developed into a national committee at the end of 1956.
[31] Dept.J. (1974) 71.
[32] Ibid., 70.
[33] Engel (1977) 66. Engel feels that the case was probably that of Te Rongopatahi, but gives no reason for her belief. To me it seems more likely that Barnett was referring to young Eddie Te Whiu, whose case had caused such public concern and whose family Haywood knew so well.
[34] Quoted in MacKenzie (1980) 77, my emphasis.
[35] D. MacKenzie (1980) 82.
[36] R. Brunell (pers. comm.).
[37] S. 172.
[38] The Bill was eventually passed by a free vote, with ten National members crossing the floor to join a unanimous Labour caucus. Although he was no longer in politics by this stage, even Clif Webb is reported to have later changed his stance on capital punishment (Belshaw 1979) 100.
[39] Ralph Hanan, the new Minister of Justice and Attorney-General, had long been the strongest opponent of hanging in the National party camp.
[40] The incident is reported in *Truth* 30 June 1959 and *NZH* 23 June 1959.
[41] *App.J.* H-20 (1961) 12.
[42] One of those assaulted was Charlie Lashmar, an informant of mine.
[43] *Truth* 19 July 1960.
[44] These were eventually dealt with by lost privileges, segregation, and restricted diet.
[45] Verbal accounts of the assaults and/or the subsequent strike were given to me by C. Lashmar, T. Hooper, D. F. MacKenzie, J. Moyle, and R. Maketu. Written accounts appear in the 1960 *Report on a Series of Assaults on Prison Officers at the Auckland Prison*; and in *NZH* 11, 12 April 1960 (three articles), 13, 14,

16 April 1960 (two articles), 23, 14, 16 April 1960; *Evening Post* 22 April 1960; *Truth* 19 July 1960. Accounts of the La Mattina incident appear in *Evening Post* 11 April 1960; *NZH* 12 April 1960; Morris (1975), 10-15; and Mackenzie (1980) 58-9. Having read the official report and heard eyewitness accounts of what occurred, I am firmly of the opinion that a retributive attack on inmates by four staff members did in fact take place. The inmates' account of events was given in a letter smuggled out of Mt. Eden prison and published fully in *Truth* 19 July 1960.
46 *PGM* 9 July 1960.
47 MacKenzie (1980) 26.
48 The last reported activity of the band was in November 1962, when a series of four performances was given. The recitals were produced, as before, by Mrs E. V. Haywood. (*NZH* 21 November 1962).
49 In September 1960 Mt. Eden inmates had rampaged and attacked an officer with his own truncheon. As a result of this incident the arming of staff with batons was stopped.
50 Robson (1974) 2-3.
51 Escape ratios in the early 1960s were in the region of 2.2 to 2.7 per cent per annum.
52 While in 1959-1960, the escape ratio had been 2.7 per cent of the total penal population, in the preceding decade this rate had varied from 2.89 per cent in 1955 to 5.89 per cent in 1952.
53 *NZH* 13 April 1962.
54 Officially, this was only the fourth time a lifer had escaped from Mt. Eden. The others were R. R. D. Smith in 1940, then L. S. Hannan, and E. R. Horton, both in 1955. In 1958 one of the undiscovered men who escaped with Maketu was also a lifer.
55 An earlier report said that a serrated knife had been used.
56 *PGM* 30 May 1962.
57 Ibid.
58 *EP* 20, 26 September 1962.
59 *NZH* 18 May 1962.
60 The tiny hanging yard, then as now, served as the detention unit's exercise area.
61 See *Truth* 5, 26 February 1963, 23, 30 April 1963.
62 NZPD vol. 335 (1963) 806; *AS* 18 July 1963.
63 *The Wild New Zealand Boy*, sung by the Howard Morrison Quartet, was released by La Gloria records in 1963.
64 NZPD vol. 335 (1963) 88.
65 Official statement by J. R. Hanan 14 March 1963.
66 This view was expressed to me by former prison employees D. MacKenzie, T. Hooper, J. Moyle, and E. Molloy. In a personal letter dated 12 August 1984, however, Robson assured me that the transfer had been requested by Haywood himself. Records on the matter have been destroyed, but Weiss (1973) 21 seems to suggest that Haywood had suffered a nervous breakdown.
67 *EP* 17 February 1964.

9: THE HANAN–ROBSON ERA 1960–1969

In Haywood's last years a fresh regime began at head office, and a new Minister commenced duty with the fall of Labour in 1960. The new permanent head and the incoming Minister were to reign over the Department of Justice for the next nine years, and the partnership between them produced a decade of reform which was every bit as progressive as that preceding it. By the time Haywood retired from Mt. Eden, this administration was in full swing and had already begun adding to, and altering, some of the policies of the previous order. The transfer of Haywood, and the man chosen to replace him, were also largely attributable to the new outlook.

J. L. ROBSON, SECRETARY FOR JUSTICE

Sam Barnett retired in July 1960, and he was succeeded by an officer who was not only academically qualified, but who already had considerable practical experience in the administration of Justice. John Lochiel Robson was born in Halcombe in 1909 and educated at Wairoa District High School. His public service career began in 1924 when he joined the Public Trust Office, but having studied law at Victoria University, he eventually left Wellington to take a Master of Laws degree at Canterbury. From there he travelled to England and in 1932 Robson graduated with a Ph.D. from University College, London.

When his studies were completed Robson returned to the public service, and in 1944 was promoted Assistant Inspector for the Public Services Commission. Two years later he became Superintendent of Staff Training. So, by the time he took over the Justice Department in July 1960, Robson was well versed in public service management. More importantly, he had already experienced nine years' involvement with that Department. For, in 1951, two years after Barnett became Controller-General, Robson was appointed Assistant Secretary for Justice, with responsibility for staff direction. By the end of the decade Dr John Robson, legal scholar, public servant, and high-flying Justice administrator, appeared well qualified to step into the shoes of his retiring superior. Robson took over the Department of Justice in July 1960.

Robson's experience, and his view of the progress, successes, and errors of the 1950s, proved important influences upon the style of management he adopted. Although already a senior officer, Robson had not been a member of the penal group when it initially convened in April 1954. Nevertheless, from the time his name first appeared in the group's minutes at the end of the following year, Robson played an increasing role in the formation of Justice policy. He had seen the mistakes of the previous administration and often disagreed with Barnett's ideas. Yet on coming to office, shrewd tactician that he was, Robson moved slowly. At his first meeting as chairman of the group, he announced that for the time being there would be no changes of substance. He had been fully identified with past policy, he said, and desired that members become used to his control before serious alterations were made. Robson was due to travel overseas, and with a likely change of Government only months away,[1] he no doubt also felt it prudent to wait and see what the new caucus might do.

THE 1960 ELECTION

As many expected, in November 1960, the party which only three years before had grasped power by a fingertip majority, again took the opposition benches. Because of its precarious tenure, Labour had never really come to grips with governing, and its behaviour generally had been timid and uncertain. Although he was opposed to apartheid, for example, the Labour Prime Minister had spoken in favour of the all-white All Black team, which toured South Africa in 1960.[2] In the Justice arena, while it unanimously abhorred capital punishment for murder and had itself abolished the penalty in 1941, the second Labour Government did nothing about repealing the Capital Punishment Act. Instead it commuted all death sentences to life imprisonment, and left the penalty sitting on the fence.

Walter Nash's poor relationship with his Ministers, his frequent overseas trips, his poor party organization, and inefficient preparation for the election have all been put forward as partly to blame for the defeat of 1960.[3] The most important factor which led to the Government's downfall, however, was Finance Minister Arnold Nordmeyer's 'Black Budget' of 1958. This package, which increased taxes and almost doubled duties on beer, spirits, tobacco, cars, and petrol, dominated the election and made an easy target for opposition fire. When the November returns were tallied, Labour had conceded seven seats. National coursed in with forty-six to Labour's thirty-four.

In itself, law and order was not a central issue in the campaign of 1960, although the subject did not escape attention. Rising youth crime had been a theme of general concern about the country, and one which opposition had been hammering all year. From this point of view, one of the most significant events of 1960 was the Hastings Blossom Festival, because it was this incident which really brought delinquency into focus. In September 1960, the Blossom Festival, an annual event held to celebrate the coming of spring, attracted some twenty to thirty thousand people to the small East Coast town of Hastings. Bad weather on the weekend of 17 September forced people inside, many of whom took refuge in the bars of local hotels. As an afternoon's drinking progressed, fighting broke out in pubs and streets, accompanied by destruction of property, assaults on police, and widespread disorder. Police, outnumbered and unable to control the large numbers of drunken out-of-town revellers, had their job complicated by groups of locals who attempted to enforce the law themselves. This added to the confusion, and a number of young people were arrested and charged with offences ranging from wilful damage to fighting and assaulting police. The incident made headline news[4] and became a major point in a Police Department supply debate in Parliament.[5] Law and order and the control of young offending suddenly became a pressing issue directly before the election.

National's leaders made good mileage out of the alarm about Hastings and had the weekend's events exaggerated and falsified in the press. A number of journals worked with the opposition to present the incident as dimly as possible. In its election manifesto, National decried 'hooliganism' and delinquency and pledged, if elected, to make serious efforts to control it. Although successful, the tactics were also venal and soured relations with Labour for years to come.[6]

Another question which sought immediate address at this time was capital punishment. Under Labour, no one had hanged for more than three years. However, the penalty still existed and the National Government was left to face the crucial decision of whether to recommence the executions or do away with them altogether. When the new Minister took over at the end of 1960, therefore, two essential issues faced him. One was juvenile delinquency, the other, capital punishment.

J. R. HANAN, MINISTER OF JUSTICE

Ralph Hanan had been an outspoken critic of capital punishment for almost as long as he had been in politics, and when the new Prime Minister, Keith Holyoake, selected him as Minister of Justice, Holyoake must have realized that abolition of the penalty would be high on Mr Hanan's agenda.

A nephew of the former Liberal party Justice Minister, J. A. Hanan, Josiah Ralph Hanan was born at Invercargill in 1909. He attended Southland and Waitaki Boys' High Schools before proceeding to study law at Otago University. In 1935, Hanan commenced law practice, and he entered local body politics the same year. In 1938, he became Mayor of Invercargill, where he remained until joining the army as a private soldier early in the war. Having served four years in New Zealand, the Middle East, and Italy, Hanan returned badly wounded in 1944. Two years later, responding to a request that he stand for the National party, Hanan was elected MP for Invercargill.

For the rest of his life, Hanan remained in this position. Between 1954 and 1957 he served as Minister of Health, Public Trusts, Friendly Societies, and Immigration. When the National Government was returned in 1960, Hanan was made Attorney-General, as well as Minister of Justice, Minister of Māori Affairs, and Minister in Charge of the Electoral Office. In 1963 he also took on the portfolio of Island Territories. By the time Hanan was appointed Minister of Justice, he was already one of the most influential people in the National party caucus. A man of great tactical acumen, Hanan was a consummate politician; a worldly individual who had both the nous and the mana to sell new ideas to a conservative, sceptical Government.

The Hanan-Robson Partnership

Although he wore many ministerial coats, Ralph Hanan, once labelled 'Legislator of the Decade',[7] is best known for his input to Justice. Much of the success he enjoyed in this field was due to the able support he received from his permanent head, and conversely, a lot of Robson's personal reputation is due to his harmonious relations with the Minister. The Hanan-Robson partnership, which lasted for most of the 1960s, is often held up as an example of the perfect relationship which can exist between a permanent head and a politician.

In John Robson, New Zealand found a capable academic, a thinker, a man with an organized mind and a wealth of innovative ideas. Beside him, Ralph Hanan was a person with a strong sense of humane justice, sympathetic to the concept of reform, and receptive to new methods. He was accordant, energetic, and most importantly, he had the power and respect needed to put Robson's ideas into effect. Hanan's strength was reinforced by the Prime Minister, to whom he was a friend and frequent adviser. Essentially, then, the relationship between Robson and Hanan was one of a philosopher supported by a pragmatist. Each complemented and reinforced the other.

On a personal level the new Minister and his departmental head appear to have got on very well. Later, Robson's many writings on penal policy seldom omitted a word of tribute to his deceased partner. However, the Secretary did not relate well to all of his colleagues and their opinions of him were equivocal. Robson declined to be interviewed by me, but he is known not to have enjoyed much compatibility with his own predecessor, whom he also knew closely.[8] As noted, Barnett and Robson disagreed on a number of issues.

Subordinate members of Robson's Department — all of whom requested they remain anonymous — were reserved in their praise for Robson as well. Robson was described as a sombre, solitary person, though hardworking, thorough, and capable. He was a ritualistic operator, a 'public servant of the old school', and one who, I gathered, was somewhat monocratic in the way he ran his Department. He was described by former subordinates as being inflexible in relation to the opinions of others, and was criticized for a reluctance to give them credit when their schemes were accepted. In his dealings with colleagues, Robson's approach was formal and correct. He lacked the jovial, impetuous nature for which Barnett was known. Where Robson was methodical, scientific, and thorough, Barnett had been flamboyant and bold, but also a little rash.

Both the fifties and the sixties were decades of progress, experimentation, and change, and the advancements of the former were as courageous and valuable as those which came with Robson. Yet it is the sixties which are remembered as the most momentous in penal reform, and it is easy to overlook the importance of the era which went before it. As we have seen, apart from the laws of 1954, there was little statutory revision in the penal policy of the fifties. The important changes which took place were local and administrative. Improvements were made across the board, but the attention given to prisons was weighted in favour of maximum security. The top prison, in the centre of our largest city, had become the nation's showpiece. Other institutions could do little but follow the example being set in the north.

Barnett had known three different Ministers within two changes of Government, and the longest he had served under a single Minister was five years.[9] The Government had changed for a third time after eight. Conversely, for almost the whole of their ten years at the helm, Robson and Hanan were a continuous pair. This gave stability and direction to the progress of policy. Harmony within Justice was such that Robson's ideas were readily able to gain statutory recognition and thereby to enjoy much broader and permanent application than was possible in the decade before. By the time Robson retired, penology in New Zealand had matured to a level which has remained substantially unaltered to this day.

POLICY FORMATION AND CHANGE

The nucleus of head office policy formation in the sixties was the penal group, in which Robson's relentless pursuit of new schemes gave fresh life and vigour. Its complement was increased to thirteen and each member was given assignments on specified topics, for submission as written reports. Considerable trouble was taken by Robson to ensure that these reports were discussed fully and that they were submitted by their requested dates. Thus there was created a constant inflow of research data to the meetings, from which a modern penology could be forged.

The Robson administration, like Barnett's, was deliberately speculative, but now developments were more objective, more systematic, and more soundly based than before. Early in 1960 — before the election — Hanan had called for experimentation in the prison system.[10] It was not until 1963, however, that the 'policy of experimentation' was officially adopted,[11] and announced in Wellington. Robson desired scientific feedback from his new ideas and in April 1963, in order to assess their effectiveness, Marshall's research division of 1957 took its first full-time director. The person chosen for the job was D. F. MacKenzie, who since 1954 had been welfare officer, then psychologist, at Mt. Eden.

To complement this programme, a sustained publicity campaign was planned. Its aim would be to hold penal reform in the public eye by keeping the community informed about its progress. Robson always advocated the sound policy of preparing the ground for a development before putting it together. In 1967 he would write:

> In all my ruminations upon penal policy I kept coming back to the question or problem of public attitudes. There seemed no other way but a direct, sustained attack over a wide front, before progress could be made.[12]

Robson, in general, enjoyed good press relations, and it was perhaps because of his success in this field that in May 1964 a plan was approved by Cabinet, requiring the permanent heads of all Government Departments to implement positive public relations programmes.[13] Robson cultivated the media and always kept at his disposal a list of editors who he knew would be favourable to him.[14] Thus he was able to ensure his official releases would be reported as auspiciously as possible.

Regardless of the context of social conservatism in which developments of the sixties took place, therefore, the publicity campaign and the strong voice of Hanan were able to override sporadic attacks by public and political critics. The left wing of Parliament could hardly slate a policy more adventurous by far than any it itself had dared implement. In 1959, when Mason had tried to set up a pre-release hostel in Auckland, public opposition had convinced him to abandon it. However, in 1961, the determined Hanan was able to establish a centre in his very own electorate, with hardly a murmur from local residents.

Hanan's personal kudos, together with the support of the Prime Minister, muffled the voices of conservative colleagues. The fact of the matter was that Robson's publicity campaign was so effective, and Hanan's endorsement of it so sound, that the mass of the New Zealand public not only supported the reforms, it began to identify with them. The quality of the treatment which criminals received was believed by New Zealanders to be amongst the highest in the world. Like its free education programme and its welfare services, penal policy became a source of national pride.

Public Relations in Action

The public relations programme got off to an encouraging start in 1961 and was accelerated the following year. In January 1962, Mr E. S. Hoddinott, Senior Prison Chaplain, presented his first annual report on the campaign. In 1961, the Justice Department had had a total of 740 meetings at head office, in the probation service and in prisons. There had been seventeen broadcasts on air and more than fifty-five organizations had been personally addressed. Good relations had been enjoyed, and in 1962 an even more ambitious programme was planned. There would be more lectures, more broadcasts, and a greater number of publications. Deliberate efforts would be made to keep a good rapport with the press.[15] Prison superintendents would be asked to try stimulating interest themselves, by inviting influential people to visit their institutions.[16]

The decade produced some important publications, their object being to provide information about the justice system of New Zealand: its development, its current orientation, and its gaols. In 1964, the Department published *Crime and the Community*,[17] the country's first comprehensive survey of penal policy since Webb's small brochure of 1954[18] and Marshall's supplement of 1957.[19] The 1964 title laid out the two central themes of policy in the sixties. These were the development of goodwill in the community and a 'firm resolve' to deal with offenders, wherever possible, without removing them from society. The Department's guiding principles would be:

1. To take every practical step to divert individuals from becoming criminally involved.
2. To use imprisonment only as a last resort for young or inexperienced offenders.
3. To reduce short terms of imprisonment to an absolute minimum.
4. That where imprisonment was thought necessary, to bring every possible reformative influence to bear on the prisoner.
5. That offenders who persistently failed to respond to reformative efforts should be placed in custody for a long time, 'in order that the community may be protected from them and in order that they may realise the futility of their criminal activities'.[20]

Between 1964 and 1970 the Justice Department produced six more publications of this nature.[21] The principles of 1964 remained effective for the rest of the decade.

LEGISLATIVE ACTIVITY

When they came to power, Hanan and Robson wasted no time in drawing up legislation. As indicated, one of Hanan's strongest personal desires was to see capital punishment for murder repealed. Thus, one of the new Government's first pieces of law was the Crimes Act of 1961, the most significant aspect of which withdrew the penalty of hanging for murder. Arguments over this section of the Bill were hotly mooted in the House, and it is interesting to note that the name of Tume, used effectively in 1950 in support of capital punishment, again appeared in 1961. Now the case was applied in defence of abolition. For a while in Mt. Eden prison, Tume had himself become the victim of a murder attempt by an inmate called Brooks, who had decided he wanted to die on the gallows.[22] Although the plan had been unsuccessful and Tume had suffered only a shallow wound, the anti-capital punishment lobby seized the incident in support of its cause. The existence of capital punishment, it argued, could become an incentive for murder among persons with suicidal tendencies.[23]

Overall the issue was emotional and complex, and the Brooks-Tume incident was only of passing importance. Feeling about murder was strong, although abolition probably preceded public sentiment.[24] After the Crimes Bill was passed, and as a palliative to its opponents, the non-parole period of the life sentence was increased from five to ten years.[25] It was announced, as a working formula, that murderers doing life should serve between twelve and fourteen years. The Crimes Act 1961 was more than just an abolitionist statute. It was important in its own right for amending the Crimes Act 1908 and consolidating more recent legislation which related to it. In particular, the new Act abolished Grand Juries,[26] revised maximum penalties, created new crimes (such as infanticide), redefined old ones (such as treason), and abolished obsolete offences (like sabotage). For the most part, however, the 1961 Crimes Act had little impact on prison policy.

While the Haywood era was remarkable largely for the changes it made in security administration, then, the Hanan-Robson decade is best known for its contribution to law. Of particular importance here is legislation relating to young offenders and the non-custodial penalty. It will be recalled that a promise to deal with rising juvenile crime had been included in the National party manifesto of 1960. Soon after he gained office that year, Robson travelled to the second UN Conference on Crime and the Treatment of Offenders in London, returning late in October. One of the major questions discussed by the congress was that of juvenile delinquency, which had begun to cause concern in many parts of the world.[27]

In 1954, detention centres had been provided for in the Criminal Justice Act, but, because of overcrowding and lack of finance, had never been established. After taking control of Justice, Robson expressed misgivings about them. He believed the centres should be merely incorporated within the borstal programme and that training should be no different from that of a short borstal sentence.[28] Holyoake, however, had called for the establishment of detention centres before, and again after the Hastings incident, organized along the punitive-cum-military lines adopted in Britain. Within six months of becoming Prime Minister, therefore, the first detention centre was established at Waikeria, based on the English model. Two years later the sentence of corrective training, which had also been introduced in 1954 and never really got off the ground, was withdrawn.[29]

In addition to these, a variety of other measures was taken. In 1954, provision for temporary release from prison had been made[30] and was first used in 1955. This was expanded in 1965 so that certain first offenders could become eligible for regularly spaced home leaves during the course of their sentences.[31] In 1961, the work parole scheme was added,[32] and the first releases-to-work took place in January 1962. Periodic detention, a long-contemplated system of 'weekend imprisonment',[33] was finally installed by the Criminal Justice Amendment Act of 1962 for offenders between fifteen and twenty-one years of age. The first centre was opened in Auckland in 1963, and in 1966 the scheme was expanded to apply to adults as well.[34] Classification procedures were also upgraded and a first offenders' classification unit opened at Wi Tako in 1964. By November 1967, Wi Tako had been completely rebuilt and today it stands as a 'show' prison, a monument to the work of Hanan and Robson.

These provisions had no direct effect on maximum security, applying principally as they did to first offenders, petty criminals, and juveniles. The emphasis on non-custodial penalties and open prisons, however, did divert from the security environment a large number of transient and uncriminalized inmates. While a proportion of youthful, naive, and short-term offenders still came to Mt. Eden, an insistence that maximum custody should be restricted to those really in need of it meant that many were now held only in transit. These were treated differently from the others and never really became involved with the inmate community.

There was a slight change in the place of security under the new regime. Although Barnett had been concerned that because of overcrowding, too great a proportion of offenders was being held in low security, policy now was to increase the number of open institutions. This was announced by Robson in his first report of 1961.[35] As a result, in spite of a stable overall prison population in the first half of the decade, the DAP at Mt. Eden began to drop after 1962. Privileges such as home leave, work parole, and later on, special remission,[36] now became increasingly available at (and largely restricted to) low security establishments. So at the same time as the proportion of men confined in camps began to increase, the gulf separating them from those in security gradually started to widen. There was a considerable range of measures taken in this period which like the above examples, influenced the running of penal institutions generally. The high security prison was among those affected, and in some cases the impact was direct.

In 1967, the legislation concerning preventive detention was altered. Under the 1954 Criminal Justice Act, preventive detention had been an indeterminate sentence for repeated offenders and was almost always restricted to a term of three to fourteen years.[37] The provision had never been used extensively and by the mid-1960s had almost disappeared. In the years of 1965 and 1966, for example, only five people had received the three- to fourteen-year penalty.[38] The unsettling effect that indeterminate sentences had on prisoners was now recognized and the maximum of fourteen years was in many cases excessive in relation to the offences it applied to. The criterion for release, moreover, — that the offender was thought to be unlikely to continue offending — was known to be difficult to judge in a custodial setting. It was therefore felt that the penalty should be changed.

In 1967, preventive detention was removed for all offences except repeated sexual ones.[39] Its role now became purely custodial, with a minimum term of seven years, and a non-finite maximum term. Between 1968 and 1978, only twenty people received this sentence, and all of them served substantial terms in maximum custody. The five who had been discharged up to April 1981 had a mean age of 38.8 years when sentenced; 52.4 when released. They had served an average of over thirteen years in custody.[40] Because so few preventive detainees were sentenced, the 1967 amendment had little effect on the security environment. The preventive detention legislation, however, had endorsed a penalty, the object of which was long-term confinement. In this respect it departed from mainstream thought at the time, that custody should also rehabilitate.

However, aside from this exception the campaign for reformation forged ahead. It was felt important that inmates be involved in treatment programmes from the beginning of their terms, and, in October 1961, Robson decided on a policy of granting full privileges on admission. Staff were to become part of the reformative process, and the Department had already initiated training for prison officers in group counselling.⁴¹ If effective, the scheme might not only improve post-release prospects, but could also service staff-inmate relations. In 1962, officer counselling began. By the end of 1963, discussion groups had commenced in a number of establishments, including Mt. Eden.⁴² The following year, groups, run by prison officers, were said to have become a 'part of the programme of every institution'.⁴³ The importance of these groups was not great in maximum security, and in December 1963 they only catered for ten out of a total inmate complement of about 380.⁴⁴ Groups are unlikely to have had much greater effect elsewhere.

Probably as a result of this, enthusiasm for them was subdued and at Mt. Eden they had no impact on the tensions which began building up again in the early 1960s. Although a Justice publication still claimed in 1970 that group discussions were a part of the programme of every penal institution⁴⁵ there were none at Paremoremo that year, nor were any operating at the detention centre when I was there in 1971. At most institutions, where they did exist, discussion groups only included a small percentage of the total muster. At Mt. Eden groups were never very popular and the idea burned out — along with the rest of the prison — in mid-1965.

In 1961 a new set of regulations for penal institutions was gazetted with the intention of consolidating and amending previous rules. While the new regulations brought no great changes, they did rationalize a code which had been effective since 1946.⁴⁶ Important from the point of view of administration and the health of prisoners, 1961 saw the abolition of bread and water punishment and its replacement with two grades of restricted diet. Following the example of Alcatraz in the United States the more extreme of these, no. 1 diet, provided for a daily regimen of bread, potatoes, milk, and dripping. Its alternative, no. 2 diet, added to the above a ration of oatmeal, salt, sugar, and cheese. The maximum period a person could be sentenced to a restricted diet was fifteen days.⁴⁷ In the case of no. 1 diet only, full rations were given on every fourth day.⁴⁸ Bread and water had been abolished — amid continuing opposition — because it was considered ineffective and archaic. It is doubtful whether its replacements, although slightly less antiquated, were any more effective. In 1975, during the term of the third Labour Government, restricted diets were abolished on these very same grounds. The law did not come into effect immediately, however, and it was not until 1981 that restricted diets were formally done away with.⁴⁹

Another long-established, but controversial, policy of the 1960s was that relating to firearms. Since formal justice began in this country, the bearing of arms by penal staff had been an established practice, and between 1875 and 1945 a total of four prisoners had been shot while attempting to escape.⁵⁰ Since 1945 there has been only one case of a man being hit⁵¹ but there are a number of instances of rounds being fired at absconding or rioting inmates.

The necessity of firearms had often been queried and in 1958 the rules relating to the shooting of prisoners were reviewed. Considerable debate followed, nevertheless it was eventually decided to leave the regulations as they were. In March 1961, soon after Robson took over, the question of firearms in prisons was discussed once again at head office. By this time the only institution at which officers were regularly armed was Mt. Eden, where Greener police (riot) guns were carried by the six or seven tower sentries and .38 revolvers by some of the night watch and escort details. The arming of officers on escape duty had ceased.

Until 1961, prison sentries had been instructed only to fire in the defence of other staff or to prevent escapes, and then to aim to wound. Shooting-to-wound was to be preceded by a challenge, followed by a warning shot.[52] Considering the situation in March 1961, the penal group decided that firearms should remain in prisons and that shooting to kill could be justified in some circumstances. In its draft circular to all superintendents some weeks later, the Department stated, puzzlingly, that shooting to kill could never be justified. From then on only sentries, night staff, and escorts would be armed. Staff were often reluctant to use their weapons and when Wilder, Matich, Wiwarina, and Awa escaped in 1963, a question had arisen over the logic of arming night patrols. If night staff could be overpowered so easily, issuing them with revolvers seemed pointless. When Buckley took over a few months later, therefore, the issuing of revolvers to patrols was stopped. Moreover, the Greener guns — single-shot, lever action weapons which fired a special twelve-gauge cartridge loaded with buckshot — were found to be obsolete and unsafe. A decision to replace them was made in 1963, but it was not until 1970 that the weapons were scrapped. From this time, sentries at Mt. Eden were armed with twelve-gauge, five-shot, pump action shotguns.

Interesting to note is that, although from 1965 Mt. Eden was designated only as a medium security facility, the tradition of arming sentries there remained until late in 1980. Today, the only prison which has an armoury is the new maximum security one at Paremoremo, but this armoury is kept locked, and staff do not routinely carry weapons.

The Fate of Maximum Security

For our purposes, the most important decision which came out of the sixties had to do with the future of maximum security. Since 1951 plans for a National Penal Centre had been at the forefront of discussions about replacing Mt. Eden. We have seen that reservations about the scheme were long standing, however, and, by the end of 1958, it was generally agreed by the penal group that a number of smaller prisons was preferable to a large, single, complex.[53] In the first meeting after his appointment, Robson had emphasized his opposition to large prisons, and suggested that the size of the planned centre — already reduced from 1200 inmates to 1000, then to 935 — should now be cut to 600. For not only was the project at Waikeria too big, its distance from a major city,[54] made it undesirable from a family contact point of view. Robson believed that no penal institution should exceed an optimal size of 100–150 inmates.[55] By the time his first annual report was released, Robson was able to announce that an ambitious building scheme for small prisons was already being formulated. Regardless of persistent overcrowding, no new institution had been opened since the war.[56] Now, ten new establishments were planned by 1975, including two for women and a new maximum custody unit for men.[57]

From 1962, monthly reports on the building programme were made to the penal group. The most prominent of these was the maximum security gaol, first officially announced in the Annual Report of 31 March, 1961. In August 1962, a site for the gaol was purchased, and by June the following year preliminary drawings had begun. At Mt. Eden prison little in the way of progressive development was seen after 1960. From about the time Hanan became Minister, Robson knew its function would soon be replaced. However for the next five years the surge of escapes, assaults, and disturbances continued to generate concern. Combined they fostered a regime which became less compromising and more illiberal, as time went on. While the trend to greater sophistication in penology continued, Mt. Eden was approaching a crisis point. A new command at Mt. Eden in 1963 suddenly put development programmes there into reverse. The era of conciliation in the fifties was transposed in the early 1960s to one of unrest and anarchy. Hanan and Robson were helpless to stop this and it was only a matter of twenty-seven short months before the maximum security regime fell apart.

NOTES

[1] A poll conducted in April 1959 had indicated that public opinion had swung decisively against Labour (Chapman 1981, 364). Anti-Government feeling at this time was strong.
[2] Sinclair (1976) 385.
[3] Sinclair (1976) 347-50; (1980) 294-5; Mason (1968).
[4] *NZH* 12, 13, 14 September 1960.
[5] NZPD vol. 324 (1960) 2244.
[6] The dirty tactics of National's 1960 campaign were regularly commented upon over the next ten years, viz NZPD vol. 327 (1962) 1370, 1373; vol. 335 (1963) 170, 171, 174-6; vol. 352 (1967) 2313, 2315; vol. 362 (1969) 2260; vol. 372 (1971) 814.
[7] *Thursday Magazine* 24 July 1969, 36.
[8] My informant on this point asked not to be named, but told me that Robson and Barnett were 'not particularly fond of each other'.
[9] This was Clif Webb, from 1949 to 1954.
[10] NZPD vol. 326 (1960) 726.
[11] Stace (1971) 4-5.
[12] Robson (1967) 58.
[13] Thynne (1976) 2.
[14] D. MacKenzie (pers. comm.); J. Cameron (pers. comm.).
[15] *PGM* 29 January 1962.
[16] *PGM* 4 December 1964.
[17] Dept.J. (1964).
[18] Dept.J. (1954).
[19] Marshall (1957).
[20] Dept.J. (1964: 5).
[21] Dept.J. (1966); (1968); (1974, first published 1968); (1969b); (1969c); (1970a).
[22] The assailant, Brooks, suffered delusions and had attempted to stab Tume with a steel spike. He was later removed to a mental hospital. Brooks had been sentenced

to twelve years for manslaughter in 1950 and the year before the Tume assault had attacked another prisoner in the quarry with a hammer, fracturing his skull (*NZH* 8, 9 January 1953, 13 October 1954; MacKenzie 1980, 78-9).
23 MacKenzie (1980) 79.
24 This was also the opinion of many officials researched by Engel (1977) 112.
25 Criminal Justice Amendment Act (1962, s. 26(1)). The Government's intention to take this step had been announced during Hanan's second-reading speech (Engel 1977, 91-2).
26 Grand juries were groups of sixteen people required to listen to Crown evidence against an indictably accused. Their duty was to return a verdict of 'true bill' or 'no bill'; that is, to decide whether a *prima facie* case had been established in respect of evidence against an accused. In operation, the grand jury hearing was similar to the 'preliminary' or 'depositions' hearing of modern criminal procedure. (For information on the latter see Hodge 1981, 4-5, 8).
27 Robson (1973) 12-13.
28 *PGM* 30 January, 27 February 1961.
29 Corrective training was abolished by the Criminal Justice Amendment Act 1963.
30 Penal Institutions Act 1952 s. 21.
31 This was an administrative innovation that began in March. No change to the 1954 legislation was necessary.
32 Penal Institutions Amendment Act 1961 s. 2.
33 The idea of weekend imprisonment was considered as early as 1947 when Mr J. H. Luxford SM suggested that it might be of value in New Zealand (*NZH* 5 August 1947). Mason disagreed, however. His grounds were that such a scheme would be too difficult to implement and would be a punitive concept, contrary to the current one of imprisonment for training and reformation (*NZPD* vol. 277 (1947) 463). The idea was periodically revived in various contexts throughout the 1950s.
34 Criminal Justice Amendment Act 1966 ss. 7 & 8. The history of periodic detention is dealt with by Seymour (1969) and Webb (1982) ch. 7.
35 *App.J.* H-20 (1961) 12.
36 Special remission of one-twelfth of total sentence (which increases standard remission of one-quarter to one-third) was made available in 1964 for diligent, exemplary prisoners and for those who had performed some outstanding act of service (Penal Institutions Amendment Act 1964 s. 2). Later, one-third 'camp' remission, became automatically awarded (subject to revocation) to all who served the major part of their sentences in minimum security, and to others on application. In October 1985 a new Criminal Justice Act (1985) s. 80 made one-third remission standard for all finite sentences.
37 However, it could be extended to three years life in the case of some sexual offenders.
38 Webb (1982) 72.
39 Criminal Justice Amendment Act 1967.
40 Lee (1981) 5, 6, 23.
41 Dept.J. (1961). This was a booklet that was discussed at the 1961 annual superintendents' conference (Caughley 1964, 14). See also *PGM* 17 April 1961.
42 Williamson (1965) 72-3.
43 Dept.J. (1964) 61.
44 Reported in Williamson (1965) 72.
45 Dept.J. (1970a) 10.
46 These were the Prisons Regulations 1946 and Amendments.
47 Penal Institutions Act 1954 s. 33(f).
48 Restricted diet provisions are laid out in the Penal Institutions Regulations 1961 ss. 79-83.
49 M. Finlay (pers. comm.). Between 1978 and 1981 the penalty was given eleven times at Paremoremo (Meek 1986, 161). Ss. 15(2) & 16(2) of the Penal Institutions Amendment Act 1975 were brought into force by s. 13 of the Penal Institutions Amendment Act 1980.

50 *PGM* 10 April 1961 (Report on Firearms). Two of these were killed (in 1875 and 1917) and two were wounded (in 1929 and 1945).
51 This man was shot in the leg during a riot at Mt. Eden in 1971.
52 *PGM* 10 April 1961 (Report on Firearms).
53 *PGM* 15 December 1958.
54 The closest major city to Waikeria is Auckland, some 150 kilometres distant.
55 *App.J.* H-20 (1961) 12.
56 The last was Arohata Girls' Borstal, opened in 1944.
57 *App.J.* H-20 (1961) 13. Up until 1975, only five of these had actually been built. New prisons were constructed at Waikune 1963, Rangipo 1966, Wi Tako 1967, Paremoremo 1968, and Paparua (a women's prison) in 1974. The building programme was reported in *EP* 11 June 1960.

10: THE BUCKLEY ADMINISTRATION
1963–1965

The Haywood regime lasted for less than three years under Robson and from the time he took over, Robson was unhappy with it. Although he made no specific reference to Mt. Eden, it was precisely the aspect of control which concerned him most about the institution's leadership. Escapes, rebellions, and assaults, which had started becoming frequent as the 1960s began, were becoming a political issue. The Wilder escapes had drawn particular comment[1] and election year 1963, a 'record year for prison escapes',[2] had given substance to Labour party scorn.

Before Haywood was transferred in April 1963, his position had already been advertised and a replacement for him found. Once again, the Department's choice of appointment was a reaction to its predecessor. Because of glaring deficiencies, the Justice Department felt that Mt. Eden was now in need of a firm, steadying hand. The reins of command would be tightened. Security would take a much higher priority. Abscondings would be prevented, and insurrection smothered. By force of authority, the superintendent and his men would bring Mt. Eden's community back under firm control.

E. G. BUCKLEY, SUPERINTENDENT

The person chosen for this job was one who already had a quarter of a century of prison experience behind him. For half of this time he had held the rank of superintendent. Born at Auckland in 1914 and educated at Auckland Grammar School, Edward George Buckley had joined the prison service in February 1937, at the age of twenty-two. Serving first at Wanganui then Waikeria, he had transferred to Auckland two years later, where for eight years he had been in charge of the quarry. In June 1949, Buckley was made deputy principal warder, and in 1950 was promoted to full principal warder upon transfer to Christchurch. It was only a year after this that he became superintendent of Waikune prison.

In 1955 Buckley was appointed deputy superintendent at Mt. Eden, shifting the following year as full superintendent to Paparua. He remained at Paparua until his promotion to the command of Mt. Eden in April 1963. Like Haywood, Buckley seems to have been an officer who, from the time the Barnett administration began, was earmarked for greater things. Under the old administration, Buckley had achieved no significant promotion in twelve years of service. An ordinary prison officer until almost the time Dallard retired, Buckley was promoted to deputy principal, then principal warder in the space of only a year. Skipping the rank of chief, he was made superintendent the very next year. Thus, Edward Buckley, who had been ignored for more than a decade, had ascended dizzily from senior grade warder to full superintendent in a little over three years. The 1950 appointment was important in itself, but it was made even more important for Buckley because it was here that the young officer was to establish himself as a therapist of ailing prison management.

For over twenty years, J. G. Quill, 'uncrowned King of the King Country', had been officer in charge[3] of Waikune prison. A huge, barrel-chested Irishman, Quill's notoriously unorthodox and corrupt[4] administration had for years gone unnoticed by Dallard. When the new broom began its work in 1949 and revitalized the penal system, one of its first actions was to sweep Quill out of office. Jerry Quill was transferred to Paparua as a lowly chief warder, still picking up his superintendent's pay.

It is significant that in 1950, it was Buckley who was chosen to attend to the decrepit administration at Waikune. It is noteworthy also that five years later, Buckley was chosen to deputize at Mt. Eden, before his appointment to the large medium security institution at Christchurch. Buckley's work at Waikune must have been considered satisfactory, and his duty at Mt. Eden undoubtedly favoured his chances of getting the top job in 1963. Buckley's straightforward manner was already renowned and it, too, made him a strong contender for the position.

The very same week that Buckley took office the newspaper *Truth* began a series of three weekly articles on Mt. Eden prison, describing conditions.[5] Although trouble at the institution had caused recreational facilities to become restricted, evening classes still took place regularly in a variety of subjects. Basketball teams visited the gaol to play against inmates, and all prisoners serving twelve months or more were allowed to see one movie in the institution every month. At the time Buckley took over, Mt. Eden held 400 prisoners, with seventy-three officers and twenty-two instructors taking care of them. By April 1963, Eddie Buckley was already well known to inmates. 'Foxy' Townes, a prisoner under Buckley at Paparua, remembered him as 'scrupulously honest', a man who, he recalled, would not have his car filled with Government oil without paying the oil back; who once forbade a prison officer to collect pine cones on prison property. The cones, he said, belonged to the Justice Department.

Mt. Eden inmates were familiar with Buckley too, and many of them worked for him during his years as quarry boss. There his dark look and stern countenance had already earned him the nickname of 'Black Buckley', and 'The Black Dog'. Buckley was a hard taskmaster and inmates who knew him dreaded the day he would return. They feared when Buckley took over the superintendency of Mt. Eden that the equity they had known under Haywood would vanish.

When Buckley came to Mt. Eden he had three main goals in mind. The first was to take control of the staff. Buckley's most important job was to try to rebuild morale and discipline within the uniformed cohort, and re-establish the rule of authority. This would require rigid recognition of status differentials, and the devolution of responsibility through officially recognized channels. Secondly, Buckley intended to dissolve the inmate hierarchy. From then on, all prisoners would be treated the same and none would have greater status, privilege, or power than any other. Thenceforth all authority would reside within the official structure. Informants would be unwelcome. In fact, Buckley told me he actively discouraged informants and would throw them out of his office. 'Any inmate who would betray his mates would betray me as well, and I didn't want that,' he said. Finally, Buckley took steps to strengthen security. Barbed wire was installed, locks were replaced, an electric trip wire was put around the perimeter wall. Prison instructors now had to assist with the duties of disciplinary staff. Instructors were told they were prison officers first and foremost, and that as such, their principal duty was custody. An all-out effort was made to stop further escapes.

From the time he assumed command, Mt. Eden prison was run the way a maximum security prison ought to be run, Buckley told me. Taking over from Haywood required the same sort of reorganization as had been necessary when he replaced Quill at Waikune. 'I didn't really tighten Mt. Eden up after I arrived', Buckley said, 'I just ran it to the book.' Head office's response to Buckley's initiatives was positive. In June 1963, Deputy Secretary Mr L. C. Cutler visited the prison and reported on the new administration. Mr Buckley was getting a grip on his job and had staff co-operation, he said. 'He is giving his attention to security and aims to achieve it to the greatest degree possible.' Mr Buckley, it was noted, had 'accepted the challenge' of Mt. Eden.[6]

When information about security improvements reached the media it made headline news. Within a few months, *The New Zealand Herald* reported, new barbed wire entanglements had been established at key points. Sentry boxes had been raised to improve their surveillance potential. Extra staff had been employed. Floodlighting had been increased. Routine had been tightened, and electronic warning devices were to be installed on the walls.[7] Soon after Cutler's visit, Ian MacKay, Director of Prisons, also visited. He too was suitably impressed. Having spent an afternoon interviewing staff and inmates, MacKay came to the conclusion that (in spite of security measures) 'the era of suppression at Mt. Eden was over'. Staff were now more alert than before, and their morale was higher since the new administration had taken over.[8]

It was not more than a few weeks after Buckley arrived at Mt. Eden that the first trouble broke out. On 23 May 1963, ninety to a hundred prisoners assembled in the main yard at the institution, refusing to go to work. Departmental officials declined to discuss the reasons for the strike and Buckley refused to bargain. Within two hours the men had given in. In his official comment on the event, MacKay described it as without reason of major significance, and as having been merely a 'try-on' for the new administration. Head office was no doubt pleased with the expeditious way in which the bluff had been called and was not slow to commend Mr Buckley's strategy.

Buckley's penal philosophy was traditional and conservative and diametrically opposite to that of Haywood. Although Haywood's officers had been dissatisfied with some of his methods, a large proportion was none the less committed to him and to his ideals. Not a single prison officer I spoke to disliked Haywood. He was regarded as a decent, humane person, and one who accepted ultimate responsibility for the mistakes of junior staff. While his informality may have been criticized, Haywood still enjoyed the respect and loyalty of his men. When he was transferred, ailing, to Paparua, it was with a genuine feeling of sadness that the bulk of staff bade him farewell.

Buckley was a different person altogether. Like the older criminals, higher employees already knew him from his days as a junior officer in the quarry. He was vociferous, brash, and uncompromising. When he arrived at Mt. Eden he loudly advertised the changes he intended to make. This irritated staff and alienated them from his goals at the outset. Tom Hooper, an instructor who was especially fond of Haywood, described his early memory of Buckley like this:

> He had all the officers lined up on parade in the front entrance and Haywood was still there in the superintendent's office, hadn't left, and Buckley had come to take over. And Haywood was upset. He was bloody upset about having to leave . . . That was his bloody prison and that's it and he was upset about the whole thing and this Buckley stood in front of the whole bloody parade and he started with a loud mouth, shouting at the top of his voice like he always does, yelling about the changes he was going to make and this was going to be a different place in future and . . . 'The administration that's been here before is no bloody good and everything's going to change now.'
> A real standover bastard . . . That Buckley really got my back up at the start and it never came down.

The prison psychologist described Buckley's first days at the institution in a similar way:

> He put on this frightful show of strength . . . he put on all this braggadocio and show that *he* was going to clean this place up and *he* would put the officers back in their correct status and he would show these bastards, the prisoners, where they belonged.

Hooper said that Buckley's attitude was one of, 'You'll do *what* I say, *when* I say it, and that's it!' His arm-waving, desk-thumping diatribes became legendary, and the way he rough-handled his staff became known, even to the prisoners. One former inmate told me:

> Even the screws were jumpy. The screws were looking behind them all the time and the screws had to yell and scream because Buckley — you know how they come in in the morning and they line up? The office boys, prisoners, used to come and tell us. 'Fuck, he lines them up and he fucking yells and fucking screams at them!' And the screws were saying 'Fucking Black Buckley Bastard!', and all this, eh.

Officers and inmates who disagreed with Buckley were shouted into silence and often ordered from his office. Buckley's style was to intimidate his critics by overpowering tirade.

The aspect of Buckley's administration which officers disliked particularly was that although he was quite ready to accept praise for his work, they felt he was reluctant to take responsibility for mistakes. Several officers

Arthur Hume, Inspector-General of Prisons 1880-1909. *Department of Justice*

Sir John Findlay, Minister of Justice 1909-1912. *Alexander Turnbull Library*

Charles Matthews, Controller-General of Prisons 1912-1924. *Alexander Turnbull Library*

B.L.S. Dallard, Controller-General of Prisons 1925-1949. *Department of Justice*

Rex Mason, Minister of Justice 1935-1949, 1957-1960. *Alexander Turnbull Library*

Ormond Burton, former soldier turned pacifist and prison reform campaigner. *Alexander Turnbull Library*

Clif Webb, Minister of Justice 1949-1954. *Alexander Turnbull Library*

Sam Barnett, Secretary for Justice 1949-1960. *Department of Justice*

Senior staff, Mt. Eden Prison 1950. Left to right: John Lauder, Superintendent; Horace Haywood, Deputy Superintendent; F.T. 'Scarface' McKenzie, chief warder; and Captain Stanley Banyard, welfare officer. *Department of Justice*

Mt. Eden Prison, late 1965. The old women's quarters are visible bottom right. *Department of Justice*

Entrance to Mt. Eden Prison 1950, with a Black Maria in the foreground. *Department of Justice*

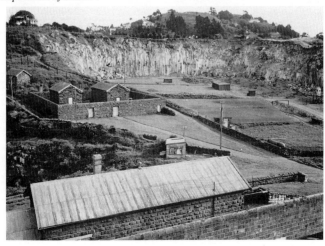

Mt. Eden Prison Quarry, early 1950s. *Department of Justice*

Remand and segregated exercise yards, Mt. Eden Prison. *Department of Justice*

Wire and perimeter wall surrounding the main exercise yard at Mt. Eden. *Department of Justice*

Armed sentry tower, Mt. Eden Prison. *Department of Justice*

The 'Dome', central hub of Mt. Eden Prison. *Department of Justice*

East block, Mt. Eden Prison, c.1950. The recess leads to the segregation cells. *Department of Justice*

Segregation cells, east block basement, Mt. Eden Prison. The wooden door leads to the hanging yard. *Department of Justice*

Three-tiered cellblock, Mt. Eden Prison.
Department of Justice

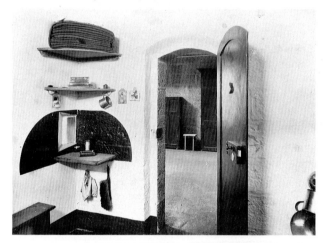

View from inside a Mt. Eden Prison cell 1950. *Department of Justice*

Cells at Mt. Eden showing parade lines on the floor. Fresh blood stains on the floor and doorway of cell 20 are from a recent fight between inmates. *Department of Justice*

Lifer's cell, Mt. Eden Prison c.1954. *Department of Justice*

Association cell, Mt. Eden Prison mid-1950s. It was from such a cell that 'Maketu' and his associates escaped in 1958. *Department of Justice*

Prison officer Dan Cavanagh (centre), c.1965. *Department of Justice*

Mt. Eden Prison staff, 1960. H.V. Haywood, Superintendent, is in the centre of the front row. *Department of Justice*

Prison Superintendents' Conference 1961. Front row, left to right: J.R. Hanan and H.V. Haywood; back row, third and fourth from the left: J. Hobson and E.G. Buckley. *Department of Justice*

remarked on this without being prompted. One, who had initially described Buckley as 'the finest man (he had) ever worked under', went on to say:

> The thing about Buckley, though, was that if anything went wrong it was always everyone else's fault but his. But he was always first to take the credit when it was right.

Another commented, 'Well if anything went wrong, we were to blame. And if it went right, he was the centre of attraction.'

Because of this, there was a general feeling among staff that they could not rely on Buckley's backing them in performance of their duty. Hargrave remembers how he was once called by Buckley and ordered to remove a man who had barricaded himself in a cell, armed with a milk bottle. It was obvious that the situation was hazardous. When Hargrave pointed this out to Buckley, Buckley said, 'You're the Judo expert, go in and get him out', adding that if the inmate got injured, it would be Hargrave's responsibility.

This attitude was a direct contrast to Haywood's. Under Haywood, officers knew that the superintendent stood behind them. Of the two, E. Molloy said:

> Mr Haywood was a *great* humanitarian. If you were in trouble, or you had a serious problem dealing with your job, you never had any hesitation approaching him and he would assist you. He would. He would assist you. But not Mr Buckley.
> 'That's yours! That's yours, now you can stew in it. You got yourself into it, now get yourself out of it.' That is a comparison between the two men.

Inmates who knew Buckley spoke about him in a similar way. Almost to a man, they had a low opinion of Buckley as a person and as an administrator. Several of both groups harboured a deep hatred of him.

Buckley disagreed with Barnett's new method, and when I spoke to him in 1983 he still believed that prisons were too easy. The way to reduce recidivism, he told me, is to make prisons more punitive. He believed that privileges should be earned, not given, and then kept within strict limits. Prisons, he said, should not be made pleasant places to be in. Unlike Haywood, Buckley would not walk freely about the institution and talk to inmates. Wherever he went he took other staff with him. If approached by an inmate with a request, Buckley's normal response was to tell him to follow official procedure and submit an interview slip. Privileges became difficult to obtain. An inmate who knew him in his later years commented:

> Buckley was a strict disciplinarian. He kept well away, there was no contact at all with the crims. You could see him by applying and go and look at him in his office. He wouldn't listen to anything. I was interested in electronics, I asked him if I could get in a few books. He said, 'No electronics books here.' He said, 'Lots of people have asked for electronics books. The next thing we'll know they'll find out how to open the electronically controlled gates at central control.' Well, that was his very narrow attitude.

Weekly interviews became formal affairs, with an inmate standing behind a steel grille, and flanked by prison officers. There was little room for discussion.

If you went and got an interview slip and you wrote one in, and straight away, like you weren't allowed to say anything, he'd look at what you requested, he'd read it through, just put it down and shout, 'Get out!'

Staff told me that Buckley's manner was but a flimsy facade, which fell apart when tested. Behind the aggressive exterior lay a weak, insecure man, forced to govern autocratically because, for him, there was no other way. One inmate believed that the reason Buckley would not walk about the prison alone was that he feared the inmates. A prison psychologist, Dr Gordon Parker, reinforced this view:

> Oh, he was absolutely frightened of the men. It always is the way with the man of power. Just the same as Stalin was afraid of his men.

Another said, 'Buckley was the weakest thing you've ever met. . . . He was like a bubble. You prick it and there's nothing there.' Dan Cavanagh's response was similar: 'No substance. Empty. No substance whatsoever,' he said.

When I attempted to interview Mr Buckley, I found him unstable and erratic. He was bellicose and vitriolic, waving his arms about with such vigour that I feared I might actually be hit. His mood swung from wildly aggressive to mild and conciliatory in a single sentence. His conversation was disjointed, repetitive, and at times contradictory. He asked me questions that he then answered himself. He ignored anything I asked, and spoke testily but aimlessly over an unconnected range of topics. He did not respond to any comment I made about any of the points he had raised. In fact, it seemed as if I was not there at all; as if Buckley was talking to himself, or to a larger audience, unknown to us both.

This conversation took place on the doorstep of Mr Buckley's house, on a dark and drizzly night. After about forty-five minutes he told me that he had visitors and had to go inside. Although he promised to let me interview him 'in a couple of months', this was the longest talk I ever had with him. He refused to give me his (unlisted) telephone number. When I tried to contact him again he was either not at home, or made lengthy postponements. I soon realized that a useful interview would not be possible.

In the light of his personal philosophy, how Buckley came to be favoured as superintendent is puzzling. As noted, he disagreed with Haywood's liberal policies and believed in discipline and deterrence. He seems to have been a man of classic 'authoritarian' personality: intolerant, insecure, highly conformist, overbearing towards subordinates, and yet submissive in his dealings with superiors.[9] Perhaps this last quality holds the key to Buckley's early acceptance — he apparently appeased his bosses and made efforts to show support, even for ideas he disagreed with.[10]

Buckley's style of management had been successful in the first two institutions he controlled. When he took command of Mt. Eden, he was quickly able to re-establish authority, routine, and regularity. In the interests of security, inmate activities were either stopped or strictly supervised. In addition to this, he completely refurbished the gaol's interior. Paint was renewed and brass was polished. Buckley may not have been the most popular of men, but he ran a tight ship and order, for a time, prevailed.

Trouble Begins

The twelve months from mid-1963 were not eventful. At Mt. Eden, plans for the new security block were proceeding. Great attention was focused on this, and on the progress of plans for the new maximum custody complex. The public wanted security, and it was security which Buckley delivered. Mt. Eden was quiet — perhaps ominously so — and remained largely out of the public eye. Emphasis was placed on the buildings of the future and an effort was made to keep publicity about the old gaol down. At the end of 1963, when three inmates were charged with assaulting two officers at the institution, the Magistrates' Court was convened inside the prison walls. It was the first time such a Court had been held there in forty years.

It was not until the middle of 1964 that evidence of rising tensions again began to show. In June, a paedophiliac inmate was stabbed in the back. No culprit was charged. That same month, a mysterious fire broke out amongst scrap timber by one of the shops. Another mystery fire broke out a month later in the tinsmiths' shop, followed by a third in August, this time in the chapel. In late August, a piece appeared in *The Sunday News*, which was to create the first of several controversies that Buckley would face. Titled, 'The Black Shame of Mt. Eden', this long article, based on information given by an ex-prisoner, described a wide variety of improper practices which it said were well established at Mt. Eden. At the 'school for scandal', the paper reported, inmate violence was ignored or condoned by prison officers, rapes occurred, and homosexuality was accepted. Bookmaking was allowed, corruption of staff was frequent, and consumption of liquor by inmates, common. Drugging of uncontrollable prisoners was a part of normal procedures.[11] Food and medical services were poor.[12]

The allegations, extravagant though they seemed, were of sufficient gravity to warrant a magisterial inquiry. However, before the inquiry began — the very next week in fact — another piece was published, this time with information from a former prison officer. 'Prison Scandal Explodes' added to, and affirmed, the material of the week before.[13] The allegations were serious and two days after the second publication, Buckley wrote to head office, vindicating his administration and dealing with the contentions point by point. Every accusation made by *The Sunday News* was denied or defended by the superintendent.[14] The weaknesses of the articles were several. Some of the examples given were patently and demonstrably incorrect. Morever, when interviewed, the men who had made the charges claimed they had been misrepresented. This allowed the report of the inquiry to assert that some of the accusations may have been valid in the past, but that the situation was now different. Indeed, referring to the rape of young prisoners by older men, one inmate deponent stated that such incidents had taken place once, but could not happen now under the more rigid system.[15]

The report of Coates SM into allegations at Mt. Eden prison was presented to the House of Representatives on 9 December. Not one of the press accusations could be justified. Claims were found to be either totally false, unfounded, exaggerated beyond credibility, or to have referred to conditions which had since been changed. Of Buckley the Magistrate wrote:

> [He] is a capable and efficient officer who is strict but fair. To the members of his staff he has made known his policy to which he always adheres. I think it can fairly be said that he has the confidence and full co-operation of his staff . . . both the inmates and the staff know exactly where they stand and what to expect. During my inquiry no serious complaint was made about the present administration. Therefore I accept that his administration is sound.[16]

The endorsement was publicly satisfying, if not somewhat predictable. Robson was pleased with the result, and he ordered immediate public release. The Coates report was covered by *The Evening Post* the same night.[17]

However, at Mt. Eden all was not well, and emotions were beginning to rise. By Buckley's own calculation, between December 1962 and September 1964 almost fifty officers — that is more than half of his total uniformed complement — had resigned.[18] In the following twelve months another eighteen left their jobs. In September 1964, a bomb was lobbed at Buckley's house. Although no damage was caused, the message in the explosion was clear. That same month, more barbed wire appeared on the walls and new locks were fitted to the doors. So while routine was now organized and ostensibly more efficient, morale had not improved. On the contrary, growing discontent was creating difficulties that security could not address.

Despite its menacing prospect it was not until the following year that serious disturbance broke. When it did, its violence was on a larger scale than anything the penal system had known. For the first time firearms were used, and for the first time since the Smith escape in 1940, there was a deliberate threat to officers' lives.

THE GENESIS OF DISCONTENT

For some who knew Buckley, such events were predictable. Like a confined explosion, violence, when it occurred, was amplified by the restrictiveness of Buckley's leadership. To one inmate it was the atmosphere of oppression which built up that made violent outbreak inevitable.

> If you start putting the pressure on people, you're asking for them to react. Violence breeds violence. That was Buckley's doing. You put a weak man in charge of a place like that and that was his first line of defence. Security and lock-up. Pressure. From the first day up there he was a cunt. Black Buckley.
> See, nobody's going to put up with that . . . 'cause the screws would come out and yell at the crims and the crims were getting all steamed up. And at that time they were tightening things up that much that the crims were resorting to anything to get out. That's why they got that fucking gun in.

However, it would be wrong to view the unrest at Mt. Eden in isolation. The effects that the new regime produced arose directly out of its predecessor. Had Buckley taken over directly from Lauder and under the traditional regime of Dallard, for example, it is unlikely that the same trouble would have occurred. Where revolutions are concerned, uprising is said to be likely when a period of progressive development is followed by one of sharp reversal.[19] Thus, at Mt. Eden, it was the *contrast* between Buckley's approach and that of the previous ten years which, more than anything, brought about the impending crisis. The changes of the fifties and the publicity campaigns of the sixties had altered people's attitudes to prisons. An emphasis on humanitarian treatment had taught the public to demand more than just custody and punishment. Likewise, those who went to gaol in the mid-1960s no longer expected to be deprived of all status and rights. Haywood's progressive style had given prisoners an appetite for better things.

To some, the troubles of 1965 were, therefore, not entirely unexpected. The prison psychologist commented:

> I felt very keenly, when I hadn't been long in head office and the big riot and fire took place, that this had been brewing. I *knew* that there would be fun and games before long, after Buckley took over. . . . Oh yes, I was sure — you could feel it in the place. I was only there a few months under Buckley. But he was full of bombast and braggadocio. And I knew there would be trouble. Because of the way Haywood had run the gaol really caused the rioting after he had left.

Under Haywood, Cavanagh told me, security had been deliberately relaxed. The prison was overcrowded and understaffed, and was in such a condition that Haywood knew antagonisms would build up. Haywood felt, acutely, the need for a safety valve through which overcharged tensions could be released. The existence of relatively easy escape routes cut the likelihood of desperate measures. On the other hand, as seen with Horton, it also allowed dangerous prisoners to abscond. The problem was that Mt. Eden had been built for nineteenth century custodial purposes, and by the 1950s it was obsolete. The only way to compensate for its physical deficiencies, Buckley felt, was to increase the presence of lock-up control. Escape outlets were sealed off. Of this effect, Cavanagh said:

> He screwed the place up so bloody tight that it became a challenge. Instead of being a gaol, it became a fortress. You know, *steel* doors. Grilles. Different locks. . . He made it into a fortress, until instead of an escape — as it was before, really just a runaway — it then became a real escape. And the only way that it could be done without blowing the place up was by firearms and taking hostages.

Cavanagh's interest in such matters is more than casual. Not only had he been for four years a prisoner of war and escapee in Germany during the war, he also had over twenty years of prison service behind him. Nearly all of this was spent in maximum security. In 1965, only twenty-two months after Buckley took over, Cavanagh himself was the victim of an armed escape.

The incident was prelude to a disaster, which later in the year, spelt the finish of Mt. Eden as a maximum security institution.

THE 1965 HOSTAGE CRISIS

For most of the time since recapture from his escape in 1963, George Wilder had been held in the security division at Mt. Eden prison. The security division was situated in the basement of the east wing, and was an especially tight 'prison within a prison' for men for whom it was felt that very close confinement was needed. Other inmates in the block included Leonard Evans, serving eleven years for escape and burglary, and John 'Dirk' Gillies, who was doing life for a double machine-gun slaying at Auckland in December 1963.

Inmates in security all had separate cells, and although they sometimes exercised together or worked with other inmates in the open wings, security men were always unlocked separately. Whenever they left the prison buildings, these men went individually, with a staff escort, and they were handcuffed beforehand. Security block prisoners had their own little exercise yard, but apart from a few hours in the morning and a few in the afternoon, all of their time was spent in their cells.

The only work available to these men was domestic and on the afternoon of 4 February 1965, Leonard Evans, employed as a cleaner in the west wing, was let out and escorted to his place of work. Wilder and Gillies, who would have gone to their yard to exercise at this time, complained of intestinal trouble and instead elected to stay locked in. Gillies said he had diarrhoea and shortly after 2.00 p.m. he attracted a screw's attention and requested to be unlocked to empty his piss pot. This was done. Gillies took his pot into the toilet and sat down.

At about the same time, in the west wing, Evans told an officer he was not feeling well and requested to be locked up again. Evans had only that morning been sentenced to a further five years imprisonment on the six he was already serving, so the request did not seem odd. Evans was escorted back to the security division, carrying his jacket over his shoulder. Hidden under the jacket was a loaded, sawn-off shotgun. At the entrance of the block, Evans was met by an officer called Daniel who unlocked the grille and admitted him. Instead of proceeding to his cell, however, the tall and powerful Evans turned and walked directly towards the ablutions section, where Gillies was now waiting. At the approach of Evans, Gillies came out towards him and the two met, then sprang apart, with Gillies brandishing the gun. Officers Daniel, Wilson, and Cavanagh were threatened, then taken at gunpoint to Evans' unlocked cell. A key was taken and Wilder was released. An officer called Holt soon entered the block and he joined the other three officers and one inmate wing cleaner, who were still in Evans' cell. Officer Jack Moyle came into the block soon after and he too was forced to join the five hostages.

Moyle and Daniel were bound hand and foot, and left in the cell while the three inmates escorted Wilson, Cavanagh, and Holt out of the block and toward the stoneyard at the rear of the prison. Cavanagh, with the gun at his back, was forced to open the grilles and having reached the final, stoneyard gate, Cavanagh ordered Percy Syddell, the officer in charge, to open it. He and carpentry instructor Tom Hooper, who had inadvertently walked on to the scene, were then made to join the party which passed through the door to the outside of the gaol. Beyond the walls, but still on Justice property an old prison truck was commandeered and the prisoners abandoned their hostages, save Dan Cavanagh, who was told to enter the cab of the truck with the escapees. The truck was then driven through a crash barrier and off down the road, followed (to Cavanagh's dismay) by a load of Greener buckshot from no. 1 sentry tower.

Up until now the plan had been meticulously thought out. The gun and its ammunition had been smuggled in, in a dismantled state, only a day or two beforehand, probably inside hobby material. The co-ordination of Gillies' and Evans' rendezvous in the toilet had been precise. However, from that point on the men appeared to have no strategy. The slow old truck rumbled into the suburb of Mt. Eden in a directionless search for somewhere to hole up. Eventually an address at 33 Horoeka Avenue was selected, crashed into, and the two occupants — an elderly woman and her adult son — were taken hostage.

It was only a matter of minutes before the house was surrounded by regular police and members of the armed offenders squad. Realizing fairly quickly that from here there would be no escape, the three prisoners became resigned to recapture. After about half an hour the old woman was released. The rest remained barricaded in the house until shortly after 6.00 p.m., when they threw out their weapon and surrendered.

As reward for giving up without causing injury, the trio negotiated a bottle of whisky. This no doubt eased the pain of failure. The five-year extensions to their sentences, awarded a few days later by the Auckland Supreme Court, were not negotiable. All three were returned to the security division at Mt. Eden, with Gillies still serving life, Evans serving sixteen years, and Wilder now carrying the heaviest finite penalty ever given in this country, a term of eighteen years, ten months.[20]

The hostage event was the first major operation for the recently formed armed offenders squad, and is still the most sensational escape in New Zealand prison history. It received maximum coverage in national newspapers as well as on television. In his report on the matter Mr A. A. Coates SM was unable to attribute blame for the escape on any person or persons, and could not explain the entry of the gun or ammunition.[21] An inmate informant[22] named a recently discharged convict[23] as the source of the weapon. Police investigations, however, were fruitless, and both the origin of the gun and the manner by which it came into the prison remain uncertain to this day.

THE AFTERMATH

The 1965 escape was a significant event in the annals of maximum security. First of all, it highlighted more than ever before the gross deficiencies inherent in the old gaol. It was made abundantly clear too that repressive control and tight organization would not deter desperate men. Indeed, a determined desire to get away might lead to dangerous measures. In his report on the incident, Mr Coates was able to make only minor recommendations for change. The inadequacy of the old prison was now obvious, and Coates urged the completion of the planned maximum security block without delay. The obvious and urgent need for the new top security institution at Paremoremo was underlined once again.[24]

The building of a new security block at Mt. Eden had been contemplated since the escape of Wilder and company in January 1963. It was officially announced in March. Although the preliminary design had been completed before September 1963, by May the following year, drawings had already fallen behind schedule. Plans were prepared before the end of the year, but in December the scheme was temporarily shelved because the estimated cost of 48,000 pounds exceeded the 40,000 pounds limit placed on it by the Cabinet Works Committee. Three days before the hostage crisis the situation had not changed, but it was this which forced the Government's hand. Although in March it had been reported that cost estimates of the unit could not be brought below 48,000 pounds, the Cabinet Works Committee was urged to give its approval. A week later, the very day on which the report on the escape was made public, Cabinet announced the calling of tenders for a new 45,000 pound security unit at Mt. Eden. By the beginning of July work on it had begun and after a number of delays, it was finally completed in March 1966.

Although Buckley had been completely exonerated by the inquiry of 1964, head office knew by this stage that his administration was faulty. Don MacKenzie, formerly the prison's psychologist, had firm opinions about Buckley and he was now the senior departmental research official. When Buckley overrode an instruction regarding access for official visitors, a disgruntled PARS[25] representative was counselled by MacKay to be patient. Mr Buckley was a 'difficult man to handle', the Director of Prisons said.

Although instability at Mt. Eden was accepted, one wonders how deep concern was at this stage. Evidence is, that compared with later years, it was still not very great. Buckley remained at his station. A replacement building was under way and it appears that policy for the moment would be to mark time with Mt. Eden, and try to avoid further crisis. While some — like MacKenzie — expected trouble to continue, it is doubtful that anyone envisaged a calamity as great as that which took place only five and a half months after the hostage incident was over.

NOTES

[1] NZPD vol. 331 (1962) 1520-1; vol. 335 (1963) 88, 806.

[2] NZPD vol. 335 (1963) 104, 282. The allegation was not challenged by National members. Forty-nine people escaped from penal institutions in 1963, a figure that

was high, but lower than in 1958 and 1959, when there were sixty-three escapes in each year.
3 Quill was controlling officer for all his time at Waikune, but was not promoted superintendent until 1951.
4 This judgement is based on the testimony of a number of prisoners who served under Quill. It is reinforced by Hamilton (1953), who devotes a large section to 'Wekunai' (Waikune) prison and its governor 'Obadayah Dowell' (Jerry Quill).
5 These articles appear in *Truth* 16, 23, 30 April 1963.
6 *PGM* 11 June 1963.
7 *NZH* 20, 22 July 1963.
8 *PGM* 1 July 1963.
9 Adorno *et al.* (1950).
10 Dr A. M. Finlay, Minister of Justice 1972–1975, for example, said that he found Buckley to be a man with a 'big front' of *bonhomie* and geniality, which he found difficult to penetrate. It also seems to have been typical of Buckley to present a much milder image to the public than that which I observed. Speaking to a conference of Catholic chaplains in1964, for instance, Buckley came out strongly against repressive prison administration and acclaimed the value of rehabilitating criminals by changing their points of view (*NZH* 10 September 1964). An *Evening Post* article once described Buckley as 'the big, genial caretaker of Mr Eden' [*sic*], a man who was 'more of a humanitarian than a prison superintendent' (*EP* 24 September 1968). Public statements by Buckley also appear in *NHZ* 9 March 1965; *AS* 8 November 1968; and *EP* 12 June 1974.
11 While attempting to research this matter at head office, I chanced upon a file that listed doses of psychotropic drugs such as Largactil (Modecate) that had been given to inmates during the period in question. 'Inmate went berserk' was given several times as the reason for medication being prescribed. I am told the drug Paraldehyde was also used. Unfortunately the Justice Department's drugs register was taken away from me before I could make any notes from it.
For a comment on the use of these substances in a forensic psychiatric ward see Newbold (1982a). The legality of administering drugs to mentally disordered offenders is discussed in Newbold (1984).
12 *Sunday News* 30 August 1964.
13 *Sunday News* 6 September 1964.
14 Memo from Buckley to Robson, 8 September 1964.
15 *Report of Inquiry into Alleged Happenings at Auckland Prison* (1964) 2.
16 *Report of Inquiry into Alleged Happenings at Auckland Prison* (1964) 14.
17 *EP* 21 December 1964; *NZH* 22 December 1964.
18 Memo from Buckley to Robson, 8 September 1964.
19 Davies (1962).
20 It should be noted that reports of the men's overall sentences vary slightly. It could be that the extra ten months on Wilder's term was a result of the time he had spent at large, when his sentence ceased to run. Unfortunately, the men's files are confidential.
21 *Report of Inquiry into the Escape of Gillies, Evans, and Wilder* (1965) 13–18.
22 The informant was a passive homosexual who, until recently, had been the lover of one of the escapees. According to Buckley's policy (unlike Haywood's) the relationship had been broken up by staff. After remonstrating with officials over that event, the aggressive partner had been confined in the security section, where he was able to take an active part in the escape and its planning (Cavanagh pers. comm., and notes of evidence of J. W. Cross and J. J. Connor, February 1965 in *Report of Inquiry into the Escape of Gillies, Evans, and Wilder* (1965) 13–18).
23 This man is now serving a long sentence in New Zealand for drugs offences.
24 *Report of Inquiry into the Escapes of Gillies, Evans, and Wilder* (1965) 22, 23.
25 Prisoners Aid and Rehabilitation Society.

11: MT. EDEN PRISON IN CRISIS 1965

Between February and July 1965, the maximum security prison of New Zealand was concealed from publicity. In December 1964 it had been suggested in the penal group that the whole public relations programme of 1965 needed toning down. Three and a half months later the group resolved that the public speaking campaign should be eased as well, and that in 1965, penal policy needed a much lower profile. After the ignominy of the hostage incident, it is easy to understand the penal group's feeling that the less said about the decrepit old gaol the better. So at Mt. Eden the policy of publicity moderation, which had begun when Buckley took over, continued into its third year.

Interest in the progress of the new institution and in the new security block at Mt. Eden redoubled. In April 1965 the Department had acknowledged that there were delays in the earthworks at Paremoremo, where the replacement gaol was to be built. The hostage incident had underlined the need for an immediate remedy, however, so the former superintendent's quarters at Mt. Eden were quickly converted to a dormitory. Female inmates were transferred there from their old wooden building within the walls on the weekend of 1 and 2 May. Tenders for the new block were called the following week. By the beginning of July, work on the unit, to be erected on the site of the old women's division, had begun.

Decrepit buildings were not the only problem facing prison authorities; there were staffing concerns as well. Turnover was high, and the Department had been losing personnel at an average of over 16 per cent, per year.[1] Despite rising establishments, which had gone from 634 in 1961-1962 to 682 in 1963-1964, there was still an acute staff deficiency, which was nowhere more serious than in maximum security. As we have seen, over half of the total complement of Mt. Eden had resigned between December 1962 and September 1964. In early February 1965, Mt. Eden was nineteen officers short. By 22 February, the situation had worsened and the prison was down by twenty-two. The March inquiry[2] had commented on this, on the excessive overtime being worked, and on the desirability of reaching establishments as soon as possible. No effective action was taken. By mid-1965 Mt. Eden was twenty-four short of its full staff establishment of 177 officers and instructors.

In the past, disturbances at the prison had more frequently been blamed on understaffing and overcrowding than on mismanagement, and this had especially been the case after Buckley arrived. However, a rising emphasis on non-custodial penalties had meant that since 1960, overall prison populations had held steady. Additionally, many offenders had been steered away from maximum security to the open institutions, so the explanation of overcrowding there was less convincing than before. DAPs at M. Eden had dropped from around 400 in 1960 to 350 five years later. In the middle of 1965 the muster at Mt. Eden was lower than it had been for some time. On 19 July, for example, the population stood at only 293.[3] So if overcrowding had indeeed been a significant cause of trouble, there can have been few occasions in the previous decade when upheaval was less expected than in July 1965.

THE ESCAPE ATTEMPT

In the early hours of 10 June 1965, two armed men broke into the Bank of New Zealand at Avondale and hid there. When two employees arrived for work later that morning, they were surprised by the intruders who beat, bound, and gagged them before making off with over 15,000 pounds in cash. One of the robbery victims was New Zealand Olympic marathon representative, Jeff Julian, but of greater moment than this, the crime was at the time the biggest of its type ever committed in the country's history.[4] The event attracted huge publicity, and it was not long before two suspects were apprehended. They appeared in court on 25 June. The accused were David Harley Western, twenty, a law student at Auckland University, and Daniel Huntwell MacMillan, twenty-eight, unemployed, alias Philip Linwood Western, alias Philip Archibald Linwood Western, who was David Western's older brother. Also charged was MacMillan's common-law wife, who also appeared under the name of Western.

The woman was granted bail, but the brothers were denied it, and on 25 June the two were admitted to the remand section at Mt. Eden prison to await their lower court hearing. From the time he entered Mt. Eden, MacMillan had his mind on escape. The charges of robbery were serious, and MacMillan was determined to be out of the prison before his scheduled depositions hearing on 20 July. One of his plans was to rob another bank as soon as he got out, and with the money to buy a boat and sail away in it. Little is known, and there is disagreement among those who knew him, about the type of person MacMillan was. However, whatever he was — philosopher, sage, or madman — his actions were to have an effect on maximum security prisons that is still visible today.

MacMillan was an intelligent man, but he was of unsound mind. At his trial in October, the evidence of his brother, his mother, and two psychiatrists was that he was insane. MacMillan felt himself to be in every way superior to the common people, and he delved deeply into the writings of Nietzsche. At times he believed he was the reincarnation of Rasputin and Mephistopheles. A tall, gaunt figure with a long red beard and fiery, unkempt hair, he provided amusing novelty to the remand wing's dull routine. Before long, MacMillan's pious ideology and his open contempt of authority began to appeal to these disgruntled outcasts of society.

In the mid-1960s challenging the *legitimacy* of official power in New Zealand was still comparatively rare. For MacMillan its rejection was automatic. Conventional society was a travesty and he saw its acolytes as vacuous fools. Prison officers were amongst the worst, and at Mt. Eden he refused to address them at all, except in the most contumely fashion. 'How *dare* you speak to me, you filthy little pleb!' he would exclaim, indignantly, when spoken to by a member of staff. 'How *dare* you feed your wife and kids on my misfortunes! Begone!' During his trial, MacMillan's behaviour was similarly impertinent. He refused to plead. He considered proceedings 'boring' and accused the Judge of 'speaking out of turn'. 'You have no knowledge!' he screamed. Later, MacMillan asked to be excused from the court. 'I don't think I can stand all this blasted nonsense today!' he said.

Although at that time he wore a long overcoat, this was later swapped for a blanket with a hole cut in the middle. He did not like prison dress. The drab, blue garb held no appeal for the messiah MacMillan now thought himself to be. In accordance with his new identity, he changed his name by deed poll from Daniel Huntwell MacMillan to Leonatus Trajax Globeous Aureolus Philotus Augustus Cosmopolitus. In gaol they just called him 'Cosmo'.

Today, some inmates still talk about MacMillan with a kind of reverent fascination. Although they caricature him, some have been introduced to Nietzsche and see much of value beneath the eccentricity of his thinking. It is in the light of this impact that MacMillan's influence at Mt. Eden must be understood. During his trial, there was argument among specialists about the state of MacMillan's mind. Two eminent psychiatrists[5] described him as a paranoid schizophrenic. A third[6] dissented, however, and MacMillan was convicted of bank robbery, attempted escape, and a number of other charges. He was sentenced to a total of nine years imprisonment on 12 November 1965. David Western and MacMillan's wife were acquitted of robbery, but David Western received two years' imprisonment for being an accessory after the fact. He was killed in a car accident in 1973.

Daniel MacMillan only served a portion of his sentence. A few weeks after his return to maximum security late in 1965, he was certified insane and transferred to a mental hospital.[7] In 1972 MacMillan was considered to have recovered sufficiently to be released on parole, and he subsequently travelled to Australia where he committed a series of bank hold-ups. Arrested, but released on bail, he then committed another robbery, shooting two bank officials in the process. One of the victims died. Now considered highly dangerous, MacMillan was pursued by the authorities to a house in Avoca on the central New South Wales coast. There a siege took place and on 29 June 1976, Daniel MacMillan was shot dead by police.

However, in mid-1965 MacMillan's madness was only beginning to surface and the brothers seemed a harmless, if somewhat eccentric pair. Prisoners laughed as MacMillan watched the screws' movements and drew maps of sentry positions on scraps of paper. It was only a matter of days before it became generally known in the yard that MacMillan was planning to escape. It was MacMillan's strange countenance and his irreverent behaviour which immediately set him apart from the remand men, and it was these qualities too which attracted the attention of other prisoners. One whose interest was particularly intense was remand inmate Jonassen Sadaraka, a powerfully built part-Samoan who had considerable criminal experience and contacts in the sentenced section. Within a few days of Western and MacMillan's arrival,

Sadaraka approached MacMillan and the two began talking together about how they might break out. Western wanted no part of it and refused to become involved, but between his older brother and Sadaraka, a number of schemes was discussed.

Security in the remand wing was lax. In the chapel, where visits took place, only one officer supervised the whole room and there was no barrier between visitors and inmates. Objects could be passed across the tables separating them without being seen. Strip searches, which were infrequent and irregular even after visits, were not otherwise customary. Thus, a person who managed to smuggle contraband out of the visiting area had little chance of being caught. Early in the week of 12–16 July, Sadaraka was seen by prisoners to have a silver hand-gun stuck in the waistband of his trousers, beneath his overcoat. The pistol was a chrome-plated six-shot revolver made by the Bernadelli-Gardone Company of Italy. This gun and about fifty rounds of .22 ammunition had apparently been smuggled into the institution by a visitor of Sadaraka's and/or a prison officer[8] on two consecutive days.

Home-made cell keys were, and had been for some time, easily procurable at Mt. Eden prison. The new locks which Buckley had had fitted proved no more difficult to open than the old ones, and it was a relatively simple task for Sadaraka to procure a key from his friends in the sentenced wings. Most of the cells in the division had solid doors and hence could not be unlocked from the inside. However, due to population pressure, there were a number of association cells still in use. Known as the 'tower cells', some of these had barred frontages. A person with long arms was just able to reach the outside lock on these grilles and thus, with the use of a key, could lock and unlock his own door. Neither MacMillan nor Sadaraka was held in these association slots, but the men who were were approached by Sadaraka, and one of them agreed to try to open his own cell.

Although it was well known in the remand section that Sadaraka and MacMillan were planning to escape, and although several inmates and possibly a member of staff knew that Sadaraka had a gun, other staff were either not aware of it or did not give the information any credence. The existence of home-made keys was general knowledge and top officials also knew of the suggestion that tower cells could be opened by their occupants. Deputy superintendent D. W. Byrne had once attempted to prove this by attempting to unlock such a cell. However, finding himself unable to do so, he had assumed the rumour was unfounded.

Two days before the escape attempt, and after evening lock-up, an inmate in one of the tower cells asked a prison officer to pass a newspaper to Sadaraka. The officer searched the newspaper, and inside it he found a ciphered message. The officer's next action is puzzling. Unable to understand the code he took a copy of it and passed the message, still inside the newspaper, on to Sadaraka. The message was then filed, undeciphered, and apparently forgotten about. Why the message was passed on after it had been discovered, and why it was not decoded, has never been properly explained. It was not until after the riot some days later that police experts were asked to unravel the code.

The oversight was a costly one, since had the contents of the note been known, the escape attempt, and the riot which it led to, could have been forestalled. When deciphered, the message read:

> John regret must cancel our trip tonight. Been questioned about cell key. Have seen Mac. Our cell got special treatment keep blankets whole mate see how things go later.[9]

It was not until two days after the note was passed, the very night before MacMillan and Western were due to appear in court to hear police depositions, that the escape plan was put into action.

At about 5.30 p.m. on Monday 19 July, an inmate in a remand tower cell, by arrangement with Sadaraka, unlocked his own cell door with the home-made key he had been given. He then went and unlocked Sadaraka, gave him the key, and secured himself back in his own cell. Sadaraka then released MacMillan, who had already made a dummy and placed it in his bed. What the pair did for the following eight hours is unclear. Perhaps they tried to get out of the prison. This they would have found impossible since the key they had only opened the cells, and a different one was needed for the gates. They thus had no means of getting through the grilles which blocked their way to other sections of the institution.

Soon after 2.00 the following morning, officer Marchant, a member of the night watch, walked into the remand section on a routine security patrol. Reaching the end of the wing he was confronted by two masked men who emerged from a lavatory recess. One of the men was holding a pistol (Sadaraka), the other (MacMillan) was armed with an iron bar. Marchant screamed and struggled with the men, who felled him with the bar. Marchant's grille key was then taken and the men proceeded towards the north wing where they encountered officer Weir, on his way to investigate Marchant's cries. Weir was taken hostage, and Sadaraka fired a shot through the barber's shop window to show he meant business.

Meanwhile, at the end of the remand wing, Marchant began to recover and blew a long, loud blast on his whistle. Using Weir as a shield, the two prisoners made their way to the 'Dome', at the centre of the gaol, where they attempted to get out through the administration block. There they found a grille, which Marchant's key did not fit. The inmates called out and were answered by officer Grubb on the far side of the grille. The prisoners demanded to be let through, but seeing the gun, Grubb dived into the adjacent office of the superintendent, followed by two more shots from Sadaraka.

At this time Mr Haines, officer in charge of the morning shift, was in the basement unlocking the prison bakers. Puzzled about the commotion in the Dome, he began to ascend the basement stairs. Half-way up them, he stopped, as he saw Weir being held at bay by two masked men. Haines quickly retraced his steps and telephoned the police. Soon after 2.10 a.m. the alert was given. Thus, within five or six minutes of the escape beginning in earnest, it was already on the road to failure. At 2.15 a.m., Inspectors K. G. Sykes and E. G. Perry were informed at their homes that there was trouble at the prison and that shots had been heard. Sykes arrived at the gaol at 2.30 a.m. and Perry, in charge of the armed offenders squad, arrived at 2.45.

Marchant, his head covered in blood, had meanwhile been let through the administration hall grille by Grubb, and he and a sentry, called Gillard, drew revolvers from the armoury. However, although the two prisoners were in sight, the officers were unable to fire for fear of imperilling the lives of Weir and Haines, both of whom were now hostage. By this stage both MacMillan and Sadaraka knew their chances of escape had foundered so they retreated, with their two captives, to the basement of the east wing. Here, among others, Gillies, Evans, and Wilder were held and it was hoped that if released, these seasoned and resourceful escapers might be able to solve the predicament.

The unlocking of the security prisoners is crucial in understanding what took place next. For it was the liberating of these men which turned what began as nothing more than a crazy escape attempt into the most destructive prison riot in the history of New Zealand.

THE RIOT

Since their escape in February, the eight top security risks at Mt. Eden prison had been held under the tightest of possible conditions. The March inquiry had recommended that all hobby materials be removed from the cells[10] and men in the security block were now locked in their 'peters', virtually unoccupied, for twenty-two hours of every day. Each door was bolted and double-locked with an extra padlock, which latter was changed daily. Men were kept in strict solitude. For an hour in the morning and an hour in the evening, each was unlocked for exercise, alone, in the tiny, mesh-covered yard at the end of the block. This was the hanging yard, where between 1952 and 1957 eight men had met their deaths.[11] At the time of the July escape attempt, prisoners like Evans, Gillies, and Wilder had all endured these mind-destroying conditions for almost six months. It is in the light of this that their violent reactions must be read.

With the use of the bar carried by MacMillan, the padlocks on the doors of Gillies, Evans, Wilder, and a man called Sutcliffe were forced. The key was then used to release the bolt. This done, the six inmates returned to the Dome. At this point the aspiring escapers seemed at a loss, and the incident might easily have been over. Evans' statement to the investigating Commission was that he believed from the start that the situation was hopeless, and that he had explained this to Sadaraka and MacMillan. However, before any thought of surrender could be entertained the destruction suddenly began. Gillies took the iron bar and without warning, smashed all the windows in the Dome office, including the electrical switchboard. Thus, at 2.35 a.m., the whole institution was plunged into darkness. Then Matich set the office on fire.

While this was going on, other prominent men in the gaol were being unlocked. Among them were Twist, Maketu, and Cecil Te Whiu, elder brother of Eddie Te Whiu. More inmates were released and the two hostages were taken to the separates division, adjoining the security block in the east wing, where they were locked up. The fire in the Dome began to grow and prisoners in cells close by were in peril of suffocation. After about half an hour, Weir and Haines were let out, therefore, and told to call out for more cell keys. This they did, but their pleas were ignored. The chapel was burning too by now, and the whole prison began to fill with fumes. Officers at the front gate still refused to hand out more cell keys, so a few inmates, the most prominent of whom was Twist, began moving about the institution unlocking inmates as quickly as they could. The home-made key was rough and had little grip, and although Twist's hands were soon raw and bleeding, he continued with his task.[12]

Mt. Eden prison was made of stone, but its walls were covered in layers of oil-based paint, over a centimeter thick in parts. This burned rapidly and produced volumes of thick, black smoke. Even more volatile was the ceiling sarking, made of 9 by 1 inch kauri planks. These were old, dry, and soaked with tar. When the heat and flames reached the roof area, therefore, a searing blaze swept through the upper levels of the institution. Few realized that the old stone gaol would burn so well and it was not until 4.44 a.m. — over two hours after the first fire was lit, that the fire brigade was called. By 5.30 a.m., about 70 per cent of the prisoners had been unlocked by other inmates, and the chapel too was well alight. At 6.00 a.m., Haines and Weir, standing in the south wing, saw firemen and police enter the Dome via the front entrance and begin dousing the flames. The two officers, unimpeded, made a dash for freedom.

Perhaps in retaliation against the loss of their hostages, or perhaps in anger at the authorities' entrance, prisoners on the upper landings began to pelt firemen attempting to extinguish flames in the Dome area. For a few minutes the firemen stood their ground, but when one of them was knocked out by a fire extinguisher they retreated with their police escort to the outside of the gaol. From this point, fire-fighting was restricted to the largely ineffective method of playing deliveries on to the building's tile roof.

By this stage rioters were divided into three distinct camps. The first comprised mainly older men and uncriminalized prisoners. These took no active part in the uprising and generally kept out of the way, waiting for the destruction to end. This 'passive' group was a large one, and sixty to seventy of them answered calls to surrender at intervals during the first day of rioting. They were held temporarily in the tailors' and tinsmiths' shops. The second group was the 'hard core'. This comprised ten or eleven desperadoes who were intent on using the incident as a means to escape or secure better conditions.

The third category was the most important, and it consisted of a significant number of principally younger inmates. These were the men who continued the riot, for, having been released and having seen the damage started by the security prisoners, they excitedly joined the rebellion. It was they who sacked the institution. They lit huge bonfires, they burned indiscriminately. In an orgy of wanton destruction, these prisoners wrecked whatever they could

Jack Marshall, Minister of Justice 1954-1957. *Alexander Turnbull Library*

Ralph Hanan, Minister of Justice 1960-1969. *Alexander Turnbull Library*

John Robson, Secretary for Justice 1960-1970. *Department of Justice*

Former prison officer Percy Anstiss in 1983. *Department of Justice*

Minister of Justice J.R. Hanan examines damage caused by the 1965 Mt. Eden riot. *Department of Justice*

Paremoremo under construction, late 1967. *Department of Justice*

Paremoremo completed, December 1968. *Department of Justice*

Front gate, Paremoremo Prison. *Department of Justice*

Entrance, Paremoremo Prison. *Department of Justice*

Paremoremo visiting room. *Department of Justice*

Paremoremo workshop. *Department of Justice*

Paremoremo gymnasium. *Department of Justice*

Paremoremo weightroom. *Department of Justice*

A. block exercise yard, showing the toilet recess from which Merv Rich tried to escape in 1972. *Department of Justice*

The 'Airstrip' — main corridor — of Paremoremo Prison. Cellblocks lead off to the left and right. *Department of Justice*

Cell landing, D. block, Paremoremo Prison. *Department of Justice*

Decorated cell landing, C. block, Paremoremo. *Department of Justice*

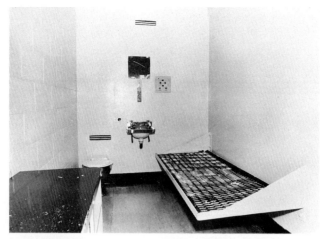

Empty cell, D. block, Paremoremo. *Department of Justice*

Long-term inmate's cell, C. block, Paremoremo. *Department of Justice*

Central control, Paremoremo Prison. Officer Smith, attacked by Dempsey Roberts in July 1970, is seated on the right. *Department of Justice*

Jack Hobson, Superintendent of Paremoremo Prison 1972-1984. *Department of Justice*

Sid Ward, Superintendent of Paremoremo Prison 1984-1985. *Department of Justice*

Les Hine, Superintendent of Paremoremo Prison 1985-1987. *Department of Justice*

Max Hindmarsh, Superintendent of Paremoremo Prison 19 *Department of Justice*

Four former Justice Secretaries, left to right: Gordon Orr 1974-1978; Jim Callahan 1982-1986; John Robson 1960-1970; Eric Missen 1970-1974. *Department of Justice*

John Robertson, Secretary for Justice 1978-1982.
Department of Justice

David Oughton, Secretary for Justice 1986-
Department of Justice

Dan Riddiford, Minister of Justice 1969-1972. *Alexander Turnbull Library*

Roy Jack, Minister of Justice 1972. *Alexander Turnbull Library*

Martyn Finlay, Minister of Justice 1972-1975. *Alexander Turnbull Library*

David Thomson, Minister of Justice 1975-1978. *Alexander Turnbull Library*

Jim McLay, Minister of Justice 1978-1984. *Department of Justice*

Geoffrey Palmer, Minister of Justice 1984- *Department of Justice*

find, including the personal property of other inmates.

As this activity intensified and the smoke became oppressive, other released inmates sensed danger and began desperately seeking an escape. Using a steel table, a number finally managed to smash through a grille and exit to the safety of the yard. Later, at about 5.45 a.m., officers D. C. Taylor and H. Stroud entered and opened another grille, which allowed more to get out.[13] They also went around the rest of the gaol, unlocking the fifty or so prisoners who had not yet been liberated by the others.

As noted, the alarm had first been raised at about 2.15 a.m., and by 2.45 the police and armed offenders squad had begun to surround the prison. Buckley had arrived at 2.30., five minutes before the lights went out. It was the opinion of inmates I spoke to, and many who gave evidence to the Commission, that, had the authorities entered the prison at any time after this point the rebellion would have collapsed immediately. Although the Commission rejected any such notion[14] the circumstances of the riot, with its lack of leadership, planning, and direction, make it difficult to believe that an early surrender could not have been easily secured. Haines and Weir were unharmed, and had been permitted to depart as soon as the administration grille was opened. Stroud, Taylor, and a number of other officers had wandered about the institution in the early hours of 20 July. They were not seriously threatened or molested by anyone.

On the other hand, at about this same time, considerable resistance had been encountered by police and fire-fighting personnel as they were driven from the prison. According to his own written evidence, it was Detective Inspector Perry of the CIB who made the decision not to enter it again, and from the time he arrived, Perry supplanted Buckley's authority. Initially, Perry's reasons for not attempting an assault were several. These can be summarized as follows:

1. The prisoners were armed and had fired shots.
2. Nobody knew where the main body of inmates was.
3. It was unknown how many men had been let out of their cells.
4. The prisoners had two officers hostage, whose lives could have been endangered.
5. There was no light in the building, and auxiliary floodlighting was not established until 4.10 a.m.
6. With only about 100 men present at this stage, there were insufficient personnel both to maintain a secure perimeter and detail a phalanx to enter the prison. To have attempted an entry would have seriously weakened the outer security cordon.
7. If armed men had entered the prison in the darkness and been overpowered, their weapons would have become available to prisoners.
8. Buckley had assured Perry that the prison would not burn.

Perry felt that to have entered the prison at its early stages would have resulted in considerable violence and the possibility of someone being shot.

Tear-gas too was considered, but decided against because:

1. Nobody knew where the loose prisoners were.
2. Many men were still locked in their cells.
3. The presence of tear-gas in the prison could have hampered any attempts made to enter it.

By 6.00 a.m. on 20 July, this situation had changed. Weir and Haines had escaped and Taylor and Stroud were inside the gaol unlocking doors. When the latter returned they must have reported on conditions inside. Parts of the institution were now burning fiercely. With the escape of the hostages and the coming of daylight, some of the problems envisaged by Perry in the early stages disappeared. Yet still he decided against a forcible entry. It was 7.30 before it was light enough to see inside and some inmates had already begun to surrender. By 8.00 a.m. the prison was surrounded by 150 armed police and prison staff, plus thirty regular army troops with automatic rifles and fixed bayonets. With the SAS first ranger squadron soon to reinforce the cordon, the perimeter was considered secure.

At 8.15 a.m. Detective Chief Inspector R. T. Walton arrived to take over operations and a meeting of senior police personnel was held. Mr Buckley was also in attendance. The prison was virtually gutted by now, and it was decided (again incorrectly) that little more damage could be done. It was considered that the prison would settle down quickly and that nothing further could be gained by force. To enter would only endanger the safety of police and inmates.

As observed, a limited response to calls for surrender was achieved during the day of 20 July. Nothing was done by those continuing the siege to prevent those who wanted to from giving themselves up. All that day and the rest of the night, however, the prison remained in the hands of the prisoners. It was midwinter and the morning which had dawned fine, later became cold and drizzly. Prisoners huddled in groups and attempted to warm themselves with blankets, but everything was smoke-ridden and wet. The excitement of the previous night soon dissolved as hunger added to the discomfort and misery of the day.

The kitchen was raided, and prisoners barbecued sausages and steaks over an open fire in the yard. Although the prison was largely quiet, a number of small incidents took place. In the afternoon, a fight broke out in the yard. A man was later admitted to hospital with facial injuries and a ruptured spleen. Sometime during the day, two young men were raped by groups of aggressive prisoners. At about 10.00 p.m. a number attempted to scale the walls, but dropped back after a warning shot. A second group tried to force a grille to a remand yard, but they too scattered when a shot was fired. Others received a gunshot warning when they tried to smash security lighting with stones.

During the night of 20–21 July, prisoners spent several hours stacking everything which was still flammable into the prison kitchen. At 6.15 a.m. on Wednesday, 21 July, assisted by some form of accelerant, this was set alight, and smoke and flames began to pour out of the kitchen's windows. As the fire took place in a part of the prison still occupied by the 200 or so remaining prisoners, no attempt was made to extinguish it. The fire continued until, like the rest of the institution, the area was little more than a smoking, blackened shell. When morning dawned on 21 July, without food, fuel, or adequate shelter, the last of the Mt. Eden rioters were ready to give in. Ronald Jorgensen — John Gillies' co-offender — had attempted to negotiate terms, but to no avail. Nothing less than unconditional surrender was acceptable to the authorities. In the end, inmate Jorgensen managed to acquire a vague verbal assurance that personal effects would not be destroyed, but in the event, no attention was paid to it. At around 7.15 a.m. prisoners began filing out into the yard. By 11.00 a.m., thirty-three hours after the riot had begun, all 293 men had been accounted for and were back in safe custody.

The gaol was now completely uninhabitable, and from the first day of the trouble, frantic efforts had been made to find alternative accommodation. Indeed, if the riot had collapsed earlier, one wonders what would have been done with the prisoners. As it was, finding beds for 293 maximum security criminals was daunting enough. When the prison surrendered, thirty-five men due for discharge by the end of July were released immediately. MacMillan and Sadaraka, along with eight other top security risks, were kept in the detention cells at Central Police Station. Meanwhile, transfers southwards had begun. Those designated for Christchurch and Wellington flew by Bristol Freighter. The remainder, who went to prisons in the North Island, went by bus.

Of the 218 transferred, the majority — eighty-nine in all — were sent to the east wing at Waikeria. This was a secure section of the borstal, which was cleared of trainees to make room for the Mt. Eden rioters. Accordingly, sixty borstal boys were shifted to Invercargill. Of the eighty-nine destined for Waikeria, the ten most 'dangerous' of them were put into the borstal's new security unit. This group consisted of MacMillan, Sadaraka, and the rest of those who had been held at police headquarters. Thirty-one men went to the newly finished secure psychiatric hospital at Lake Alice. Opening of this facility was delayed for some months by the Health Department, and the institution was temporarily renamed Rangitikei Prison. Thirty men went to Paparua and thirty to Mt. Crawford. Thirty-seven were reclassified to minimum security, and sent to Tongariro prison farm and Waikune. The forty short-termers who stayed at Mt. Eden assisted with cleaning-up operations and were housed in the less-damaged areas of the institution.

Destruction in the prison was considerable, and early estimates put the cost of repair at 250,000 pounds. Apart from total loss of most of the gaol's contents, the roof was almost completely destroyed. The fire service reported 90 to 100 per cent destruction of the roof in most of the wings except the west and remand wings, 40 to 80 per cent of which was damaged.

The transfer of potentially dangerous prisoners to less secure facilities had obvious risks, and placed considerable pressure on their administrations. At Mt. Crawford, transfers from Mt. Eden, still fired by the excitement of the rebellion, assaulted and obstructed prison officers trying to search them on arrival. Some days later, two of the Mt. Eden transfers were charged with attempting to set fire to their cell. At the end of July coshes, knuckle-dusters, and two bottles of petrol were discovered at Invercargill borstal, after a plot among prisoners to burn the prison was revealed. Of more serious concern was an incident at the medium security gaol at Paparua, which had taken thirty of the Mt. Eden men. Here, at 5.30 p.m. on 25 July, the maximum security transfers began an organized rebellion in the chapel, where they were joined by numerous of the local prisoners. In all, forty-three took part in the insurrection. The east wing was routed and set on fire, and prisoners engaged in a running battle with baton-wielding staff.

Three ambulances, eight fire engines, and fifty police, including the armed offenders squad, turned out to the incident, but twenty-three prisoners barricaded themselves in the east wing and refused to surrender. At 10.00 p.m., police fired six tear-gas canisters through cell block windows. By 10.30, the last of the men, forced to break cover and badly affected by the gas, gave themselves up to the authorities. Damage was estimated at 100,000 pounds.

The devastation of the Mt. Eden debacle and the threat of a riotous epidemic in the week following it, caused distress to administrating authorities. A Commission of Inquiry was ordered into the Mt. Eden disturbance, and a similar one began at Paparua. On 22 November the taking of evidence started and the report on Mt. Eden was presented before Christmas. Search procedures were criticized and visiting security was deemed inadequate, but otherwise staff and administration at Mt. Eden were absolved. 'Apart from the matters which I have mentioned', the Commission chairman wrote, 'I found no evidence to establish that the practices in force and the procedures followed in Auckland Prison contributed to the commencement or continuation of the riot.'[15] 'I firmly believe that no fault can be found with the appreciation of the position made by senior and experienced police and prison officers on the spot, nor with the measures which were taken to deal with the situation which had arisen. . . . In my view the measures which they took were proper, adequate and effective.'[16] Once again the rapid construction of a replacement prison at Paremoremo was called for.[17]

Of the 293 prisoners in the institution on 19 July, only sixty-nine gave evidence to the Commission. As previously mentioned, many of them insisted that the riot could have been forcibly quelled at any stage, had the initiative been taken. Buckley himself had been subordinated fairly early in the piece and thereafter took his instructions either from the police or by head office telephone. MacKay had flown to Auckland on the morning of the very first day and became the senior Justice official at the site. Buckley's role in the Mt. Eden riot is significant, therefore, in its very insignificance. He did not emerge as a man of leadership or initiative at a time when possession of such faculties might have been a special asset. Buckley was an avid rule-book man and like many of his type, he found difficulty making decisions when routine was awash.

Of course the riot at Mt. Eden, so swiftly followed by rebellions at Mt. Crawford and Paparua, stunned the aspirations of Robson. Dissatisfaction with what had been happening at the institution since 1958 was catalysed by the disaster of 1965. Nothing like this had ever happened before, and to many, the 'mollycoddling' of prisoners was to blame. Letters to newspapers accused the Government of running a 'five-star-plus hotel' at Mt. Eden and demanded a return to punitive discipline. Hanan and the Prime Minister received a wad of correspondence, all of which was replied to, criticizing the 'soft' policy on prisons and even calling for Hanan's resignation. When it was revealed that after their surrender inmates had breakfasted on bacon and eggs, the Minister was forced once again to justify the whole dietary scale of prisons.

In the House, conservative MPs demanded a toughening up of the country's gaols and an end to the 'play-way' method of dealing with convicts.[18] Their calls were greeted by applause from both sides. Arnold Nordmeyer, now Leader of the Opposition, called for 'steps to restore discipline' in the prisons. Hanan replied that superintendents had been instructed hereafter to take stern and severe measures against recalcitrants.[19] However, a reaction of this type was predictable, and the Minister of Justice was no doubt prepared for it. Hanan had for years cherished the idea of having Mt. Eden pulled down, and with the building now almost empty, it seemed for a while that his hopes might be realized. Yet, there had been talk of this since 1950, and so far not a stone had been shifted.

THE FATE OF MT. EDEN

As far as the public was aware, the old prison was to be demolished and a new one built some time before the end of 1968.[20] What was not known, or ever fully validated for that matter, was that from the early months of 1965 there had been rumours in Parliament, and about the Department of Justice, that the planned building at Paremoremo would be postponed. Hanan certainly wanted a new prison, but in Treasury and some sectors of Government, opposition to it was growing.[21] Had this movement held its impetus, the construction of Paremoremo might well have been shelved. The Mt. Eden riot put a decisive stop to any deferment plans and although it made efforts to do so, the opposition was unable to identify anyone who had been behind the movement.[22] So, for Hanan and Robson, news that Mt. Eden had been burnt out was not entirely unwelcome. At head office, tidings of the riot were received with muffled levity. One senior official told me, 'Well, we did in the Department suggest, somewhat unkindly, that (the riot) might have been "managed"!' Later, Hanan himself 'denied' that he had supplied the box of matches used to start the fire.[23]

With the extent of the damage and the estimated quarter-million pound restoration bill, Hanan can be excused for believing that Mt. Eden would go. On 21 July, before the riot was even over, he announced that he thought and hoped the cost of rebuilding the prison would be too great, and that it would be removed. However, if Mt. Eden was demolished, Hanan knew he would be faced with a difficult situation: apart from the fourteen who could be contained in the soon-to-open security block, there would be 300 fewer security beds.

By the beginning of August, most of the work of cleaning up the debris of the riot was over. Before 31 July, nearly all of the forty short-termers who had been retained there had been released or transferred. On 29 July, there were only thirteen prisoners left at Mt. Eden, the lowest muster the site had known since its opening 109 years earlier.[24] This situation was short-lived. Two weeks later it was announced that the population had grown to 100. Some of the men were in leaky cells, which took in water when it rained, but the superintendent, without sympathy, told the press there was nothing he could do about it. 'They will just have to put up with it,' he said.[25]

By the middle of August, the kitchen was operating again. The more distant the riot became, so did the prison's demolition appear more remote. On 21 July Hanan had suggested that *if* the prison was dismantled, the new security block and a new remand section would have to remain on the site. On 9 August he announced that Mt. Eden would be partially rebuilt, to house 150 men in separate cells. Fifty of these would be remand units, forty would be for short-termers and another sixty-three, for medium security prisoners. The proposition was received critically by the Mayor of Auckland and other public figures, but the dissenters were assured that measures were only temporary. The Mayor of Auckland, thus convinced, now attempted to assuage the fears of critics by telling them that no vestige of the old gaol would be left standing in a few years' time.

However, the steel beams which were used to replace the burnt-out roof supports were anything but temporary, and the considerable (albeit underestimated) cost of 95,000 pounds for partial renovation may have seemed a rather extravagant stop gap. The kitchen and some of the wings were rebuilt, but the north wing and most of the east were left unrepaired and sealed off. These areas were now marked for demolition.

By 12 October 1965, re-roofing was well under way. The chapel was being recommissioned and the south, west, remand, and administration wings were being made serviceable. On October 15, Hanan, under pressure from opposition to plans for the prison's future, made a definitive statement about it. The work on Mt. Eden prison was only a temporary expedient, he promised.

> Within a year or so of the completion of Paremoremo, Mt. Eden will be completely demolished, and not a vestige of the existing institution will remain, except for the new security unit being built now, for fourteen men. The firm decision made now is that the existing unit as we know it will be completely demolished.[26]

Hanan's statement could hardly have been less ambiguous, but the opposition was not easily convinced. Bob Tizard commented that 125,000 pounds[27] (an increase on the earlier estimate) seemed a lot of money to spend on a temporary measure, and questioned the Government's sincerity.[28] Opposition Leader Norman Kirk observed prophetically that there is 'nothing quite as permanent as temporary work'. He implied that the Government was stalling for time.[29]

By 19 November, it was reported that the institution at Mt. Eden, still unfinished, but being rebuilt for 150, already contained 200. To forestall any alarm, therefore, Hanan announced on 27 December that demolition of parts of the north and east wings would begin soon. 'I have directed there should be no delay in starting the demolition,' Hanan said.[30] Speed was apparently of low priority, and it was three months before this work began. By the end of March 1966, 6 feet of stone had been removed by inmate labour from a tower on the north wing. However, the 'demolition' only involved a small section of ornate battlement on the crest of the tower, and had no effect at all on the prison's structure. Moreover, almost as soon as it was started, demolition stopped. Stone blocks were being damaged as they fell, and besides, said Mr Hanan later, the stones were too large and heavy for work to be done by inmates.[31] No one reminded him that inmate labour had laid the blocks in the first place.

In July 1966, Marshall announced that when restoration work was finished, Mt. Eden would hold 200 inmates (not 150 as originally planned). On 9 August, it was declared that a decision would be made soon on the demolition of the north and east wings, and from this point demolition plans were forgotten. A few days earlier, MacKay had revealed that Mt. Eden now held 250 men, not many less than it had at the time of the riot (although they were now held in a smaller space).

This same year there was a huge leap in prison populations, as large as any in the fifties. The leap was repeated the following year and rises continued, virtually unabated, until 1971. These huge population shocks, involving DAPs which in five years increased by over 50 per cent,[32] effectively sealed the fate of Mt. Eden. After 1966, the Department of Justice was again confronting a crisis as dramatic as any that had faced Barnett. From here on, any serious talk of getting rid of Mt. Eden vanished, although in 1969, soon after the opening of Paremoremo, Hanan said that now the institution would definitely be pulled down by outside contractors. MacKay thought inmate labour might be used. Nobody really seemed to know, and in the end, nothing at all was done. Populations continued to rise and the accommodation situation got worse.

In March 1971, there was another riot at the prison. Cabinet immediately approved the renovation of those parts of the institution which had been left unused since the 1965 fire. The east and north wings were reopened later in the year at a raw cost of over 32,000 dollars. Finally, in January 1977, the battlements which Hanan had ordered to be pulled down eleven years before were repaired on the orders of another administration.[33] The institution still stands today, and there are no plans for its removal.

Post-Riot Conditions

While the old gaol was being repaired and a new one constructed, Buckley remained at Mt. Eden. In 1966 he was awarded the MBE, a distinction which led to facetious speculation that he was being rewarded for burning down his prison. In fact it was given largely in recognition of his service in helping to quell the riot. However, the fire was a painful blow for Buckley. He was deeply disturbed by the event and the way things had broken so violently from his control. To Buckley, the riot had been totally unexpected and he was anxious to prevent a repetition. So was the Department of Justice. Head office felt that a contributing cause was that there had still been too much liberalism. With populations rising, it was resolved that the safe custody of prisoners would now be paramount. Superintendents were therefore instructed to take stern and severe measures against recalcitrants.

Although Buckley had been exonerated by the Commission of Inquiry his security procedures had been questioned.[34] Further tightening was endorsed and Buckley quickly set about his task. This job was assisted by the fact that, since the majority of facilities had been destroyed anyway, the gaol's conditions were completely spartan. Now, inmates were locked up for between sixteen and a half and seventeen and a half hours every day. No newspapers, no hobbies, and no recreation materials were permitted, except for library books. On weekends it was compulsory for prisoners to go into the yard, and if it rained there was only one leaky shelter for refuge. Weekly visits were now tightly monitored and no physical contact at all was permitted. Showers were limited to two a week. For a long while, Buckley banned Prisoners' Aid visitors from the institution, and when he finally permitted them entry two years later, it was under the most restricted conditions that they worked. One officer who remained at the prison after the fire said of Buckley's reaction:

> ... he turned into almost a bloody sadist, where he had all the inmates running at double-time and snapping to attention like bloody soldiers. Almost a sadist. Almost as if they'd burnt the bloody prison down and it was their fault and he was going to make them pay for it. And he had the officers jumping around like bloody sergeant majors and inmates snapping to attention and everything on the double.

Because, in the early stages at least, the prison was almost empty, most of the staff were now forced to perform duty at other institutions. The majority worked either in the east wing of Waikeria or at Rangitikei. Some transferred permanently, others — twenty-one at Rangitikei and fourteen at Waikeria — were rostered to one of these locations for two weeks out of every seven.

After the trouble at Auckland and Christchurch, demands by staff for better pay and more personnel became more strident. Constant delays followed by inadequate concessions forced the prison officers' subgroup[35] at Mt. Eden to threaten a work-to-rule. In response, on 9 November, Hanan ordered the beginning of a campaign to recruit eighty more prison officers and thereby reduce overtime demands. The effort brought small, if any, relief. For one thing, in August it was revealed that the Department was ninety, not eighty, men short.[36] For another, recruitment was slow and unable to keep up with resignations. Although by July 1966, fifty to sixty new personnel had signed on, in November the chairman of the Government State Services Tribunal had to declare that prisoners were now 100 men short of establishment. Hoping to attract more interest, further pay rises were ordered.

In spite of these moves, morale at Mt. Eden remained low. The riot had destroyed in many the last vestige of confidence in the existing regime. While only a small number of officers believed that Buckley's administration had caused the riot, a much larger proportion thought his indecisiveness had worsened it. It was widely believed, moreover, and deeply resented, that Buckley had tried to blame the disturbance on his deputy, who was a popular and highly respected officer. On top of this, the superintendent's redoubling of disciplinary fervour ran counter to the personal philosophies of many and created a depressing environment. So, although disenchantment and resignations had been high before 1965, these problems were compounded. Many who had previously suffered conditions in the hope of some improvement, now found their situation intolerable. Between 20 July and Christmas 1965, Mt. Eden lost forty-four officers, a third of its total complement. Reasons for the spate of resignations were general dissatisfaction with pay and career prospects, together with the requirement that staff serve so much time away from their homes.

By Christmas, Lake Alice Hospital (Rangitikei Prison) was already being vacated, and Mt. Eden was starting to fill up again. However, it was a different type of inmate who now entered. Because from the time of the riot Mt. Eden was no longer considered maximum security, the old lags, the men with whom the senior screws still enjoyed some rapport, were gone. Most of them were now at Waikeria where, as we shall see, conditions and staff–inmate relations were poor. Even after duties at Mt. Eden began returning to normal, therefore, worker satisfaction among Auckland's prison officers never returned to its pre-riot standard. It seemed that as long as Waikeria continued in its interim capacity, this situation would remain. For Justice officials and prison staff alike, the only hope for real relief appeared to lie in the new institution being built twenty miles to the north at Paremoremo.

THE RIOT: ANALYSIS AND SIGNIFICANCE

In his report on the Mt. Eden riot, Mr A. A. Coates SM, chairman of the Commission of Inquiry, found the disturbance to have been caused by an escape attempt which, when frustrated, became transformed into an orgy of senseless destruction. As noted, the chairman made the observation, doubtless correct, that had search procedures been more stringent — that is, if either the pistol had not been brought in or cell keys not been easily available — the riot might never have begun.[37] However, to analyse only the immediate causes of the 1965 rebellion is potentially misleading and of little real value, as it provides only a superficial picture of the event and the context in which it took place. For this crisis had been looming for a considerable period. Its likelihood had been increasing from the time Edward Buckley took over as superintendent.

The violence of the Mt. Eden disturbance, in fact, exhibited a variety of characteristics which are visible in the general pattern of riots in other parts of the world. Concerning the wave of insurrections which swept the United States in 1951-1953, for example, Schrag has written: 'Correctional administrations generally contend that certain forms of institutional maladministration are responsible for the recent riots.'[38] Martin, in a similar study, observes that although these riots were in many cases linked, general mismanagement was a factor common to all of them.[39] Hartung and Floch concur that long-standing grievances in institutions have usually been at the heart of America's more brutal and destructive rebellions.[40] Analysing the epidemiology of riots, a 1973 American Select Committee on Crime found that, 'Prison riots, save for isolated cases, are indications of long-standing problems in our correctional institutions.' Substandard conditions and repressive administration create an environment that is easily triggered into extreme and often bloody uprisings.[41] Similar conclusions are drawn by Fox in his well-read book *Violence Behind Bars*,[42] by Martin in *Break Down the Walls*,[43] and lay behind the political types of resistance discussed by Jessica Mitford in *The American Prison Business*.[44] In the United Kingdom the pattern is similar.[45]

Because each institution differs so much from every other, the causes of riots can only be discussed generally. It is broadly true, however, that prisoners lose a great deal more than they gain by rioting, a fact which is often overlooked in official reports. Although immediate causes of rebellions — as we have seen at Mt. Eden — often seem trivial and insignificant, violent outbreaks frequently point to deep-seated and long-standing deficiencies which inhere in custodial systems. It will be seen at Paremoremo, and it is generally the case, that riots are often effective in drawing deficiencies in prison management to official attention. The disturbances recorded in the English prisons at Hull[46] and Albany[47] in 1976 are examples in point, although without outside support, winning any sympathy can be difficult.

In 1965 the underlying causes of the riot at Mt. Eden drew little constructive comment, and it was not until after the Commission's report that members of the penal group attempted to analyse it from the viewpoint of its less obvious factors. The majority of these had been either neglected or rejected in the inquiry. As its principal information source, the group used a booklet on riots which had been published by the American Prison Association in 1953.[48] Published after the American riots of the early fifties, the Prison Association booklet was in broad agreement with informed opinion on the subject. Events such as riots, strikes, and mutinies, the Association wrote,

> ... are almost always the direct result of the shortsighted neglect of penal and correctional institutions, amounting to almost criminal negligence in view of the costly results, by (those) responsible for the administration and management of institutions.
> Prison riots should be looked upon as costly and dramatic symptoms of faulty prison administration.[49]

The American Prison Association (APA) listed seven conditions commonly associated with prison riots. These conditions were:

1. Inadequate financial support, and official and public indifference.
2. Substandard personnel.
3. Enforced idleness.
4. Lack of professional leadership and professional programmes.
5. Excessive size and overcrowding of institutions.
6. Political domination and motivation of management.
7. Unwise sentencing and parole practices.[50]

Each of these conditions was discussed by the penal group and rejected as inapplicable or insignificant at Mt. Eden. It was agreed that, apart from inadequate searching procedures, faulty administration had not been a factor in the trouble. There had been no recent complaints about food and in July 1965, the prison was less crowded than usual. Violence had been unplanned and that when it came had been from the younger, irresponsible element in the gaol, for which the prison executive could not be held accountable.

Point by point the head office penal group dismissed or disclaimed responsibility for conditions which might have been decisive in the destruction of Mt. Eden. While the rise in tension after tightening up of the new regime was acknowledged, the Secretary concluded that the riot had been caused not by any kind of mismanagement, but by an 'unfortunate chain of circumstances'.[51] In this way, the fundamental role of administration was dismissed, and the Mt. Eden riot's most important factors were ignored. However, to even the most unsophisticated examiner, it should have been plain that the effects of overcrowding at Mt. Eden could never have been removed by an incidental dip in daily musters. Mt. Eden had been overcrowded for years and instability that resulted from that situation was profound. It is naive to suggest that a sudden and temporary drop in a prison's muster would be accompanied by an immediate reduction in tension. The discontentment which had built up at Mt. Eden was residual and, inured as it was by a repressive management, could have precipitated rebellion at any time. This remained the case, irrespective of day-to-day fluctuations in numbers.

All of the conditions suggested by the APA, with the possible exception of item six, could be argued to have had a part in creating the trouble which eventually broke out. Factors three and seven, for example, relating to enforced idleness and unwise sentencing practices, impress as having immediate and indisputable pertinence to the case. It will be recalled that it was the security block inmates, not the escapees, who ignited the 1965 rebellion. These men played as important a role as the pistol in the prison's eventual fate. Locked in their basement cells for twenty-two hours of every day, without work or recreation, their terms of solitude were indefinite. Many had been confined in this way for months. Their explosive behaviour when suddenly liberated requires no other explanation.

Item seven may be considered similarly. Gillies of course was doing life. Evans and Wilder, who had entered prison with relatively short terms, were, after the February incident, carrying extraordinarily heavy finite sentences. Wilder's term of almost nineteen years was, and is still, the longest determinate sentence ever given in this country. Both he and Evans could have expected to spend a similar length of time in prison to that of a person convicted of murder. There was little that could be taken from these men who were confined for indefinite periods in the bleakest corner of an eighty year old edifice.

However, while this small group became the catalyst of the riot, its behaviour does not explain the actions of the many who, once released, hastened to join the rampage. In chapter eight we considered the élitist system which Haywood established at Mt. Eden. At Oahu, Hawaii, in 1960, dissolution of a similar order resulted in a destructive rebellion.[52] Hartung and Floch have described an almost identical situation where dismantling of the semi-official, informal inmate structure produced the same effect.[53] Schrag too has commented that managerial adjustment in any institution — even where change is necessary and beneficial — is frequently associated with a rise in tension and anxiety among prison staff and inmates.[54] In the bloodiest riot in prison history, at New Mexico in 1980, a contributing cause was held to be the dissolution of informal control.[55] It is generally true that when self-control mechanisms are dismantled, the interest of leaders in propping up officaldom disappears. Prisoner bosses become alienated from goals they once supported and, unless neutralized or appeased, they can develop into a reactive force.

So it happened at Mt. Eden. From the time Buckley took over, the prison's inmate hierarchy was ignored. Powerful men now had nothing significant to gain from co-operating with the authorities. The rapport they enjoyed with senior staff diminished and their special privileges vanished. In addition to this, the new repressive philosophy was producing tensions of its own. Staff, brow-beaten and harried by their new boss, worked their frustrations out on the men. As a result, antipathy generated by the superintendent diffused downwards. Dissatisfaction and tension in the gaol ascended sharply after mid-1963.

When they were unlocked in the early hours of 20 July, inmates found the prison in their hands and parts of it on fire. The men who were largely responsible for the liberation — Evans, Gillies, and Wilder — were folk heroes to the prisoners. They were dangerous men, infamous men, courageous and very daring men. As such they were revered by the others. It was the younger, more impressionable inmates who were influenced by the notoriety of the first escapees and who, when freed, contributed so energetically to the destruction. In a frenzy of nihilistic zeal, the young men at Mt. Eden joined their heroes in ransacking the prison; source and symbol of their loathing and discontent.

The serious trouble which took place at Mt. Eden in February and July 1965 is not, to this author, particularly surprising or perplexing. Like that at New Mexico, its origin was spontaneous, its course directionless, and its motives irrational, but the warning signs of trouble had been visible for some time. The precarious instability of the new regime made some form of outburst likely, if not inevitable, after the middle of 1963. When trouble did occur, it was the extent of the violence and the vigour of the reaction, rather than the occasion of rebellion itself, which surprised even the most prescient observers.

The 1965 riot at Mt. Eden was a highly significant event, which had a lasting effect on things to come. Firstly, it crushed, suddenly and decisively, any notion that a new maximum security gaol might not be essential. The building of such an amenity now became an immediate priority. Secondly, it encouraged those in charge of providing this facility to spare little expense in creating an institution which would rank among the most advanced and secure in the world. Work on construction was approved without delay. Thirdly, the riot highlighted, more distinctly than ever before, the incompatibility of dictatorial action and liberal philosophy. Notwithstanding, the painful lessons of 1965 were to be repeated several times before Buckley was removed in 1972.

The final important effect of the Mt. Eden riot was that many who had taken part in it, and all 'dangerous' receptions subsequently, were confined for the next four years in the extremely tight, spartan, and forbidding recesses of Waikeria's security wing. The resentment, the *esprit de corps*, and the callous fatalism which grew so abundantly among those held in this unit had a powerful impact on subsequent relations. It was these relations — forged, wrought, and tempered in the east wing at Waikeria borstal — which travelled with the men when they moved back to Auckland, setting the tone for interaction in Paremoremo.

NOTES

[1] Of a total personnel complement of about 650, the Department had lost about 110–120 officers every year. Staff loss in the penal division was higher than anywhere else in the Justice Department.
[2] Report of Inquiry into the Escape of Gillies, Evans and Wilder 1965.
[3] Of these, twenty-two were lifers and thirty-four were preventive detainees.
[4] Nash's 19,000 pound heist in 1956 was a crime of theft. The largest robbery committed in New Zealand to date was the 285,000 dollar robbery of an Armourguard security van on 19 October 1984.
[5] These were consulting psychiatrist Dr G. D. Tetro and the superintendent of Tokanui psychiatric hospital Dr H. R. Bennett.
[6] The dissenter was Dr P. P. E. Savage, superintendent of Oakley psychiatric hospital.
[7] The Court of Appeal had quashed the convictions on the ground that MacMillan was insane at the time of the offences.
[8] The tentative conclusion of the 1965 Commission of Inquiry was that the weapon had been brought into the prison by a visitor (pp.18, 31). Evidence was given by an inmate, however, that an officer on duty had been seen being handed a parcel outside the prison wall, during the period in question. This officer, who resigned his position immediately before the riot and left the day following it, was visited several times by police and questioned about the incident. The police accused the officer of smuggling the pistol into the prison (*Truth* 1 March 1966), but were unable to make any case.
[9] *Report of Commission of Inquiry into Auckland Prison* (1965) 22–33.
[10] Ibid., 21.
[11] This yard is still used for the exercising of men on punitive segregation.
[12] Twist's level-headedness in unlocking endangered prisoners was acknowledged on p. 13 of the Commission's report.

13 Officers who entered the prison were commended by the Commission of Inquiry into Auckland Prison 1965, 32.
14 *Report of the Commission of Inquiry into Auckland Prison* (1965) 34.
15 Ibid., 31.
16 Ibid., 34–5.
17 Ibid., 38.
18 *NZH* 27 July 1965.
19 Ibid.
20 Official estimates of the prison's completion date varied from 1966 (*NZH* 17 October 1963) to early 1967 (*NZH* 10 April 1965) to some time in 1968 (*NZH* 1 August 1962, 11 February 1964, 10 April 1965).
21 Robson (1974) 9.
22 NZPD vol. 343 (1965) 1416–33. See also *The Christchurch Star* 24 July 1965.
23 *N.Z. Listener* vol. 61 (1969) 11.
24 When the Stockade — as the site was then known — was first opened in 1856, sixteen prisoners were transferred there from the old Auckland Gaol in Victoria Street. Interestingly, the year of the riot was the Mt. Eden site's centenary as Auckland's principal place of penal confinement. Auckland Gaol, which had opened in 1842, was closed in 1865 and its inmates were transferred to the Stockade.
25 *NZH* 9 August 1965.
26 NZPD vol. 345 (1965) 3563.
27 Hanan later confessed that the estimated cost of partially repairing Mt. Eden was 125,000 pounds, not 95,000 pounds as earlier reported (NZPD vol. 345, 1965, 3557). In a lively exchange over Justice estimates, the Government was vilified for attempting to conceal information about the Department's finances (NZPD vol.345 (1965) 3556–60). The eventual cost of repairs after the fire was later announced to have risen to 150,000 pounds (*NZH* 16 June 1966).
28 NZPD vol. 345 (1965) 3565.
29 Ibid.
30 *NZH* 28 December 1965.
31 *NZH* 10 February 1969.
32 The actual increase in national DAPs was from 1741 in 1966 to 2639 in 1971.
33 This was Jack 'Bugs' Rogers, who took over in 1972.
34 *PGM* 22 August 1966.
35 Prison officers' subgroups are local union subcommittees, ancillary to the Public Service Association.
36 *NZH* 9 August 1965.
37 *Report of Commission of Inquiry into Auckland Prison* (1965) 7–8.
38 Schrag (1960) 137.
39 Martin (1955) 203ff.
40 Hartung & Floch (1956–1957) 51.
41 Select Committee on Crime (1973) 3.
42 Fox (1956).
43 Martin (1955) 203–10.
44 Mitford (1977) 227–45.
45 Thomas (1972); Fitzgerald (1977) 119–35.
46 Report of Inquiry into Hull Riot 1976.
47 Amnesty International (c.1977).
48 American Prison Association (1953).
49 Quoted in *PGM* 15 August 1966.
50 Reproduced in Martin (1955) 207.
51 *PGM* 22 August 1966.
52 McCleery (1961b; 1969).
53 Hartung & Floch (1956–1957) 53–6.
54 Schrag (1960) 138.
55 Colvin (1982).

12: INTERIM SECURITY 1965-1969

Between 20 July 1965 and the middle of March 1969, New Zealand was without an operating maximum security prison. The redesignation of Mt. Eden to medium security followed hard on the heels of the riot and was officially confirmed in August 1965. For over three and a half years, high-risk prisoners were detained in the security section of the east wing at Waikeria borstal and, when it opened in 1966, the small security block at Mt. Eden.

THE EAST WING, WAIKERIA

From 23 July, a total of 100 prisoners travelled under heavy police escort to Waikeria. Some were shifted almost immediately on to other institutions, so that by 26 July, eighty-nine prisoners remained in the wing at Waikeria. Because of fear that the trouble at Mt. Eden might be repeated and because the institution was not really designed for prisoners of this type, conditions were grim. Initially there were no exercise yards. Prisoners were kept locked in their cells without access to newspapers or radio and with nothing else to occupy them for twenty-four hours a day.

Security in the half-century old structure was as tight as it could be. The escorts which transported the prisoners were armed, and the sentries guarding the ten most 'dangerous' of them carried loaded weapons. When the block was finally vacated in 1969, the special armoury at the institution consisted of five Greener riot guns, two SMLE .303 rifles, one .32, and four .38 revolvers. There were nearly 900 rounds of ammunition for these weapons as well as extra magazines, pouches, clips, and tear-gas apparatus.

In the event of an escape, instructions were that a warning round was to be fired, followed by a shot to hit. Accordingly, when, on his way to court, MacMillan attempted to escape by scaling a drainpipe, a warning was discharged. Realizing his danger, MacMillan climbed down again before another shot was necessary. Officers were told that defiance was not to be tolerated. Prisoners were forced to accept conditions as they were and had no recourse to complaint. One officer said to me:

> Naturally, they didn't get exercise. I remember when we put them in there we said, 'That's it!' and they shut up, mate. I think even the guys who'd done a while were a bit frightened. They were a bit upset. They didn't know what was coming next. Because there was no compassion shown to them in any shape or form. If they stepped out of line, they were for it. And that's how it worked.

One inmate who had tried to escape was knocked to the ground and sat upon while a doctor stitched a laceration on his scalp. No anaesthetic was administered.

In the early phases, security in the east wing was the sole priority. The Government could not have stood the embarrassment of any more trouble, so when the disturbances broke out at Mt. Crawford and Paparua, a convoy of police cars rushed from Auckland to Waikeria to guard against any insurrection there as well. However, apart from precautions against violence from within, the Justice Department also insured against attack from outside. Quarter-inch steel plate was bolted over the main doors, and exercise was taken in steel mesh cages, especially designed for the purpose. By early August, the east wing was considered the most secure facility in the country.[1]

Such a spartan lack of amenities meant that inmate tolerance would be short. This the authorities knew and they worked rapidly to provide some basic freedoms. By mid-August a secure visiting area was half completed, as were the cages. However, because the yards were still insecure, inmates were given only twenty to forty minutes' exercise a day, walking in small groups inside the building. In mid-September it was announced that a six-booth visiting bay with concrete walls and floors, an all-steel ceiling, and shatterproof-glass dividing panels was almost completed. Early the next month, Waikeria's exercise yards became operational. Meanwhile, the population was being reduced to a minimum and only the most serious risks were retained. By March 1966, numbers in the east wing had been cut from eighty-nine to sixty. Four months later, after the top security men had gone to the new block at Mt. Eden, the muster was down to fifty.

From the time prisoners arrived, the east wing was divided into two sections. The lower left landing was walled off from the rest of the unit. This was the maximum security division and its dozen cells were reserved for isolation of the most 'dangerous' of the inmates. These were known as the 'top twelve'. The top twelve were locked in isolation for upwards of twenty-two hours every twenty-four, and each day they were let out individually and exercised for a short period in small enclosures outside. Sometimes termed 'rabbit hutches', these facilities were empty cages of about three by six metres in dimension, which gave prisoners their only contact with direct sunlight. Like the exercise compounds, cells of the top twelve were almost bare. They had a double-locked door and inside, a basin, running water, and a flush toilet. As there were no beds, inmates slept on mattresses on the floor. There was no other furniture. Prisoners ate either on the floor, or sitting on their toilet seats. Because of the complexity of security arrangements between the borstal kitchen and the wing, by the time meals arrived in the top twelve they were almost cold. This was a continuing feature of life in that division and became a source of aggravation to its inhabitants.

For eight months, the leaders of the riot and other 'dangerous' men were held in the top twelve cells at Waikeria. These prisoners were moved to the Mt. Eden security block in March 1966. After that, the security cells served as a detention unit and for long-term segregation of troublesome inmates.

The top twelve cells were seldom fully occupied during the 1965-1969 period, and the majority of inmates lived in slightly more open conditions in an area known as the 'Wing'. The Wing consisted of two tiers of cells facing inwards on either side of a central corridor. Wing cells had beds in them, and after a time prisoners in this section were also permitted to have transistor radios. Here conditions were thus more relaxed than in the walled-off sector, but by any standards they were still bleak. Inmates were exercised in the cages outside for two or three hours every day.

Apart from the few hours when they were unlocked, and apart from regular film showings, there was little to occupy the occupants of the security wing. Work was almost unobtainable. A few prisoners were employed 'feeding out' (distributing rations) or cleaning, and eventually a small carpentry shop for around half a dozen was opened at the end of one of the yards. However, less than 10 per cent of the men could be employed at any one time, and one, who at the age of nineteen, had started a life term there, told me that for the first two and a half years of his sentence he did not do a scrap of work. Although the prison became virtually escape-proof, the impact of these conditions on the thinking of inmates was costly.

SOCIAL RELATIONS AT WAIKERIA

In 1942 Polansky observed that in autocratic situations such as in prisons, there are two likely responses to naked power. Firstly, as coercion intensifies, inmate organization may collapse under its pressure, or alternatively, an increase in the use of force may result in systematic resistance.[2] As a rule, a large, culturally diverse population in loose custody, which is controlled by a powerful convict hierarchy, tends towards atomization. Unity is lost and there is little overt hostility towards staff. Leaders tend to identify with officialdom. This system is best illustrated by the concentration camp,[3] but has also been observed in, for example, the New Zealand detention centre,[4] in the Southern United States,[5] in contemporary South Africa,[6] and in Soviet labour camps.[7]

On the other hand, conditions which encourage cohesiveness are largely the converse of those producing division. Where a captive population is small and held in close custody, with members whose background and culture are similar, where official power is dominant and there is no co-opted élite, prisoners tend to become mutually supportive. There is hostility towards or rejection of authority, and inmate organization waxes tight and cohesive. Systems of this type have been described by McCleery in segregation facilities in Hawaii and North Carolina,[8] and by Cohen and Taylor in the United Kingdom.[9]

McCleery points out the high levels of inmate aggressiveness in both mass and close custody situations.[10] However, if we enquire more deeply, we find that the quality of aggression in either context is different. Coercion by leaders in the concentration camp tends to have a highly alienative, disruptive purpose and effect. In the segregation unit, displays of aggression can become the only effective means of maintaining group integrity. Threat and violence become necessary to enforce commitment to inmate interests. For like the underground resistance organization, the ability of members to stand firmly together is the best self-defence weapon they have.[11]

The powerful emotional bonding that appears in groups of this type is difficult to portray, but is well depicted in the outstanding firsthand accounts of men like Jimmy Boyle[12] and John McVicar.[13] After twenty-four years behind bars in the United States, maximum security prisoner Jack Henry Abbott writes:

> Myself and my fellow prisoners lived a hard code, but it was one of survival. . . .
> [The convict] has no 'revolutionary ideology,' true. But eventually he'll run into me in the hole and I'll tell him things that will clear this confusion and give his rebellion a cause . . . And when he rebels alone, if I see him fighting a squad of pigs in the yard or in the hole, I will never hesitate to dive in. We are brothers under the skin. His fight is my fight. If I pay the highest price and he later cops out, it doesn't bother *me*. I've done the right thing and I have no hard feelings for him. We got no one but each other, and I learned that a long time ago.[14]

The importance of such understanding will become clearer when the subject is discussed in chapters 15 and 16. For the moment it concerns us because the solidarity which Paremoremo knew in its early years was conceived and given birth to in the confines of the east wing.

Life in the wing at Waikeria was similar in its routine and organization to that of segregation units overseas. The reference group was small; facilities, recreation, and work were almost non-existent, routine was predictable from one week to the next. Physical security was high and contact with staff minimal. The men detained in the wing had no idea of how long they might be kept there, except that it would be for a very long time. They knew there was little chance for improvement and they also knew that even in the top twelve, things could not be much worse. The result of this situation was a changed outlook among prisoners. A new — or at least, more severe — vision developed, and a different set of social relations grew with it.

The men at Waikeria were rallied by their poverty, bonded by their bitterness. The community saw itself as remote from and opposed to an authority structure from which it now stood little to gain. Lacking any real physical or statutory power, the group did the only thing it could: it assembled its own defence mechanisms and closed itself behind a barrier of extremism. Inmates knew that their best weapon was their unity. Amiable conversation with staff was now censured: the screw you shared a joke with today might baton you down tomorrow. Those who seemed to be on amicable terms with staff could be suspected of informing and called 'policemen'. The 'policeman' ('copper'/ 'grasshopper'/'grass' in the prison's rhyming slang) is still the basest pariah in the security prison environment.

Complementary to these defensive mechanisms were others which protected inmate integrity. Exploitation and bullying of fellows were condemned. Imposing on the freedom of others ('acting like a screw') became a form of conduct which met with the strongest disapproval. As there was no room for differential privilege under these conditions the society became markedly egalitarian. The new emphasis was on fortitude where, at a group or a personal level, 'staunchness', that is, courage in the face of adversity and a refusal to compromise against group interests, became more important than it ever had before.

The puritan atmosphere which arose from these conditions meant that deviation from the prisoners' social code could no longer be tolerated. Any who transgressed could be summarily dealt with. The story is told of one man, a known 'tealeaf',[16] who stole a bottle of shampoo, which had been deliberately left out where he would see it. Unknown to the thief, the soap in the bottle had been laced with battery acid and when he used it to wash with, all his hair fell out.

Another to run foul of the system was Maketu. At Mt. Eden he had been a recognized informant, but under its loose moral structure, his behaviour had been accepted. During the riot, Maketu was known to have given himself up early and volunteered information to the police. He also co-operated with authorities during the Commission of Inquiry and provided damning evidence against some of the riot's leaders. In the changed situation at Waikeria, Maketu found himself an outcast. Others refused to speak to him, and they booed loudly when he was unlocked to testify to the Commission. During the preliminary court hearing of riot charges, Maketu was set upon by his fellows and stabbed several times in the back of the head. Twenty prison and police officers were rushed to the block to prevent trouble spreading, and immediately afterward Maketu was transferred to Mt. Crawford.

However, overall, violence among the prisoners was rare. Consensus was high and the community looked after its own. Weaker inmates could get support from the stronger men. The young Paul Morrison, sentenced to life imprisonment for killing a policeman during a gaol break in Dunedin,[17] later described his reception in the wing:

> I ran into Gary Mulholland up there. Gary sort of took charge over me because when I came in I'd been pretty knocked around, you know, by the police. My nerves were shattered and that, so Gary used to give me a boot up the arse to keep me going three miles around the yard. . . . Gary straightened me out in a hurry. Exercising. Sparring. Kept my sanity.

Another lifer said:

> It was pretty close down there, with the boobheads down Waikeria. . . . I found the boobheads totally different from Mt Eden and Hautu. Down there they seemed to have a sense of unity amongst themselves.

Although enmities inevitably developed, these were largely overridden by the strong friendships forged among men facing prolonged adversity. Many of those who met in Waikeria still remain the closest of friends. Paul Morrison said to me:

> There wasn't much fighting. A few differences of opinion — it would happen anywhere — but considering it was so confined and the amount of violence which could have been generated, it was amazing how close those guys became. Like I've still got a lot of affinity with the guys I was in with there. . . .
> It was a whole different social structure than what emerged later. It was uncanny really because there was so much empathy between us. We *knew* each other, and so much more personally than later on. It was the closeness, and the barbaric conditions we were basically living in. It just bonded you together.

Staff Relations

Contrasting with these circumstances were those at Mt. Eden in the 1950s, which it will be remembered involved a large, highly stratified community organized around the power of a few key members. The co-operativeness of the leaders had made employment at Mt. Eden in the fifties relatively easy and pleasant. A number of officers I spoke to mentioned this and lamented the passing of a time when staff-inmate relations were governed by a code of familiar ethics. Under the previous system, they felt, individual dignity had been preserved. A sort of social mutualism had softened antagonisms and permitted an amount of good-natured harmony.

Although this situation had already begun to alter by the time Eddie Buckley took over, it was at Waikeria that the most dramatic changes took place. The aggravations which had started after 1963 were one component of the new outlook, and the riot definitely added another. As already recorded, staff were demoralized by the rebellion and hated having to work away from their families. Inmates had been vexed by the destruction of their personal effects after the riot, and many felt that conditions in the east wing were unjust. This added to their solidarity as well as to their rancour.[18] This antipathy was unnerving, even for experienced staff. It made their jobs difficult and it alienated them, miles away from home as they were, from their working environment. One Mt. Eden officer who was sent to Waikeria at this time eventually served almost thirty years in the prison service. Asked about working conditions he replied:

> Oh Jesus, tough down there, boy. That was really crucial, that one was. . . . I think that was about as tough a situation as I've ever handled anywhere.

I asked whether it was tough for inmates or staff.

> For staff. From a staff point of view. The tension. The tension between inmates and staff. You know, there were chaps down there that I knew, that I used to talk freely to and that, and it was as if I didn't know them. I couldn't break them. Couldn't break them . . .
>
> And believe me, we were glad to get into that bloody hotel at the end of the day. Tension. I used to see officers that I was associated with, come in at lunchtime, you know, and they just couldn't eat their lunch. Tensed up inside. Well you never knew what was going to bloody happen there from one minute to another . . . you don't know. Never knew what the bloody hell was going on.

Although violence between staff and inmates at Waikeria was rare, it was the threat of violence and the unified opposition to their authority which officers found disconcerting. At Waikeria, a new era in prison management had been born, and it was a tougher, more determined, and more calculating brand of criminal that came out of it. As far as maximum security was concerned, the days of compromise were over. From now on the best type of relationship that staff could hope to have with their charges was a fragile, armed truce.

The Mt. Eden Security Block

Waikeria was not the only security facility in use in the 1965–1969 period. From early 1966, the security block began functioning at Mt. Eden, and henceforth all of the very top risks were held there. It will be recalled that the idea of building a new block at Mt. Eden had arisen out of the escapes of Wilder, Matich, Wiwarina, and Awa in January 1963. Progress on the unit had been slow and was finally held up by financial matters. However, with the headlines of the hostage crisis early in 1965, the calling of tenders had soon followed. The tender for the new block to hold a maximum of fourteen men, was won by L. J. Heffer Ltd. in May 1965.

At the time of the Mt. Eden riot, work on construction had only just begun and the crisis awarded the project a sudden new urgency. Concrete for the floors was poured a week after the fire, and by the end of July labour on the walls had started. Work on the block proceeded quickly, and in early November it was predicted that construction would be finished before the end of the month. Then the finishing date was postponed until before Christmas. Two weeks later, Hanan announced the facility would not be finished until January 1966. More delays followed and it was not until 10 and 11 March 1966 that the Mt. Eden security block finally received its first batches of prisoners from the east wing. There were fourteen cells in the new unit, and all of the twelve top security men were transferred there. This included Wilder, Evans, Gillies, and Gillies' co-offender, Ron Jorgensen.

Conditions in the security block were similar to, if marginally better than, in the top twelve division at Waikeria. Cells measured 2 by 3 metres each and they had a steel bed and a seat bolted to the floor. There were no toilets or radios in the cells, but doors had a small mesh grille so that communication with others in the division was easy. Security was very tight. Strategic points were controlled by electronic devices and visits took place in locked booths, where inmates were divided from their visitors by bars and thick, shatter-proof glass. The unit itself was split into two wings of seven cells, each wing being separate from the other. This allowed inmates who might constitute a threat to one another or to the order of the block, to be kept apart.

The men were locked up for around eighteen and a half hours of every day. All meals were eaten in cells. The block was equipped with dayrooms, and for two and a half hours in the morning and one in the afternoon, the men worked in groups of three on braille programmes. Every afternoon from 2.00 p.m. to 4.00 p.m. exercise was allowed in one of the small yards. These were tiny enclosures, 4 by 4 metres in size, with bars and two layers of heavy-duty mesh on top of them. At 4.00 p.m. inmates were locked up for the night. Like Waikeria, therefore, conditions in the security block were extreme. Contact with the outside was heavily veiled. Visits were electronically monitored and to begin with, little information filtered to the media. It was not until August 1967 that controversy over the new facility broke out.

On 13 August 1967, author Maurice Shadbolt, opening the ninth New Zealand University Arts Festival in Christchurch, commented on the absurdity of regulations surrounding George Wilder, at that time still resident in the block. Strenuous pressure from Shadbolt had resulted in permission for Wilder to be given a pencil and crayons, but paints of any kind had been vetoed. Moreover, for security reasons, Wilder had been prohibited from drawing other prisoners and was not allowed to have a mirror so that he could draw himself. In Parliament the case was taken up by the opposition, which managed to badger Hanan into agreeing that an easel, brushes, and oil paints should be permitted. However, Hanan stated that glass containers, mirrors, or pallet knives would still be unavailable.[19]

The victory was minor, but it was significant because the pettiness of the restrictions drew attention not only to the paucity of facilities in the block, but also to the narrowness of Buckley's thinking. The solution of the dispute was enervated by the fact that, having been forced to allow the paints, the superintendent then decided not to permit them in the cells. So 'Catch-22' was that Wilder, although granted access to materials, had almost no time in which he could use them. Buckley refused to compromise, and on 10 September all twelve inmates went on hunger strike and refused to leave their cells. Although MacKay claimed not to know what it was about, he was aware that the source of friction was general routine and the issue of Wilder's paints. A list of grievances had been presented, but the following day the Assistant Secretary still claimed to be ignorant of them and to have received no written complaints.[20] Buckley refused to let the Secretary of the Howard League into the institution to discuss any of the contentious matters.

On 13 September, nine of the twelve prisoners broke their fast. Another succumbed the following day. On Saturday 16 September, after a visit to the gaol by opposition Member Dr A. M. Finlay, the final two strikers agreed to eat. The grievances were then made public by Dr Finlay.

As soon as the strike was over, a magisterial inquiry was begun, and results were presented about two and a half months later. Principal complaints revolved around uncertainty among prisoners as to why they were in the block and for how long they would remain there, combined with the absence of any effective grievance procedure. Wilder was still not allowed to have paints in his cell, and Jorgensen had been refused permission to have an art book on the ground that part of it was 'obscene'. Beds, set alongside a 2-metre long wall, were too short for tall men to lie straight, cells were stuffy in summer, draughty in winter, and the layer of mesh over the yards meant that there was hardly any sun. To compensate for the latter, inmates now received a supplement of vitamin pills.

The magistrate's report had some effect, and minor adjustments were made. Food was served on a table instead of the floor, and Hanan promised he would consider the report's suggestion that a television set be provided. Eventually he decided against it. In future, inmates would be told why they were in the block and a local magistrate, Mr L. G. H. Sinclair, was appointed to advise the Justice Department on transfers in and out of segregation.

The improvements made to living conditions in the security block were thus mainly cosmetic. The need for custody, the physical limitations of the unit, and the obstinate opposition of the superintendent to most forms of change, made a relaxation of conditions impossible. Policy of the Department towards the security prisoners was dominated, not by enlightened humanitarianism, but by an overriding fear that concessions might become tools in another disturbance.

Conditions in the block themselves had little to commend them, which is why MacKay was at such pains to obscure the causes of protest. When, in January the following year, three more prisoners went on strike, the reasons for their actions were not disclosed either.[21] From that point on, in fact, almost no publicity about the unit reached the press, and media attention was directed away from the existing situation, towards the future. When the new institution at Paremoremo was opened, it was hoped, problems associated with close confinement would disappear.

SOME IMPLICATIONS FOR THE FUTURE

The social impact of the security block on later relations at Paremoremo was similar to, but less important than, that at Waikeria. There were only twelve men in the Mt. Eden block and two of them, Wilder and La Mattina, had been transferred before the new institution opened.[22] Of the rest the majority went straight to Paremoremo's D. block, a long-term isolation facility in which conditions were not much different from what they had been in security. In D. block these prisoners met up with old friends from Mt. Eden and from the early days at Waikeria.

In the new institution near Albany there were new faces as well as old, but lack of familiarity mattered little. The men soon grew acquainted with one another, and in spite of their differences, D. block men found they had one thing in common: an 'intractable' label, the meaning of, and reasons for which, they did not fully comprehend. Once again, deprivation, repression, and a deep sense of personal wrong became the grist of inmate relations. As we shall see, although, in its early years, Paremoremo experienced difficulty which was not restricted to its segregated population, it was in this division that the core of resistance formed. The energy and bitterness of the conflict which developed here was to have a lasting effect on the institution even after the original regime was replaced.

NOTES

[1] *NZH* 2 August 1965. It should be noted that Waikeria at this time still consisted of two other institutions. These were what remained of the borstal and the detention centre. Although under the command of a single superintendent, inmates from the three units were kept strictly apart from one another.
[2] Polansky (1942) 20–1.
[3] See, for example, Bettelheim (1943) 447–9; Bloch (1947) 338; Cohen (1953) 177–9; and Kogon (1958) 274–7.

4 I noticed this tendency when incarcerated at Waikeria detention centre in 1971.
5 Dollard (1957) 124, 174-5.
6 Etzioni (1968) 362.
7 Cressey & Krassowski (1957-1958) 227-8.
8 McCleery (1961a).
9 Cohen & Taylor (1972).
10 McCleery (1961a) 302-3.
11 For a broader discussion see Newbold (1978) 213-17.
12 Boyle (1977).
13 McVicar (1974).
14 Abbott (1982) 79-80.
15 The rhyming slang of prisoners derives from cockney. A glossary of the vernacular of modern New Zealand prisons appears in Newbold (1982b).
16 Thief.
17 The case I refer to is that of Paul Morrison, an accomplice to a murder committed by one J. T. Wilson. *Morrison* is now a leading authority in New Zealand case law (accomplices). See *R v Morrison* (1968) NZLR 156. The case also drew comment in Parliament (NZPD vol. 348 (1968) 2077).
18 One who was vocal in this regard was Angelo La Mattina. La Mattina had been transferred to Waikeria, even though at the time of the riot he had been bedded in the prison's medical section.
19 NZPD vol. 348 (1967) 2309; *NZH* 16 August 1967.
20 *NZH* 14 September 1967.
21 *NZH* 15, 16 January 1968.
22 George Wilder won a correspondence art scholarship in March 1968 and was transferred to Paparua prison. He was released in June 1969 after serving eight years of his nineteen year sentence (*NZH* 30 March, 8 April 1968, 21 June 1969).

La Mattina, who had been transferred to the block after a strike at Waikeria in August 1967, was released in May the following year. Deported to his native Sicily, La Mattina was re-charged with his New Zealand murder (in accordance with Sicilian law) on arrival at Palermo Airport. In 1973, then aged thirty-nine, La Mattina was sentenced to a further twenty-one years' imprisonment for the crime he had committed in Wellington some sixteen years before.

13: THE BUILDING OF PAREMOREMO
1960-1969

Development and Delay

Paremoremo prison was the brain-child of John Robson. Within a few weeks of his appointment to Justice in July 1960, Secretary Robson had embarked on a trip to Europe, to study overseas systems. Upon his return, the Department's building programme was completely reviewed, and an area which received great attention was that of prison security. On 19 December 1960, Robson presented to Hanan his views on accommodation policy, and included in his memorandum was an argument for a new top custody prison at Auckland.

The Minister agreed with Robson, and by the end of March the following year, the Department was able publicly to announce plans for a large rebuilding programme, with a new security prison as an urgency. Four months later, the Department sent a representative to Auckland to look at land for the prison, which it was anticipated would cost around 935,000 pounds to construct. Meanwhile, a variety of designs was being evaluated.

The early 1960s was a period of experimentation in prison security, and both New Zealand and Britain attended to the results of new systems. In the United States, Alcatraz had been emptied in 1963 and an early improvement in the conduct of inmates transferred to its replacement at Marion, Illinois, was noted.[1] By this time, a British building programme was well under way, and after the escapes of train robber Ronald Biggs in 1965 and spy George Blake in 1966, the whole of England's security programme was revised. Although the resulting investigations — the Mountbatten Inquiry and the Radzinowicz Report[2] — arrived too late to influence the design of Paremoremo, they were noted with interest in New Zealand. Likewise, the building of the world's most modern security prison in the South Pacific was a development which British criminologists watched keenly.

In 1959, the Prison Commission of England had undertaken its largest building programme for a century, and, in order to stimulate thought on the subject, four papers on penal architecture were published in a special edition of the *British Journal of Criminology*.[3] Their content was discussed at a special meeting of the penal group in July 1961. Although (in an unusual departure from tradition) the Department was not heavily influenced by British developments, the material did have a bearing on the institution's eventual design because among the projects dealt with in the papers were those at Blundeston in Suffolk and at Marion, Illinois. Both of these were eventually visited by the New Zealand Government's architect, before the drawing of plans.

Building a prison of the size and importance contemplated proved complex. In March 1962, twelve months after the official announcement had been made, a site still had not been found. The following month, when the Cabinet Works

Committee approved in principle the construction of the institution, it had nowhere to build it. The escape of La Mattina and Tell in 1962 had accelerated that search, and in May Robson announced to the Penal Group that a block at Paremoremo, north of Auckland, was being studied, as well as one at Manurewa to the south. At a meeting on 28 May, the Paremoremo option won the favour of the Department's Buildings Committee, and having conferred with the Justice Department, money to buy it was requested from Cabinet Works. The requested was granted on 18 July 1962.

To begin with, the decision was kept secret, even from members of the Government.[4] The press was kept in the dark as well, and not until 7 August did Hanan confirm that 276 acres were being purchased at Paremoremo for a new maximum security gaol, with room for a less-secure institution there in the future.[5] For the next few months, speculation about the institution abounded. The Department was undecided about whether radar and closed-circuit television should be used, and looked elsewhere — particularly to Sweden — for ideas. Eventually it was resolved that no policy decisions would be made until the architect had returned from overseas and reported on recent developments.

In March 1963, the assistant Government architect Mr J. R. B. Blake-Kelly had been booked to travel abroad to study concepts of modern prison design in England, Sweden, and the United States. In the meantime, plans of some modern gaols (among them Blundeston, and Kumla prison in Sweden) had been received and were being studied by the Department. Later, the plans for Marion were also acquired. Site surveys were now conducted so that by the time Blake-Kelly returned early in June, all that would be left undone would be the prison's actual design.

Within a month of Blake-Kelly's return, work on preliminary plans was under way, with a first draft scheme expected by about mid-July. On 31 July, the Cabinet Works Committee approved increasing the projected size of the prison from 200 to 250 cells, in order to include a special unit for classification. The institution would consist of five wings of fifty cells each, and it was hoped that construction would start quickly. At the end of August approval for working drawings was given. By the beginning of October preliminary investigations by the Ministry of Works were complete and it was announced that construction would start at the end of 1964. The prison could be ready for habitation, it was felt, by the close of 1966.

At the beginning of February 1964, however, final sketch plans had still not been completed. A press conference was called then, and a detailed statement was issued on the progress of the new gaol and the form it would take. This statement is important because it marked the beginning of a deliberate campaign, which continued right to the end of Hanan's life, to 'sell' the idea of the prison to the New Zealand public. Characteristic of this campaign were elaborations on the secure, but aesthetically pleasant, nature of the gaol, augmented by grossly deficient estimates of its cost.

The prison, which Hanan said would now be opened by the beginning of 1968, would be modelled on the 'school house' system. To be built for a bill of one million pounds, the institution would be totally self sufficient in terms of amenities. Although security would be high, efforts would be made to rid it of an 'institutional' character. Corridors would be lined with pot plants and ornate pools. 'Painted vistas' and a 'parkland setting' outside the walls would remove the prison's fortress appearance. In spite of this, security would be tight. Key areas would be surrounded by a bullet-proof casing. There would be closed-circuit televison at strategic points and bars would be of hardened steel. Dining rooms and the visiting area, both potential trouble spots, would be under close surveillance and control.

As we have seen, moves to hasten the building of the new facility as well as the new block at Mt. Eden had followed closely on the heels of Wilder's and his associates' escape in January 1963. The building of Paremoremo had then become a 'main priority', and although in 1961 the annual report had said the security prison was one of the most urgent of all,[6] by 1964 it had been relegated to third place. Then in May of that year, with working drawings ahead of schedule and due by the end of December, the erection of a secure prison was revived as the first imperative.[7]

Privately it was known that earthworks would be delayed for at least seven months. An impatient press had been informed that work would begin 'soon',[8] but tenders for it were not actually called until the end of October, and the contract was not let until December. That month it was announced again that the commencement of work on the land was imminent.

Seven months before the Mt. Eden riot, press comment about the slow progress had already begun,[9] but despite criticism, the delays continued. Final plans, first scheduled for December 1964 then June 1965, were now put off until October. On 10 April 1965, it was announced that the opening of Paremoremo, targeted in February 1964 for the end of 1967, would now be a year late under a 'new programme' worked out by the Department. The end cost of the institution — originally estimated at less than one million pounds — was no longer known and final drawings were still incomplete. By early July 1965, the excavations, which in March the Department had said could be done by April, were also unfinished. By the beginning of July 1965, therefore, excitement about the rapid construction of the prison had eased. We already know that in Parliament at this time there were moves afoot to have the institution postponed indefinitely. Even after the riot, when building a new institution suddenly developed true urgency, the Department remained uncertain about how long it would take. As recently as 9 April 1965, Hanan had suggested construction would take three years. Now he predicted four. A few days after the riot it was announced that under an accelerated programme, the buildings could be completed in two and a half years and be ready by early 1968.

However, no sooner had the Ministry of Works agreed to step up its pace, than heavy rain turned the site into a quagmire. *The New Zealand Herald* noted that the Māori translation of Paremoremo was 'to be slippery',[10] and that the site appeared to be living up to its name. Earthmoving equipment was bogged down and draglines were being used to remove masses of viscous clay. Within three weeks, heavy rain had forced the cessation of works altogether, and they were not resumed until September. On 13 October a contract was let to Tileman (NZ) Limited, for construction of the institution to start.

Weekly reports on the projects at Paremoremo and Mt. Eden were now supplied by the Department, with a more extensive assessment being prepared every month. By early December, earthworks were completed and by February 1966, building had commenced. Like the earthworks, this progress was erratic. First of all there was trouble with the foundations, then workers walked off because of the mud. In July the scheduled completion date of 1 March 1968 was extended to 4 June. However, mud was still a problem and in November 1966, a sea of sludge forced another postponement, this time to 25 June.

The building of the prison was followed closely by *The New Zealand Herald* and regular columns, accompanied by photographs, appeared in the paper every few weeks. Although progress was reported to be good, postponements continued; completion finally being promised in mid-November with a commitment to hand over the prison to the Justice Department on 10 December. This date would remain fixed, irrespective of whether or not the gaol was finished. The institution would not, as it happened, be ready for its first batch of receptions until late January or early February 1969. In February, Hanan announced that the prison would receive its first prisoners on 1 March. On 13 March a small group of short-term inmates was admitted to Paremoremo. These men were only a test-group, however, and it was not until the end of the month that the first loads of permanent detainees were delivered. By 29 March, fifty-nine maximum security men were contained in the new gaol.

Although the new prison had been opened more than a year later than projected, it was still incomplete, only partially staffed, and only three of the five cell blocks were in use. Its construction, originally estimated at less than one million pounds, was finally completed at a total cost of almost seven million dollars.[11]

ARCHITECTURE AND DESIGN

When it was opened in 1968 — despite the controversy it provoked — Paremoremo was arguably the most modern and technologically sophisticated gaol in the world. The Justice Department avoided describing the prison as 'escape-proof',[12] but within the limitations of humane custody it was as close to being escape-proof as could reasonably have been hoped for.

As noted, the design of Paremoremo was derived largely from studies made by Government officials on modern custodial facilities operating or being built overseas. Of these, Blundeston was already in use, Kumla was under construction, and Marion was as that time nearing its completion. In varying degrees, all three institutions contributed towards the planning and development of Paremoremo, but it was Marion which proved the most instructive.

Marion had opened as a federal prison camp in June 1963 and was designated a penitentiary the following year. Originally built as a partial replacement for Alcatraz, Marion was designed to contain some of the more troublesome elements in the federal system. The smallest of more than thirty such prisons, Marion has a capacity for 700, but now holds less than 350 men. It is the only federal correctional facility constructed primarily for single cell accommodation. Within its nine cell blocks, Marion originally kept maximum, medium, and minimum security prisoners for whom work was provided both inside and outside its perimeter. Additionally, within the double-mesh fence enclosing it were full-sized football and baseball pitches, as well as a gymnasium, an auditorium, and a chapel. After trouble in the federal gaols in the late 1970s, Marion's role was changed to one of top custody for extremely dangerous prisoners. Accordingly, in 1979 it became the Bureau of Prisons' only 'level 6' penetentiary. A series of fifteen murders (with two of the victims being staff members) after this point resulted in a regime of total lock-up in the institution since October 1983 and a closure of all its facilities.[13]

The New Zealand Department of Justice in the early sixties favoured some of the features of overseas institutions, others it did not. Solid concrete perimeter walls — as seen at Kumla, for example — were rejected. However, the Swedish idea of curving the tops of internal walls to make them grapple proof, was one which did appeal.[14] The perimeter of Marion is secured by heavily armed security personnel. This was contrary to New Zealand policy and to the recommendations of the Mountbatten Inquiry, which the penal group discussed in February 1967. Consequently, the Department insisted that Paremoremo staff should not be armed. On the other hand, Blundeston, which also had no armed sentries, was felt not to be secure enough, and when plans for Paremoremo were drafted, playing fields were excluded. The only facilities for physical activity were the gymnasium and the small, high-walled concrete enclosures which served as exercise yards.

Where recreation was concerned it was perhaps the Marion gymnasium which impressed the New Zealanders most. Likewise the Department liked the internal central control system in use here and at Kumla. The workshops at Kumla also caught the eye of the New Zealand officials[15] and, when their prison was designed, provision of a worthwhile array of industries was a matter for some attention.

So, although small aspects of several overseas institutions were incorporated at Paremoremo, it was North America which drew the New Zealanders' closest attention.[16] Design of the sentry towers, cell layout, and the overall plan of the buildings were unmistakeably of the North American mould.[17] The remote door-locking mechanisms were made in Joliet, Illinois, while the hardened steel grilles were freighted from Canada. In 1978, the superintendent of Paremoremo even had a copy of the Marion penitentiary rule book shelved in his office.[18]

Like Marion, the New Zealand prison was formatted on the 'telegraph pole' pattern, a style of construction unambiguously identified with the United States, originating in Fresnes prison, France (1898)[19] and used widely in the States after 1932.[20] Cell blocks of the telegraph pole pattern do not emanate from any common point, but extend like cross arms, in parallel fashion off a central corridor. The advantage of the 'telegraph pole' over the older radial design of prisons like Mt. Eden is that it allows for freer mobility within a secure interior. However, at the same time as it does this, it permits specific areas to be closed off from different classes of convict, thus reducing problems of supervision. Thus, it was the tradition begun in France last century, spreading to the Americas[21] and in some instances as far afield as Japan,[22] which eventually found its way to Auckland in 1968.

The site that the Justice Department had chosen for its new prison was on the upper reaches of the Waitemata Harbour, 30 kilometres north of Auckland. Set in a district near the township of Albany, the prison was built to accommodate 248 inmates in five, three-storeyed cell blocks. Adjacent to each block was constructed one or more exercise yards. Structurally, the prison now is unchanged from when it first opened. Internal amenities are entirely self-contained. There is a well-equipped gymnasium with adjoining weight room, a workshop complex, a library, a study room, visiting room, chapel, kitchen, laundry, administration section, and a 'central control' unit, all under the same roof. In addition, there is a small psychiatric division, a detention unit, and a fully equipped hospital. The emphasis of the whole concept lies heavily upon security. On top of the initial investment of seven million dollars, by 1976 a further two million had to be spent every year in maintenance.

Internally the prison is divided into separate housing units, which are known as A., B., C., D., and classification blocks. The first three of these are the so-called 'open' or 'standard' blocks, the inmates of which all enjoy full privileges. Each block has an exercise yard and most of the open-block inmates have access to the workshops, the gymnasium, the chapel, the library, and open visiting at designated periods. These prisoners spend the greater part of the day unlocked — from 7.00 a.m. to 8.30 p.m. during the week, and 8.00 to 11.40 a.m., 1.00 to 4.40 p.m. on weekends. Classification and D. blocks are closed areas and their inhabitants are entirely segregated from the rest of the population. They never leave their wings unless under officer escort (three officers per man in the case of D. block) and they spend a great deal of their day — between eighteen and twenty-one hours — in their cells.

All of the blocks at Paremoremo have cellular accommodation on their second and third levels, with recreation in the basement. The standard blocks, and the lower landing of classification have dining-room facilities. The rest are 'fed in', that is, they are delivered meals in their cells. A., B., C., and D. blocks all contain forty-eight cells, in four rows of twelve. Classification houses fifty-six. Each cell-row may be isolated from the others by a manually operated steel grille at the end of the corridor. All cell doors are opened and closed from behind this grille.

Cells are not large, measuring approximately 3.2 by 1.8 metres, but are designed only for single occupation. They are well equipped with hot and cold running water, a flush toilet, three channel radio-cum-PA speaker, a chair, a fixed steel writing desk and good lighting with interior switches. The entire frontage of the cell is barred. This is an American idea and is one of the greatest drawbacks of Paremoremo, eliminating any privacy for prisoners.

The most important feature of the organization and structure of Paremoremo is that it is almost totally secure. In fact, the whole complex is obtrusively designed to fulfil this essential function. For the aspiring escapee, the minimum physical barrier between himself and freedom is always at least one, 7-metre concrete wall with a trip-wire on top of it and two, 7-metre close-mesh perimeter fences, also equipped with trip-wires. The windows of the institution are of shatter-proof glass in steel frames and are guarded by steel-reinforced concrete mullions. The bars of the cells and grilles are made of tool-resistant manganese steel, each reputed to be able to resist the action of 1000 hacksaw blades or a press of several tonnes.

In his movements about the prison, it is never necessary or possible for an inmate to leave the confines of the buildings. The whole institution is constructed under a single roof; a maze of long concrete corridors, steel grilles and barred windows. Indeed, except when he enters the high-walled exercise yards, the inmate is never without a roof over his head. In the standard blocks, the yards are larger than those of the closed divisions. These former measure 35 by 15 metres and are bordered by buildings or walls which range from 7 to 10 metres in height.

The physical design of Paremoremo is one that affords a high level of control over movement, and it is possible for large sections of the community to be isolated at the flick of a switch. This is permitted by the electronically controlled sally port system and the internal complex of microphones, speakers and closed-circuit television. Most of these are linked to central control, a bullet-proof unit set in the prison's middle, with armour-plated windows looking out upon the four main corridors. It is always manned by two or more personnel. The sophistication of the central control system is such that monitoring inmate movements is easy. Theoretically, at any time of the day, the exact whereabouts of any specified prisoner should be immediately available to staff manning the control office. More important for our purposes is the fact that this control is impersonal and mechanical, and continues without the necessity of face-to-face or verbal contact between prisoners and staff.

Such a situation has consequences for the development of inmate–staff relationships and, we shall see, indirectly affects inmate belief systems and interaction.

PAREMOREMO AND ITS CRITICS

Inevitably a project as expensive and as heavily publicized as that of Paremoremo drew comment from many quarters. Although critics were often unfamiliar with the prison's structure and with the principles to which it conformed, there were few without an opinion on it. The majority welcomed Paremoremo. No one denied that Mt. Eden had outlived its usefulness, and its replacement with a safe and, hopefully, humane alternative was applauded.

On the other hand, Paremoremo was attacked for his inhumanity. The gaol was felt to be 'escape proof'[23] and this, for some, was condemnation enough. Some believed its austere concrete interior would be psychologically destructive and that the lack of privacy in cells would debase their inhabitants. For many, the building of Paremoremo signified the end of the reformist era in New Zealand penology. Its construction heralded the reincarnation of the primitive spirit of retribution.

In its early days, though, comment was only sporadic. To most the gaol was still an idea, a cardboard model at the Easter Show,[24] a drawing in the local newspaper.[25] It was not until after July 1965, with the need for a new facility unquestioned, that comment gathered momentum. Even so, it took another three years for the general reaction to crystallize.

Robson of course was still highly media conscious, and he hungered for public acceptance. Initially Hanan had talked about pot plants, ornate pools, vistas, and the gaol's 'parkland setting'. Then Blake-Kelly prepared a statement which extolled its bright, airy atmosphere and the 'interesting plays of light and shade', which the concrete pillars guarding the windows would cast upon the interior.[26] The press, especially in Auckland, was interested in all aspects of the project and it was not until early in 1968 that the first real challenge to official claims was levelled.

On 16 April and 6 May 1968, *The Evening Post* and *The Auckland Star* carried identical articles on Paremoremo prison. Entitled 'The "Intractables" Will Live Alone', the columns predicted a gloomy style of life in the new institution, with fifty-four prisoners locked in semi-permanent solitary confinement and exercising in small, high-walled yards. Watched in their cells twenty-four hours a day by closed-circuit television, the men would be guarded at all times by armed sentries and surrounded by twin fences of barbed wire.

Robson lost no time in reacting to this, and on 20 April he approached Mr Hardingham of *The New Zealand Herald* to publish a response. Accordingly, a little over two weeks later, 'New Prison Has Luxury Touch' underlined the modern amenities and humanitarian conditions that life in the institution would offer.[27] The piece was a counterview rather than a specific negation of the offending allegations, and it is not surprising that *The Evening Post* and *The Auckland Star* versions retained some credibility. Thus, rather than being dampened by rebuttal, the passion of the battle increased. Rejection of the Government's multi-million-dollar baby was becoming fashionable; and before the first fires of contention could die, another log was added. On 23 June, *The Sunday Times* exposed a new angle to the problem. Not only would the prison be inhumane, the paper said, but the 'solitary confinement' of intractables might also be illegal. In addition to the myths about barbed-wire fences, armed guards, and television cameras in cells, *The Sunday Times* now introduced the spectre of cell doors, which were really 'remote-controlled electric gates', allowing staff to keep well away from prisoners. 'SOLITARY PRISONERS WILL NOT EVEN HAVE CONTACT WITH PRISON OFFICERS LOCKING AND UNLOCKING THEIR DOORS AND UNDER THE TV CAMERA'S EYE WILL BE DENIED PRIVACY EVEN FOR THE MOST INTIMATE OF PERSONAL ACTS', the article cried. Men would be driven close to insanity by such conditions. They would become like animals. The institution would produce 'robot prisoners' and stand as a 'terrible monument to a lost battle in the war for prison reform in New Zealand'.

The New Zealand Council for Civil Liberties was disturbed by the claims, and requested Government clarification. On 24 June, a ministerial statement criticized *The Sunday Times*' inaccuracies, pointing out that there would be no solitary confinement in the segregation block, and no cameras in the cells either. 'Intractable' prisoners would not be under punishment, and there would be frequent contact between prisoners and staff, with liberal access to a library, hobby work, instruction classes, and general education as well. The 'windowless' cells, in fact, had barred fronts looking on to an airy corridor, many with windows facing towards the open countryside, the statement said.

Although somewhat exaggerated, Hanan's reply was substantially correct,[28] and he slated *The Sunday Times* for a 'deplorable piece of journalism'.[29] A meeting with the publisher's general manager achieved conciliation and there the matter would then have rested. However, not two days later, and to Hanan's great astonishment, another article appeared in *The Sunday Times*, which reiterated each of the claims that Hanan had so explicitly denied.[30]

The Minister of Justice was furious. A terse letter was drafted to the publisher Burnet, who did his best to placate the Minister. The second piece had been published in error, he explained, but with the paper's editor refusing to withdraw, Burnet could only apologize and suggest the matter be dropped. Bitterly, Hanan agreed, although with the unveiling date so close he knew a media black-out was impossible. For in September 1968, two eminent criminologists were scheduled to visit New Zealand on a speech and lecture tour, and they would inspect the new institution. One was Sir John Barry, a world authority on crime, the other was Mr C. H. Rolph, a former Chief Inspector of the City of London Police and latterly legal editor of the magazine *New Statesman*. Hanan hoped that their reactions would be favourable, as both were recognized specialists in their fields. Perhaps more importantly, Rolph's comments were due to appear in the December issue of *New Statesman*, the release of which had been timed to coincide with the prison's opening. Although Barry's visit was forestalled by terminal illness,[31] Rolph's visit proceeded. Having toured Paremoremo, Rolph remarked on the 'world recognition' it had earned from overseas officials. Paremoremo, he announced, was a 'masterpiece of designing and construction'.

Australian observers were also favourable. Their only scepticism was over the (now) stated four million dollar cost of the project,[32] which the Department already knew was incorrect.[33] Unwilling to respond at so critical a point, however, Hanan was concealing Paremoremo's seven million dollar bill until opening day, when a pamphlet[34] containing the true figure would be distributed.[35] In the meantime the publicity campaign continued, and now everyone seemed impressed. Secure in the visitors' responses, and with opening publicity now poised for action, it seemed that the stage was set for a trouble-free transfer.

However, not five days before the opening date, another scandal broke. I. M. F. McDonald was an English-trained lawyer and a criminologist, teaching at Auckland University. While most of his work had been at Edinburgh, McDonald had been at Auckland for three years and over a two-year period claimed to have visited every penal institution in New Zealand. He also said he was a doctor — although he possessed no such letters[36] — and the report he said he was writing in fact never appeared. None the less, the critic's credentials looked important, and when, on 4 December, he released a statement on the new prison, detailed accounts of it appeared in all the major tabloids in the country.

Paremoremo, McDonald said, was a 'monument to the spirit of vindictive retribution', with a concept based on 'nothing more enlightened or progressive than society's age-old desire for revenge and punishment of the criminal'. He criticized its 'zoo-like' atmosphere. The electronic controls and televison cameras, he said, 'add to the general atmosphere of inhuman gimmickry'. The prison was based on a view of prisoners as wild animals, and if a man was not a wild animal when he entered, he almost certainly would be by the time he got out. The eighteenth century penal philosophy still prevails in New Zealand, said 'Dr' McDonald, and 'accords perfectly with the public service mentality which is responsible for penal policy in this country'.[37]

The statement brought swift support from other quarters, and in sympathy with McDonald, they now choroused with him in decrying its 'Orwellian' overtones.[38] In Wellington, Hanan sensed the beginning of another onslaught and sprang to the counter-offensive. He sharply reproved McDonald for his ignorance of the New Zealand penal system and his unfamiliarity with the major policy declarations of the Department. However, practised though Hanan was becoming at dealing with his detractors, this time the best reply was non-official. For the very day after the McDonald release, Rolph's comments in *New Statesman* appeared. In direct rebuttal of his countryman, Rolph heaped praise upon the nascent institution. Paremoremo, Rolph wrote, 'does everything a maximum security prison ought to do in the way of imagining and forestalling escape methods . . . but it's a perfect illustration of how much freedom and open training there can be inside a really secure perimeter — the best I've seen anywhere in the world'.[39]

Like the McDonald statement, the contrary view of Rolph received a wide press and when he opened the institution on 10 December, Hanan compared the comments of the two. Speaking to a guest audience in the prison gymnasium — one of the members of which was McDonald himself — Hanan said the lecturer's impressions were representative of minority opinion only. He then attacked the criminologist for his false claim to qualifications, and said that McDonald's statements about Paremoremo constituted a 'shocking abuse of academic freedom'.[40] Hanan's treatment of McDonald was ungracious, but it should be understood that for this Minister, Paremoremo had become a personal and sensitive concern. The whole project, nine years of careful thought and research, had been the product of him and his permanent head. Every resource the Department had — technical, academic, and financial — had been poured into its planning. Much of the feedback the Government had received was emotive, spurious, and sensationalistic, and goes far towards explaining Hanan's energetic response to the attacks. It also helps in understanding why, when the institution was ready for handing over, it was in a 'gala atmosphere'[41] of triumph that the ceremony was conducted.

The Justice Department was fortunate that it held favour with the editors of *The New Zealand Herald*, New Zealand's largest daily, and the day after opening, in a large feature, *The New Zealand Herald* eulogized the prison as 'a five-star, inside-out fortress', from which 'only a genius could escape'.[42] Other pieces, some of them checked for accuracy by Robson, appeared in subsequent months. Their responses now were all positive and before long, interest began to recede. The new institution was opened. The staff had been appointed. By the end of January 1969, there was little more to do, but wait for the first prisoners to arrive.

NOTES

[1] Klare (1968) 495. Klare does not mention that it was only the younger, relatively harmless inmates who initially went to Marion. The majority of Alcatraz prisoners were sent to Leavenworth and Atlanta. In 1973 the Federal Bureau of Prisons returned to the policy abandoned in 1963, of concentrating disruptive inmates in one institution. These were sent to a special 'control unit' at Marion. It was not until 1978–1979 that Marion became a 'level 6' prison, for the containment of the most dangerous of convicts (*Consultants' Report on Marion, Illinois* (1985) 1–2).

[2] In December 1966 the 'Mountbatten Inquiry' (*Report of the Inquiry into Prison Escapes*) made numerous recommendations, many of which were challenged by the 'Radzinowicz Report' (The Regime for Long-Term Prisoners), two years later. The effects of these reports and the differences between them are discussed by Klare (1968) and McClean (1969).

[3] See the *British Journal of Criminology* vol. 1 (1961) 4, 305–75. The four contributors were A. W. Peterson (Chairman of the Prison Commission), N. Johnston, L. Fairweather, and J. Madge.

[4] NZPD vol. 331 (1962) 1217. One who was particularly interested was R. D. Muldoon, who later became Prime Minister.

[5] *NZH* 8 August 1962. The less secure institution — a medium security structure — was eventually opened in 1981. Designed to hold 144 inmates, it was constructed at a cost of eight million dollars.

[6] *App.J.* H-20 (1961) 13.

[7] *PGM* 4 May 1964.

[8] *NZH* 20 June 1964.

[9] *The Sunday News* 30 December 1964.

[10] *NZH* 25 August 1965.

[11] New Zealand changed to decimal currency on 10 July 1967. One pound was worth two dollars.

[12] *NZH* 11 February 1964.

[13] United States Bureau of Prisons (1969); *Consultants' Report on Marion, Illinois* (1985); Johnston (1961) 328–9; *US News and World Report*, July 27 (1987) 23–4.

[14] Robson (1974) 12.

[15] Robson (1974) 6.

[16] Robson (1974, 30) confirms that Paremoremo was 'substantially based' on Marion.

[17] See for example the series of photographs of American institutions reproduced in Nagel (1973) 59–64.

[18] United States Bureau of Prisons (1969).

[19] Fresnes Prison, designed by Francisque-Henri Poussin, was itself an extension of the separate block plan built under du Cane at Wormwood Scrubs (1874–1891).

[20] The first recognized American telegraph pole prison was that of Lewisburg, Pennsylvania (1932). This institution was designed by Alfred Hopkins, who visited

Fresnes before starting his work. Modified versions of the plan had appeared earlier in the states, for example, at Stillwater, Minnesota (1914) (Johnston (1973) 43-4). Material on the origins and development of telegraph pole prisons can be found in Johnston (1961) 327-31; (1973) 41-9; and Nagel (1973) 39-41.

21 Telegraph pole prisons have also been constructed in Canada and Latin America.
22 For example, Yonago prison, Honshu, opened 1923.
23 I have noted, of course, that no official ever said that the gaol would be escape-proof.
24 This model was ready by March 1965 (*PGM* 8 March 1965 and was first displayed at a press gathering in Wellington on April 9. (*PGM* 3 May 1965). See also: *The Dominion* 12 April 1965. It later appeared at the Auckland Easter Show and at other exhibitions as well (Robson (1974) 13).
25 For example Dept.J. (1964) 82-3; *NZH* 19 November 1964; *EP* 24 September 1968.
26 Reprinted in Robson (1974) 6-7.
27 *NZH* 8 May 1968.
28 In reality, little in the way of recreation was offered in the block and only half of the cells have a (restricted) view of surrounding countryside.
29 Official press release 24 June 1968; Letter to H. D. Brass, *Mirror Newspapers*, Sydney 24 June 1968.
30 *Sunday Times* 30 August 1968.
31 The series of lectures Sir John was to have delivered was later published by the Government Printer (Barry 1969).
32 *EP* 24 September 1968.
33 Memo from Hanan to Buckley 22 November 1968.
34 Dept.J. (1968a).
35 Memo from MacKay to Buckley 22 November 1968.
36 McDonald's true credentials were an LL B from London and a Diploma in Comparative Law from Luxemburg.
37 It was well covered, for example, in *The New Zealand Herald* (Auckland) 5 December 1968; *The Evening Post* (Wellington) 5 December 1968; and *The Press* (Christchurch) 5 December 1968.
38 *EP* 5 December 1968; *Press* 5 December 1968; *NZH* 5 December 1968.
39 *New Statesman* 6 December 1968, 782. Also reported in *NZH* 7 December 1968; and *AS* 7 December 1968.
40 *AS* 10 December 1968; *NZH* 11 December 1968; *EP* 11 December 1968.
41 This was the impression of author James McNeish, writing in *The Straits Times* (Singapore) 7 January 1969.
42 *NZH* 11 December 1968.

14: HABITATION AND ADJUSTMENT 1969

The Prisoners Arrive

Eddie Buckley moved to the job at Paremoremo soon before the inmates came in March 1969, and began the task of training staff. Buckley was proud of his new position. Having been consulted on aspects of the project throughout its construction, he suspected from the start that he would be favoured as superintendent.[1] Like Hanan and Robson he had a personal commitment to the gaol. An *Evening Post* report in September 1968 said that Mr Buckley 'regards the new prison as he would his first-born child'.[2] The same impression was gained by an inmate who was in the prison's initial test-group.

> Yeah, when we arrived he took us all over the nick. He was so fucking proud of the place, he was trying to bullshit us that it was part of his planning and all this crap. He was rapt in it, eh. It was his baby.

An officer on one of the first escorts had a similar experience:

> And Buckley's there and he's saying, 'This is it! This is gunna be *the* place from now on. The acme of the penal system'.
> So these blokes [the inmates] come in and they file down the kitchen. 'Hello!' he said, 'You hungry?' He said, 'It's all there! Everything's ready for you! There's more if you want it.'
> (Oh God! This is supposed to be *it*.)
> And Tuyt was there too, and Buckley says, 'Come and have a look around.'
> ... He says, '*This* is it! I've got everything right. This is the way it's going to be, and that's the way it's going to be. If there's any trouble up there', he says, 'we'll go this way, and if there's any trouble over there we'll go this way', he said. 'We've got the place sealed off! Nothing can happen here!'

Such was the effusive flush with which Buckley took on his new job. The unqualified enthusiasm which the superintendent displayed for the appointment, however, was not matched by all.

Buckley was the only penal administrator in the country who had any real maximum security experience and of those who applied, certainly he was considered the best for the position. However, in the capital, a feeling already existed that Buckley was not the 'ideal choice'.[3] After the riot, there had been concern about the way Mt. Eden was being run and some months after Buckley had gone, MacKay inspected the old prison, now in the hands of Jack Hobson from Wi Tako. MacKay returned to Wellington satisfied with the way the new superintendent had relaxed security, and how he had extended the gaol's recreational programme.

So at the time of Paremoremo's opening, head office was already aware of Buckley's zeal for discipline. Before the prison had even been occupied, steps had been taken to impress on Buckley the way it was to be organized, and that a positive reformative programme was to be installed. A report was asked for in March 1969, detailing how he had responded to this direction. In recognition of his security penchant, Buckley was also firmly admonished that under no circumstances was he to arm sentries without approval from Wellington. It had been resolved to keep the appointee on a fairly tight rein.

Robson's requirements were noted by Buckley and he confided that he was going to deal with inmates on a more personal level in future.[4] His genial reception of the very first of them might have been indicative of this resolve. The week after the first security men arrived, two visiting head office representatives were pleased by the relaxed atmosphere apparent between staff and the prisoners.[5] Members of the Parole Board who toured some months later were similarly impressed.

For inmates, arriving at Paremoremo was a novel and exciting experience. There was tension and apprehension arising out of the news reports, but after forty months in the drab austerity of Waikeria and Mt. Eden, almost any change could be welcomed. The first contingent went straight into D. block. These were the 'intractables' and they numbered about twenty prisoners. The second escort went to A. block and the third, comprising mainly lifers and older men, was taken to the cells in B. block. By 28 March, most of the prisoners from Waikeria had been transferred. On 29 March, the nine remaining in the security block at Mt. Eden were also shifted, bringing the total at the prison to fifty-nine. With the security divisions cleared, transfers from other institutions were able to commence. In the first week of March, more men from Mt. Eden arrived, followed by arrivals from Wellington and Christchurch. By early May the prison contained 118 prisoners. A. and B. blocks were now full, and the remainder were housed in the intractables unit.

The date of the first escorts at Paremoremo had been kept secret. The bus-loads of convicts, under heavy police cover, had followed a circuitous route northward, staying off the main roads as much as possible. To those from the confines of Waikeria this summer's day tour of the countryside was one they would never forget. Arriving at the prison, with its bright, open cells equipped with toilets, hot and cold water, and radios, was a rare and stimulating experience.

Food, prepared in the gleaming kitchen, was well cooked and although they were yet to open, the standard block inmates knew that the gymnasium and properly equipped workshop industries would soon be available. Here there were no guns, and the yards afforded an unimpeded view of the sky. Compared with the east wing, whose echoing steel doors and clanging grilles had been decorated only with squads of key-rattling screws, the contrast was acute. One of the Waikeria transfers described how it felt:

> To get transferred from that, Victorian age in prisons, into something complex like Paremoremo . . . it was just like stepping on to the moon. When we first got there, we'd just stand there. And there's no screws, there's just grilles. And no keys. And you stand there and a voice just comes through a box and says, 'Twenty-three, A.' Add the grille opens, you walk through, it shuts. You're looked at by a camera.

Dempsey Roberts and Stan Rangi, who I spoke to together, were among those in the first escort to D. block. Their impressions of the gaol were similar:

SR: D. block was strange, put it that way. It was strange. It was very clean. Very clean.
DR: It was much fresher than what we came from, you know, like out of the dungeons into the bloody open air.
SR: Like a five-star hotel, sort of thing.
DR: It was open. It just seemed so open until you went into the cell. And found you couldn't get out! [laughter.]
SR: I mean we had our own toilets. We had radios. You know. We had a washbasin. Oh, it was really good, eh. It was a shock. No getting away from that.
DR: Mmm.
SR: Even our first few meals were good.

However great the contrast between Waikeria and Paremoremo might have been, the feeling of surprise and contentment which placated the men when they first arrived was bound not to last.

At Waikeria there had been nothing. The prisoners had become used to deprivation and accepted it because they knew there was no alternative. They also realized that a modern facility was being prepared for them in the north, promising amenities they had never before contemplated. However, when they arrived, the new gaol was still incomplete, and like its construction, final work took a good deal longer than anybody expected. In addition, no one had told prisoners about D. block. It was a number of weeks before Paremoremo's first occupants began to realize this, and to discover that conditions and treatment in the new institution were going to be far less auspicious than early optimism had led them to expect. It was as this knowledge became established that the first seeds of rancour were sown.

Early Problems

In its early stages, Paremoremo really consisted of two administrative subsystems, both housed within a single complex. First of all there were the ninety or so men in the two standard blocks. C. block was still unopened and classification was empty also. Secondly, there were the twenty 'intractables' in D. block. These were kept completely isolated from the rest, and could be split into two divisions: upper and lower. It was some time before the gaol's two groups learned of each other's existence, and it was some time too, before the intractables discovered that their status and treatment were different from those of the rest of the prison.

The Standard Blocks. To begin with, Paremoremo was run along strictly controlled lines. Routine and movement were rigidly monitored. For several months there was no work and only limited recreation was offered. Prisoners from A. and B. blocks were kept apart from each other, and communication between the two groups was difficult.

On weekdays these men were unlocked from between 7.00 a.m. and 8.30 p.m. Every morning they filed into their block basements where they were left to occupy themselves playing cards, chess, table tennis, or carpet bowls. The top sections were closed off at these times, and during the day, prisoners had the choice of either remaining in their cells or going downstairs. Yards were open in fine weather, but they were almost bare, and in order to prevent inmates in the B. block yard talking to men in the adjacent A. block recreation rooms, those in the basement were required to vacate the recreation rooms whenever the yards were open.

Late arrival of machinery from overseas was blamed for the delay in work provision, and in August, five months after the institution had opened, there were still no workshops operating. The bulk of standard block prisoners spent their days in the poorly equipped lower section of their respective wings. Once a week they were allowed to use the library and once a week, each of the two blocks had separate access to the gymnasium. Visits were also permitted once a week, although because of an absence of public transport, visitors at first were few. A permanent bus service was organized from 17 May. Those who wished to pursue education could do so; however, there was no education officer and hence nobody to organize study courses. Moreover, the type of literature permitted was limited. Books on electronics and law were prohibited, so were magazines containing pictures of naked women. Because of the limitations surrounding them, few education courses were taken, and those which did start often lapsed.

After his initial burst of congeniality, Buckley soon returned to his familiar pattern, becoming autocratic, aloof, and inflexible. His office he located in the administration section, away from the prison proper, and he was seldom seen informally about the wings. He was unreceptive to requests for improvements. His attitude, according to one prisoner (and this was confirmed by Buckley himself when I spoke to him), was ' "What you've got, that's all you're getting" (and we had nothing, man). . . . "What you get, you earn." '[6] With most of the men unemployed, exactly how they were to earn anything was left unexplained. Buckley was not a person who discussed his policies with anyone — least of all with convicts.

Under such conditions, problems were certain to arise, and it did not take long for this to happen. Inmates from the east wing were used to deprivation, but they were hard men whose sense of group loyalty was strong. It will be recalled that at Waikeria an ethic of non-communication with screws had become established, and this was a tradition which was now transferred to Paremoremo. If anti-authoritarian mores had been relaxed when the men moved into their new homes, the shortage of amenities and intransigence of administration rapidly fixed them again. It was thus only a short while before vocal complaints commenced.

When MacKay visited the institution after it had been working for only a month, prisoners protested that the ventilators were too noisy and that they were unable to sleep. MacKay told them they would have to get used to it. Then the men objected because they got no morning or afternoon tea. They sat down and refused to move until extra tea was ordered. They went on strike because they had no work. Strikers were told work would be provided as soon as possible. Then they complained because A. and B. blocks were segregated so heavily. This was eventually relaxed as well.

Although these issues may have seemed trivial, they were in fact fundamental to a population which was already possessed of so little. Small things take on exaggerated importance in a gaol, and to the men at Paremoremo, provision of a luxury like an extra cup of tea represented a significant improvement in lifestyle. However, forcing the administration to accept minor adjustments in procedure involved more than just a betterment of conditions. It might have been taken as an indication that inmate tolerance had its bounds. Apparently it was not, and Buckley's obduracy remained. Early action by prisoners enjoyed limited success, on the other hand, and certainly showed them the value of their unity. This asset now became employed with increasing effect.

So the importance of early events was not that they brought any marked improvement in amenities, but that they gave birth to an era of conflict at Paremoremo, where circumstances taught inmates to use corporate power in the pursuit of their wants and needs. From the beginning, the necessity for demonstrative action in securing concessions reinforced the men's commitment to one another. In doing so, it also strengthened their sense of shared purpose, and alienated them further from their custodians. Staff, always in a minority, felt demoralized by the uniform rejection they received and defensively tightened ranks as well. As the cycle of conflict continued, bitterness between the two cohorts increased. Much of the early trouble was mute, however, and it was not until October 1969 that matters came to a head.

On 10 October an inmate punched an officer to the ground and beat him as he lay on the floor. He was sentenced to an extra two months' imprisonment. Then at the end of October, a dispute over Christmas parcels led to a major strike.

The amount of money inmates are allowed to spend on Christmas parcels is normally set by head office, but tradition has for many years allowed considerable local discretion. At Paremoremo, however, Buckley had decided that this flexibility should stop, and that under him, spending would be limited to the official figure of four dollars. Such an amount was considered insufficient by inmates and, in an attempt to have the quantity doubled, on 29 October nearly everyone in the standard blocks refused to work.[7] As a result, ninety were confined to their cells. The refusal continued for a number of days, and although some returned after two, sixty-one prisoners were charged. A week later, eighteen of the strikers remained uncooperative. Eventually a verbal assurance that the situation was being looked at was accepted and the remainder then returned to work. The results of negotiations were not reported in the papers and when approached, MacKay refused to talk about them.[8]

Not two weeks after the Christmas parcel problem was over, another crisis arose. In mid-November a fifty year old woman was detected as she attempted to smuggle twelve rounds of .22 ammunition into the institution during a visit. The discovery was hardly noticed in the press and the offender only received a fifty dollar fine for her action, but a search of A. block revealed a toy pistol, modified to fire .22 calibre bullets, taped behind a toilet cistern. Although it was believed officially that the pistol was brought in for an escape attempt, it may in fact have been intended for use by a prisoner in the murder of one of his rivals. However, irrespective of what the pistol was for, its appearance in the prison must have added to the anxiety which was already mounting in Wellington. For regardless of its expensive technology, the gaol had now to be seen as dangerously pregnable, and in spite of Buckley's blustering authority, prisoners had shown a pert readiness to defy.

For the standard blocks, the year of 1969 thus displayed the sort of extreme measures such as had become evident at Mt. Eden around the time of the riot. The difference now was that the men were more co-ordinated and better organized than before, and more importantly, this time they were bound in their efforts by a powerful moral code. To some degree, though, standard block men were open to conciliation. Administration reserved considerable rein, which could be extended in the form of privileges, for example, as necessity dictated. This proved an important pacifier, and it was through a sequence of procedural concessions to the open sections that much potential violence was avoided during the early years of the seventies.

D. Block. In the bare corridors of D. block, living conditions were somewhat more severe than those elsewhere in the gaol. For unlike the standard division, D. block's inmates were contained in an environment where room for latitude was narrow. These men, taken straight from security at Mt. Eden and Waikeria, were moved into a situation little different to that from which they had come. Although at first they too were exhilarated by the move, disenchantment for them followed rapidly.

It did not take long for the men in D. block to discover that it was they who were the much-publicized 'intractables'. Accompanying this came the realization that for them, the future was grim. D. block routine consisted of lock-up eighteen hours a day, meals in cells, and a few hours' token employment in unequipped basement rooms. There was nothing else. Unlike the rest of the prison, these twenty men would be denied access to the magnificent new facilities of which they had heard so much. Because they had not been informed about why they were considered intractable, nor even of the means by which this designation might be altered, resentment grew up sharp and bitter. Dempsey Roberts and Stan Rangi explained how it felt:

> DR: When we realized that out in these other blocks — which we knew existed, but we didn't know where they were, or where we were, for that matter — but we got all this feedback [via the windows of the kitchen and laundry] that they had a gymnasium and places like that.
> SR: Having films and going to church and —
> DR: And we started wondering why we were excluded from it. . . . I mean these (other) guys are using the gym and all of a sudden they're telling us we can't. And people started looking at themselves and thinking, 'Aw, what's this, "Intractable"?' you know. 'How'd they classify me?' sort of thing.
> SR: There was no classification board whatsoever. They might have had a classification board to choose us, but we *did not* have an interview.

DR: We should have gone to Class. But we didn't. We were branded straight away. . . .
SR: I tell you, I didn't actually find out what I was in D. block for until I was transferred in 1976. . . . It was bloody Hobson that told me, just before I went to Crawford. And that was seven years later!

The D. block men had been arbitrarily labelled 'intractable', mainly on the basis of their past, refractory behaviour. Many knew they were violent, but it was the manner in which they felt they had been prejudged which cemented their rebellious commitment. Had the intractables been given a chance at the outset and tested in the open blocks, the prison's history could have been different. However, they were not. The administration felt that segregation of the most dangerous men was in the interests of safe administration. These men, united now by martyrdom, hardship, and bitterness, quickly formed the vanguard of a wave of violence that began in D. block, and eventually spread to the rest of the institution.

In early April 1969, after the prison had been operating only a few weeks, Tony Reid of *The New Zealand Weekly News* reported on life in the new, seven million dollar complex. Although he was kept away from prisoners and slept in the uninhabited C. block, Reid spent twenty-four hours in Paremoremo, locked up as an inmate. By coincidence, the night Reid chose to stay in the prison was the night the D. block men first reacted against their confinement. In a seven-page article Reid expressed shock and concern as he recalled the screaming and banging from inside the unit, and the howling siren as staff rallied to stop inmates from flooding their cells.[9] Yet few could have predicted at this point that things would deteriorate as far as they did.

Had the public been aware of it, the very grimness of D. block could have seemed sufficient to account for the restiveness of its inhabitants during the initial phases. Conditions there were nothing like they had been portrayed. On the other hand, years later the block became quiet, although conditions did not change that much. So in 1969, environmental austerity only goes part of the way to explaining why tension in the unit eventually ascended so critically. It is in the nature of the gaol's organization that the other part of the answer lies.

When Paremoremo opened, there was no system of fixed cell block assignment among staff. Personnel rotated their duties to work in all sections of the institution. An officer on duty in D. block one week, for example, could be assigned to B. block the following week, and A. after that. It was impossible to identify officers according to the areas in which they operated. This meant that inmates never got to know their custodians well and likewise to staff, inmates seemed anonymous and impersonal. If an official was abused in a certain unit, his retribution might be taken the following day on prisoners of another wing. Conversely, if inmates were upset by an officer on the last day of his shift, this anger could easily be transferred to prison employees generally. Thus, when the live broadcast of the 1969 moon landing was suddenly shut off, all staff bore the brunt of prisoner chagrin. There was no procedure for conciliation and this assisted officers and inmates in maintaining an impression of one another as unreasonable and querulous.

A second factor which added to friction was that few Paremoremo staff had experience in maximum security. As mentioned earlier, a number had been drawn from other institutions, but many were raw recruits. The Department had intended signing on more personnel after the facility was opened, but response to advertisements was slow, and manning shortages soon added to the problems already present. Among existing staff little consensus could be achieved about how and to what extent regulations ought to be enforced. Managerial inconsistencies at the basic contact level were thereby compounded.

Finally, and bearing directly upon the above, was that at Paremoremo there was an apparent lack of direction from the top. As mentioned, Buckley located his office well away from the institution and he was seldom seen by prisoners. Inmates in D. block, especially, comment that they had little contact with the superintendent at all in the crucial early phases. When Buckley was not around, they generally had only a vague idea of who was in charge. This meant there was no ready mediator for grievances, and disputes between staff and prisoners had to be resolved by impromptu methods. Once again, it was in the segregation unit that this problem was most acute.

To begin with, D. block was subject to even more fastidious discipline than that which had begun in the standard wings. Bed-rolls had to be made, floors had to be polished. When doors opened in the morning, prisoners had to stand outside their cells ready for inspection. Smoking was forbidden after lock-up, and talking during working hours was discouraged. Regulations of this type seemed pointless to prison-wise criminals, and they ridiculed the decreasing number of officers who tried to enforce them. From very early in the gaol's life non-compliance, belligerence, and increasing licence thus emerged as the hallmark of D. block behaviour. As the power of authority receded, resistance to it grew, and it was Paremoremo's beleaguered personnel who bore the brunt of their charges' contempt. Of the genesis of this situation, Stan Rangi observed:

> It only took about a month, mate. Because, see, it was *mainly* the screws. It was nobody else, it was mainly the screws. These screws had no training in security work. They'd had no training whatsoever. . . . Most of them came from the prison farms, see. And in the prison farms it was sort of like — like treating kids. . . .
>
> And it was only about a month before the first screw got bowled over. Because with this new environment the old borstal attitude, the old treating-you-like-a-kid persisted. It was a sort of a contradiction to the environment itself. And that's when the guys started rebelling.

Strikes and fires became frequent, and after the first flood-out in April these too were commonplace. Pre-existing antagonisms compounded themselves and a climate of open hostility developed. Inmates baited their custodians, who in return began to needle, provoking responses which could be rewarded with lost remission and solitary confinement. Within six months of its opening, an atmosphere of stress and commitment to resistance was already palpable in D. block. Dean Wickliffe, an inmate who had been transferred from the standard blocks in October 1969, said that he could feel the tension as soon as he entered the unit. 'The crims looked different, felt different to most others. They looked like a bunch of loners thrown together, proud, stubborn, hard, yet solid too,' he later wrote.[10]

It was several months before the first dangerous incident occurred in the intractables' unit, but when it did occur, it came as no surprise to many. Late in November 1969 — the same month in which the first major problems arose in the standard wings — two officers in D. block were badly beaten up. On the morning of November 27, inmates Dempsey Roberts and Atenai Saifiti were employed in the D. block basement, where it was their job to clean the corridor. The task took only a few minutes, after which there was nothing to do, so Saifiti and Roberts had stopped, as they customarily did, to pass the time of day with some other prisoners locked behind a workshop grille. An officer approached and, seeing the men talking during their work period, ordered them to move on. What next took place is best described by Roberts:

> We just hung in there for a while . . . so he gave us a warning. We ignored it. You know, we couldn't see any harm in it all and we couldn't do that section without breaking into some sort of conversation. . . . It was just at that moment, you know, they didn't want us to talk. So he give us a warning. Said we were going to be on charge. Then they came back with five or six screws.[11] And that was it. Confrontation straight away. Just the fact that they'd brought in so many of them.

Roberts and Saifiti attacked, injuring two employees badly enough to require hospital treatment. The officers retreated and the prisoners barricaded themselves behind a grille. Two television cameras were smashed and a fire hose was turned on staff, but finally the pair was 'forcibly overpowered' by a large contingent armed with batons. In the Magistrates Court Roberts (five years) and Saifiti (twenty-one months), each received three years cumulative on the charges of assault and wilful damage.

By the end of 1969, the lines of conflict at Paremoremo had been firmly drawn. From October onwards in the standard blocks, and from November in D. block, levels of tension escalated dramatically. In late December another officer was attacked and it was not long before aggression between prisoners and their keepers became an almost daily occurrence.

THE MEDIA RESPONSE

I have noted that it was some time before the public became aware about the problems emerging at Paremoremo. There was an election in 1969 and it was also Robson's final term in office. There were two good reasons, therefore, why certain officials might have desired that negative publicity about the institution should be contained.

Contrary to the 'complete frankness' policy of earlier years, the mass media were now treated with reserve. Official information, if released at all, came only in diluted form. At the end of 1969, MacKay had refused comment on the solution to the Christmas parcel affair. And in the annual report of March 1970, this week-long strike, which had attracted the participation of almost every standard block prisoner, was dismissed as having been short-lived and created by a 'handful of trouble-makers'. '. . . the great majority of prisoners', the report claimed, 'would gladly have no part in it'.[12] Mention of the pistol/ammunition incident, and of the problems in D. block was absent.

In its initial phases, apart from Reid's report in April, the only suggestion of trouble in maximum security was when inmates appeared in outside courts.[13] However, as antagonism within the institution increased, the trickle of independently based media reports also began to grow. This was added to in September, when three former officers approached *Truth* newspaper to publicize their grievances.

Truth had only just concluded its famous 'Birch the Bashers' campaign, which in July and August 1969 had made noisy calls for a crackdown on crime, especially violent offending.[14] A statement condemning liberal penal treatment was thus well within *Truth*'s frame of focus at the time, and on 16 September, 'The Con is King in this Rest Home With Bars' attacked Robson and Buckley for defective management. Paremoremo was run on a principle of 'peace at any price' the paper said, turning it into a 'rest home for many thugs and no-hopers who don't deserve one good free meal a day, let alone three'. Prison policy was changed at the whim of inmates. Officers had no power or authority, no protection from violent prisoners, and no co-operation from the superintendent, who favoured criminals and was reluctant to post standing orders, the newspaper alleged. The result, it said, was that between mid-July and the end of September 1969, six prison officers had resigned and another twelve of the ninety-six staff complement[15] were planning to follow. In the five months since the institution had opened, twenty applications for transfer had been lodged.

A response to *Truth* was not long in coming, and three days later *The Auckland Star*, *The New Zealand Herald*, and *Truth* were all invited to visit.[16] First *The Auckland Star*,[17] then *The New Zealand Herald*,[18] published pro-Government reports. *The New Zealand Herald*'s 'Jail Security System Works Perfectly' argued, for example, that the prison had settled down, and (contrary to *Truth* claims) 'now gives the impression of running with the precision of a well-oiled machine'.

However, speaking on behalf of the Department, MacKay conceded that there had been trouble in the form of floods, fires, wilful damage, and assaults on staff, and he also admitted that resignations of almost 10 per cent of the prison's work force in the first six months, (or 20 per cent, per annum) was higher than expected.[19] In fact it was about double the turnover rate of the penal division at large.[20] All of the Paremoremo departures had occurred in the ten weeks since July. Resignations were just 'teething problems', MacKay explained. Things were improving, and there was no cause for worry.

On 23 September, *Truth* published its own report on the visit. True, said *Truth*, none of the staff approached had admitted to being dissatisfied, but during the tour the chief officer had always been nearby. The paper tried to uphold its previous allegations, observing that since opening there had been 173[21] disciplinary charges recorded at the institution.[22] This amounted to about 2.92 disciplinary reports per inmate, per year (and compares with figures of only 2.12 recorded in 1976 and 1.26 in 1981).[23] Not published was the fact that ten of these charges involved assaults on officers. The future would depend very much on how the situation was identified, and in the way in which successive difficulties were handled.

THE POLITICAL CLIMATE

In November 1969, the nation went to the polls. Following its trouncing in 1960, Labour had risen marginally in 1963 and again in 1966.[24] Socially, conditions were working against the Government. In 1967 full employment ended,[25] and by 1968, New Zealand was entering another recession. Although in 1969 the economy recovered a little, there was industrial tension, continuing high unemployment, and increased concern for community issues. The end of six o'clock closing in 1967 had created a change in drinking patterns, and with it the face of unpremeditated crime. The 'six o'clock swill' was replaced by more sophisticated habits, but the activities of evening drinkers, no longer concealed within private homes, were now more visible than ever. When the pubs closed at 10.00 p.m., discharged patrons often constituted a majority on inner city streets.

Drug use was another area that was becoming worrisome. In the year up to the end of 1968, the number of reported drug offences rose by nearly 200 per cent, from 107 in 1967 to 312. *The Auckland Star* journalist Noel Holmes campaigned for tougher penalties and punishment of people seen associating with known drug addicts,[26] and in February a Committee of Inquiry into drug offending was set up.[27]

In 1969, a twenty-page Government White Paper on crime, tabled in July, found that reported offences had almost doubled in ten years. Convictions had increased by 50 per cent.[28] These rises could not be wholly accounted for by changes in population size and structure, and the report said that public concern warranted maximum penalties for the worst types of offending.[29] Only days later, the National Party Conference called for more 'law and order', implying a hardline preference for the treatment of criminals. The following month, the Leader of the Opposition echoed the cry.

Allegations of social malaise were grist to the Labour party's election mill, and in Parliament Dr Finlay attacked the Government for its poor handling of the problem. Once again, the bones of National's promises during the law and order debate of 1960 were disinterred, and thrown back at a defensive Ministry.[30] Declining as it already was after nine years' continuous office, the apparent deterioration in community standards served the National party poorly. Although it managed to survive 1969, when the country voted that November, the Government's share of the ballot was down by 0.92 per cent. National conceded several seats, and held on to power by a margin of only six.

Many factors contribute to the outcome of an election and publicity over rising crime, once again, was probably not an important one in 1969. None the less, the problem had been identified as one of concern, and a tide of opinion favoured sterner measures. This hardening of attitudes did not augur well for Government or for prisoners, and the instalment of a new Minister after November did little to stem the attrition that was steadily taking place at Paremoremo.

D. J. RIDDIFORD, MINISTER OF JUSTICE

In July 1969, Ralph Hanan travelled to Queensland to inspect its bauxite mines, and on 24 July, in the middle of the 'Bashers' campaign, he suddenly died. Hanan's death in Australia came as a shock to the nation, but with an election looming large, little time could be lost in organizing his replacement.

Jack Marshall, former Minister of Justice and at that time Deputy Prime Minister, ran the Department until December. With the elections completed, overburdened with work, Marshall shed the weighty portfolio with the reshuffling of Cabinet. Finding someone to fill Hanan's shoes at this time was problematical, for the depleted National caucus had few lawyers with the experience and ability to run a sector of Cabinet so close to the Executive. A number of names were considered and in the end it was a novice to the management of portfolio, with nine years' parliamentary experience and a Wellington law partnership, who became the favoured choice.

The son of a prominent Wairarapa farmer, Daniel Johnston Riddiford was born in Featherston in 1914. Educated at Downside public school in England, Riddiford progressed to Oxford University from where he graduated with an MA and a degree in jurisprudence. In 1932, Riddiford returned to the Wairarapa to continue farming, but at the outbreak of war he volunteered for the army and was posted overseas. In 1941, Captain Riddiford was captured by the enemy. Two years later, however, he escaped, and, travelling through Northern Italy and Yugoslavia, managed to regain his freedom.[31] For this action he was awarded the Military Cross. On his return from the war, Riddiford, now holding the rank of Major, entered law practice in Wellington, securing a partnership in 1950. In 1957 he unsuccessfully contested the Petone seat, but entered Parliament as Member for Wellington Central in 1960.

Apart from his degree and his experience as a prisoner of war, Riddiford had no special qualification for, or interest in, the post of Minister of Justice. 'Dapper Dan', as he was known to some, was a cultured person, but he was not of great practical bent. His private English education and archaic speech patterns set him apart from the average New Zealander, and some found him difficult to understand. He also had other shortcomings which were generally known to his colleagues. Sir Jack Marshall who worked alongside him, for example, described Riddiford as a good and intelligent Minister, but one whose feet were not always on the ground. 'He was perhaps inclined to be a bit *dreamy*', Marshall explained.

Eric Missen, permanent head for most of Riddiford's office, agrees with

Marshall, and told me that Riddiford was vague and 'lived in another world for a great deal of the time'. Another former Justice official remembers Riddiford once stopping in the middle of a policy speech, which the Department had written for him, to exclaim, 'Oh, I don't think I agree with that!' The Minister had forgotten to read the address before delivering it.

Apart from, but possibly related to, Riddiford's poor attention span, was the fact that during his term he was unwell. The Minister had a serious heart ailment and had difficulty staying awake during meetings of Cabinet. His heart condition made it hard for him to discharge his duty under pressure, which did nothing for the situation at Paremoremo. He lacked the energy, temperament, and enthusiasm for the task he had been given, and did not involve himself deeply in the affairs of his Department. As far as prison policy was concerned, observed his former Director of Research, Daniel Riddiford was a nonentity.

Soon after Riddiford entered Cabinet he announced that penal reform had gone far enough. Although the policies of past years would remain, under the new Ministry there would be no vigorous changes in Justice philosophy. Now was the time to take stock, he said, and watch how it performed in practice.

The appointment of D. J. Riddiford, foreshadowing a new phase in this country's penal development, concluded what had begun some twenty years before. In 1960, Riddiford had voted against the abolition of capital punishment and his stance on social issues generally was conservative. At a time when calls for a hard line on criminals were strengthening, it was this conservatism which perhaps best fitted Riddiford to his post. For his party's small majority in the House meant that activity now would be restrained. National would have to move prudently in its fourth executive term, if it were to stand a chance of re-election in 1972.

E. A. MISSEN, SECRETARY FOR JUSTICE

Three months after Riddiford took over, and before the 1970 session could get underway, John Robson stood down. Robson had been permanent head for almost ten years and his retirement, soon after the death of Hanan, truly marked the finish of a historic era in New Zealand's correctional history. The man chosen to replace him was Eric Alderson Missen, a competent servant whose forte lay in public administration.

Born at Ohaupo in 1914, Missen was educated at Palmerston North Boys' High School and Victoria University, where he took an MA in history. At the age of twenty, Missen joined the Labour Department, before entering the State Services Commission in 1939. Like Robson, Missen became superintendent of staff training in the Commission, succeeding Robson when the latter went to Justice in 1951. Missen gained the rank of Senior Inspector in State Services, then he too joined Justice, as Assistant Secretary in 1963. In 1967, Missen rose to the position of Deputy Secretary. As such he was poised to take over from Robson when the Justice chair was vacated in March 1970.

Although Missen was an administrator more than a penologist, this proved no handicap to his performance of duty. Patricia Webb, who worked under him in the legal division, says that Missen was a methodical man, but one who recognized the limits of mechanical control. Loyal to staff and receptive to comment, Missen was easy to work with and generally enjoyed good relations with his staff. It was Missen's high level of organization which proved one of his greatest assets during his short period as head of the Department. Missen had known Robson in State Services, moreover and later, as a senior Justice official and a member of the penal group, had been involved in the formation of policy throughout the 1960s. He was well-versed in and sympathetic with the aspirations of the previous decade, and while the era of reform had lost most of its thrust by 1970, Missen's leadership brought no great changes to its course.

In his new position, Missen struggled to work under a Minister whose interest he found difficult to inspire. Riddiford lacked the kudos of his predecessor, and had neither the energy nor the virtuosity to carry on through these difficult times. Although the public was still only marginally aware of the problems, in the Department itself officials sat uneasily at their desks. The Government's seven million dollar investment was beginning to show its flaws, and it was with apprehension that developments were monitored in the hope of recovery in the new year.

NOTES

[1] The choice of the new superintendent was not officially announced until November 1968.
[2] *EP* 24 September 1968.
[3] J. Cameron (pers. comm.).
[4] G. Parker (pers. comm.).
[5] The two officials were Messrs Bateup and Missen.
[6] P. Morrison (pers. comm.).
[7] Work, by this stage, was available to all A. and B. block inmates.
[8] *NZH* 5 November, 26 December 1969.
[9] *The New Zealand Weekly News* 16 April 1965.
[10] Quoted in Bungay & Edwards (1983) 104.
[11] According to press reports, only three staff were initially involved (*NZH* 28 November, 9 December 1969).
[12] *App.J.* H-20 (1970) 11.
[13] That is, when prisoners faced charges laid under the Crimes Act as opposed to internal charges laid under the Penal Institutions Act.
[14] The campaign began on 1 July 1969 and ended on 2 September. The 2 September issue carried an article entitled, 'We Want Men with Guts' and defamed opposition MP Dr Martyn Finlay. The publication resulted in Finlay being awarded a total of 15,000 dollars in damages.
[15] This was nineteen short of the planned establishment of 115.
[16] This took place at the instruction of the Minister of Justice.
[17] *AS* 19 September 1969.
[18] *NZH* 20 September 1969.

19 Only the previous day it had been reported that at Paremoremo only married men had been employed, in an effort to keep staff turnover as low as possible (*AS* 19 September 1969).
20 In July 1970 an official publication, Department of Justice (1970) 14, gave the total number of staff working in New Zealand penal institutions as 934. MacKay had laid the total annual turnover in the penal division at eighty to one hundred, or about 10 per cent, per annum.
21 This figure is apparently in error. Official data gave the number of charges as 177 (Memo from Missen to Riddiford: 26 September 1969).
22 *Truth* 23 September 1969.
23 These percentages are based on a DAP of 200 at the institution in later years, and are drawn from figures presented by Greer (1983) 374.
24 In 1966 both Labour and National lost votes to Social Credit and non-voters, but National lost 4.59 per cent to Labour's 3.51 (Chapman 1981, 365-6).
25 Dunstall (1981) 339. The monthly average of 463 in 1966 jumped to 3852 in 1967 and 6881 the following year. (Department of Statistics, 1976, 849).
26 *AS* 21 May 1969.
27 This was the Committee of Inquiry into Drug Dependency and Drug Abuse in New Zealand. It produced two reports, dated 1970 and 1973.
28 *Report on Crime in New Zealand* (1969) 4.
29 Ibid. 10.
30 I refer to the consequences of the Hastings Blossom Festival, discussed in chapter 9.
31 An unpublished manuscript written by Riddiford of his wartime experiences exists in the Alexander Turnbull Library, Wellington.

15: PAREMOREMO IN CRISIS 1970–1971

After the trouble which had darkened the final months of 1969, 1970 got off to a shaky start. No sooner had the year commenced than a member of staff was convicted of smuggling food and money into the lifer Jorgensen. However, like the case of the woman caught with ammunition only eight weeks earlier, the event received no media sensationalism. There was small media interest and the offender was discharged without conviction, and twenty dollars court costs.

The following month a prisoner who had assaulted an officer after being ordered to tuck his shirt in was also punished lightly. His moderate four-month extension contrasted with the three-year terms handed down to Roberts and Saifiti not two months before, and also received little press coverage. It was almost as if such behaviour was becoming acceptable.

RISING TENSION

In the first half of 1970, violence continued at a similar rate as before. In Paremoremo's first six months there had been ten attacks on staff. In the ten months following, fifteen more were recorded. Perhaps in an effort to keep publicity down, prisoners who committed assaults were now more often dealt with under the restricted provisions of the Penal Institutions Act,[1] thereby avoiding appearances in outside courts. Those who were tried externally were not always harshly treated, and frustration at what they saw as a soft response to serious crimes caused staff rancour to grow. However, resentment against inmates was still largely covert early in 1970 and had yet to encounter the levels it would find later on. Although staff violence was still relatively low, animosity towards prisoners soon made itself felt in other ways. It was in D. block that this was especially noticeable.

Security inmates began to complain that their meals were lukewarm, with the (plastic) cutlery either absent or broken. Requests for hot meals and proper utensils often met with no response. Material such as pubic hair was found in food, and although staff blamed other inmates, the D. block men felt officials were responsible. To aggravate matters further in the block office, PA systems were sometimes left on while the duty shift talked about convicts, or made obscene comments about their wives and girlfriends. It was impossible to determine who was responsible for the remarks.

Prisoners in the security block were helpless to prevent these vexations from continuing. Procedure for formal complaints was obscure and in any case, few had any confidence in administration. In 1969 there had been trouble over inmates' mail not being delivered promptly, and an internal inquiry had insisted that the situation be rectified. However, the retention of mail continued. In fact, where communication was concerned, all avenues were poor. In 1970 requests to see the superintendent often failed to reach Buckley, or if they did, they were ignored, inmates said. Likewise, letters to the Secretary for Justice were sometimes never replied to, and applications to see visiting officials were received indifferently or were given no response at all.

The complaints were few, but in close confinement their combined effect was considerable. Locked inside their concrete rooms prisoners seethed in impotent rage, and as the list of dissatisfactions mounted, they dreamed of their revenge.

Limited access to grievance facilities, mystification about how they worked, and suspicion about official motives[2] meant that in the early years, formal procedures at Paremoremo were hardly used at all. Instead, inmates chose increasingly to solve their problems on individual terms. The violence which resulted served only to intensify official bitterness, and alienated staff even further from their charges. Stan Rangi comments:

> Things were getting worse by this stage. They were getting fucking worse with the screws. There was no communication whatsoever by them. You could not communicate with fucking screws and they would not communicate with you, other than to tell you what to do.

Inmates refused to compromise either, and the cycle of non-communication and antagonism which resulted inevitably led to violence. Although not all incidents were recorded, Buckley's former deputy reports that at this time it was not uncommon for two or three members of staff to be assaulted in a day.[3] It was only a matter of time, therefore, before staff reacted in kind. Wickliffe writes:

> Conditions in D Block [in 1970] deteriorated to the point where we were forced to fight to survive. Clashes between ourselves and the screws became a weekly occurrence. Indeed it was an accepted hazard of life to get beaten up by the screws, or vice-versa.[4]

While coercion developed as a basis for control, management ventured on unsure footing. Crisis became inevitable, and in the second half of the year, Paremoremo suffered the first of what was to become a series of violent exchanges.

The Saifiti Case

The Assault on Smith and Downs. In New Zealand prisons the evening meal is served between 4.30 and 5.00 p.m., and from then until breakfast next day there is nothing. Many institutions nowadays serve a cup of tea around 8.00 p.m., but in D. block there was and is still, no provision of this kind.

People get hungry when shut in a cell for hours on end, and to alleviate this it was customary for the D. block men to save some bread from their dinner rations and eat it later in the evening. D. block was locked up at

4.30 p.m. every night, and on the upper landing where Roberts and Rangi were housed, Roberts had responsibility for keeping all the extra bread in his cell. Each night he would pass it down to the other inmates on his side of the wing, although sometimes, in order to avoid having to throw the food up the corridor, Roberts sometimes asked one of the night staff to deliver it when he came around on patrol.

One night in July 1970, D. block was locked up as usual, and Roberts asked an officer called Ken Smith to deliver Rangi's bread to him, while the men listened to a rugby match on the radio. This Smith agreed to, and taking the bread, he proceeded towards Rangi's cell. He then walked back down the wing and left the block. A few minutes later Rangi called out that his bread had not arrived. This perplexed the men, but what they suspected was confirmed next morning when Rangi's ration was found dumped into a rubbish bag. Roberts described his reaction:

> You know, that really, really riled me. I felt bad because I felt, Stan'll be thinking 'Oh shit! This bastard's frozen on the scoff!' I thought I could right the situation by punching Smith in the mouth, sort of thing.

The following night when Smith did his patrol he was questioned by Roberts and admitted to having thrown the bread away. Pressed as to why, he said he had done it because the passing of objects between cells was against regulations. Roberts then told Smith that next time he saw him he would 'punch his fucking lights out'. Roberts had a long history of violence, and perhaps in an attempt to let the matter die, Smith disappeared and was not seen in the wing for another two weeks. However, on the afternoon of 27 July, as the men filed out of their cells to attend a monthly movie in the gymnasium,[5] Roberts spotted Smith standing on duty outside the block sally port. Roberts decided immediately what he would do and turning as he drew level, he punched Smith to the ground then kicked his head and body until he was unconscious. Seeing Smith in trouble, an officer called Downs tried to stop the assault, but he too was knocked to the ground and kicked by an inmate called Tamarua.

When the fight was over Roberts and Tamarua were arrested and locked up, before being taken individually to the punishment cells. Smith was admitted to Middlemore hospital with a broken cheek bone, a broken jaw, and concussion, while Downs was treated and discharged. Roberts and Tamarua were charged under section 193 of the Crimes Act with assault with intent to injure,[6] and some hours later the two were joined by Atenai Saifiti, who faced the more serious offence of injuring Smith with intent to cause him grievous bodily harm, by kicking him while he was on the ground.[7]

Roberts and Tamarua were surprised at Saifiti's appearance because Saifiti had been housed on the lower landing of D. block, and at the time of the attack was already in the sally port. Both Roberts and Tamarua, as well as other prisoners who witnessed the incident, knew this and realized that Saifiti was innocent.

The Aftermath. The Smith assault was the most serious the prison had seen to that date and reaction to it was immediate. On the evening of 28 July, a telegram was received by the Justice Secretary from the prison officers' subgroup at Paremoremo, asking him to attend an emergency staff meeting the next day. This the Secretary declined. On 29 July the staff at Paremoremo stopped work to express concern about their safety and to examine pay and working conditions. Ninety of the hundred-odd officers employed there attended the meeting, which called for all violent men to be kept on the top floor of D. block, for tighter visiting restrictions, and for even greater segregation. An increased staff–inmate ratio and easier access to batons were also requested. These suggestions were accepted in principle by the Department.

An anonymous letter listing complaints was drafted to certain newspapers and Members of Parliament,[8] in which it was alleged that the superintendent lacked autonomy and was being trammeled in his job by directives from head office. It was also suggested that the Minister was out of touch with what was going on at the prison. Increasingly severe assaults on staff and inaction in Wellington had seriously eroded staff morale, the letter said. Missen hastened to reply, insisting it was incorrect that the superintendent had no power. A policy of humanitarianism was aimed at rehabilitating prisoners — not appeasing them — and recent allegations of widespread damage were exaggerated.

The following day *The New Zealand Herald* received another anonymous letter from staff answering Missen's denials, and one of their reporters visited Paremoremo to investigate for himself. 'Smashing Time At Jail by Prisoners' confirmed the letter's insistence that damage had indeed been extensive. At least fifteen wall telephones had been destroyed, windows, window fittings, and clocks had been wrecked, and lights and cameras had been smashed. There had been fires in cupboards and in the wooden ceiling. Although most of the damage had been done by undetected offenders, between January and June 1970, fifteen prisoners had been charged with destroying prison property.[9]

The incidents and the letters caused deep concern and that same day in the House, Mr Riddiford, while denying the need for an inquiry,[10] announced new security measures.[11] In future all dangerous men in D. block would be kept separate from others[12] and would not be allowed out for evening activities. Those in this special section would no longer congregate together and would be exercised in groups of no more than three per yard. 'Open' visiting for them would cease and instead, electronically monitored booths would be used, with inmates separated from their visitors by thick panes of glass. No mention was made of destruction in the other blocks — to which most of the information applied — nor of how it could be stopped.

In August 1970 Roberts, Tamarua, and Saifiti had their depositions. Roberts pleaded guilty to a reduced charge of common assault (maximum penalty, twelve months) and was later sentenced to eight months imprisonment. Tamarua and Saifiti were committed to a higher court for trial. The hearing of evidence against the remaining two began in the Supreme Court at Auckland on 7 October, before the Chief Justice, Sir Richard Wild. Roberts, called to testify, admitted assaulting Smith and to kicking him while he was down. Tamarua admitted he had assaulted Downs, but argued that he had been provoked. Saifiti was alleged to have been involved in the kicking of Smith, after the latter had been felled by Roberts. Although evidence of both Roberts and Dean Wickliffe exonerated Saifiti completely, officer Cummings claimed to have positively identified Saifiti as Smith's second assailant. Months later, Cummings was to doubt the accuracy of his testimony,[13] but at the time of the trial it was largely upon this evidence that the verdict of the Court was based.[14]

On 6 October 1970 both Tamarua and Saifiti were found guilty as charged and subsequently sentenced to two years' imprisonment. These sentences were cumulative upon their existing terms.

The Hostage Crisis. When news of Saifiti's conviction reached D. block, it was clear that trouble was in the air. Stan Rangi became Saifiti's chief advocate and wasted no time in requesting an interview with the superintendent. This was granted and after Rangi had explained the situation to him, Buckley summoned Dr Gordon Parker, the prison psychologist, whom the men trusted. Parker genuinely believed that Saifiti had been mistried, but could do nothing other than urge the angry prisoners to sit, and await the outcome of the appeal.

Feeling was running high in the intractables' unit and the prisoners did not want to wait. The three accused, plus Wickliffe, were now in isolation on one side of the top landing where they had been since the time of the assault. When news of the conviction arrived, Wickliffe smuggled a note to the members of the block's general population, asking for their support in a strike. Although Wickliffe had no idea of what the reaction would be, co-operation was immediate. Only four days after Saifiti's trial had ended and before sentence had even been passed, a planned, co-ordinated, and desperate action took place on the top landing of D. block.

At 4.10 p.m. on Sunday 11 October, as ten inmates from the unsegregated section were unlocked to move downstairs, the five screws in charge of them were attacked. Four surrendered without much struggle, but the fifth, a tall, powerful man called Wills, attempted to break away. Stopped by a locked grille he turned and was set upon by inmate Yorker, who had only recently been moved to the block after bashing an officer at Mt. Eden. Hit in the face, Wills fell backwards, cracking his head on a grille as he did so. This caused a long, deep cut in the back of Wills' scalp and sent him into unconsciousness.

With the staff subdued and the block now under their control, the rebels retired with their hostages, dragging Wayne Wills behind them. Wills was bleeding profusely from the head and his eyes were lifeless. Believing him to be dead, the attackers left the officer lying on the top landing outside the grille, where his 'body' could be recovered by his workmates. Staff on the scene also believed Wills had been killed and it was not until they rescued him that they found he was still alive.

A feeling of total commitment now came over the prisoners and they prepared themselves for a siege. Keys were jammed into locks to prevent the grilles being opened and the four captives — officers Hindmarsh, Te Rangiita, Daniels, and Browne — were bound with insulated wire. Hindmarsh was selected as first hostage. A piece of wire was placed around his neck and he was systematically beaten up in front of staff assembled on the other side of the grille. A steel weight was then produced. The prisoners announced that if any attempt at forced entry was made, this would be used to bash Hindmarsh's brains out.

The situation could hardly have been more critical. The armed offenders squad was called and quickly took charge of negotiations. Tear-gas and cutting gear were made available. At 7.45 p.m., lawyers, Peter Williams and Kevin Ryan, two of Auckland's top criminal defence counsel, were summoned to assist in the setting of terms, and at 11.18 p.m. Riddiford, Missen, and MacKay arrived from Wellington. Justice officials were told that Saifiti was innocent and that a re-hearing of charges was being sought. Parleying continued past midnight and it was not until after 1.00 a.m. that an assurance was gained that the matter would be looked into. Having been promised that there would be no repercussions, the hostages were then freed and the prisoners returned to their cells.

The Security Clamp. Not surprisingly the hostage incident and the threats that had been presented to officers' lives were of profound concern to the authorities, and that same night Dr P. P. E. Savage, medical superintendent of Oakley psychiatric hospital, arrived at the prison to interview Rangi in his cell. Satisfied, presumably, that the spokesman's faculties were sound, a delegation of three officers then came and talked with him.

According to Rangi, the delegation told him that it realized an injustice had been perpetrated and that Saifiti was innocent of the charge. This was taken by the men as a hopeful sign and for a while their anger abated. However, the law proceeded unaffected and less than a week later, Saifiti and Tamarua each received their two-year sentences for the July assault. That same day it was announced that eight of the ten D. block rioters would be charged in relation to offences committed during the hostage crisis. Those who were found guilty subsequently received penalties of between six and eighteen months. Rangi, who was already serving life, received an eighteen-month cumulative term.[15]

The hostage affair caused an upheaval in the already tumultuous D. block. New security procedures were drafted. From then on, recommendations that the block be divided into two distinct groups would be implemented fully and the July provisions enlarged. The whole of the upper landing would now be reserved for especially dangerous inmates, who would be locked up twenty-three hours a day. Of the twenty-seven men in the block, thirteen were classified to this section. The latter group consisted of Roberts, Saifiti, Tamarua, and Wickliffe, and nine of the hostage-rebels.

Visits for such prisoners would only be taken in security booths and they would be without other privileges. Grilles would be installed half-way down the two corridors of each landing so that instead of four, the block would be divided into eight, independently manageable units. There would be an overall increase in D. block staff ratios and all D. block inmates would be unlocked individually and escorted everywhere by three prison officers. Prisoners on the top landing would not be allowed to congregate at all and would be exercised alone, for a single hour every day. In addition to this, men on the upper landing could not be classified to the standard blocks without first being 'tested' in the lower section, where conditions were more relaxed. Here prisoners would be permitted to use the recreation areas from 2.00 p.m. to 4.00 p.m. and from 6.00 p.m. to 7.45 p.m. every day. They would have open visits and a movie would be screened to them every month. The outline of these new measures was worked out at a meeting between Missen, MacKay, and members of the Paremoremo officers' subgroup on 12 October. They were announced in detail eight days later.

Staff questioned by a reporter said they were reasonably satisfied with the proposals; however, one member of the subgroup was less confident and foretold more problems in the future. Tensions were still running high, the officer claimed.[16] While new restrictions would make administration simpler, they would do nothing to ease the bitterness which under this same repressive policy had already become so intense.

Living Conditions and Social Relations. After the hostage crisis, and in spite of new security arrangements, the situation in D. block remained critical. The fact that the few privileges which had been conceded could not always be given added to the tensions being felt. Once more this was a fact that the Department hesitated to admit. A claim that D. block had access to the gymnasium[17] was false, and for those on the top landing, exercise was irregular. Wickliffe estimates that during the year of 1971, prisoners in the top division were given use of the yards for an average of only half an hour per week.[18] Thus, inmates of the intractables' unit spent nearly all of their time in the block, where facilities were limited and few.

Initially a variety of manual work was provided, but because of sabotage and the threat of tools being used as weapons, most forms of labour were withdrawn. By the end of 1971, the only employment available to segregated men consisted of the making of paper envelopes and canvas tags by hand. Absence of work incentives, general apathy, and poor relations with adminstration meant that production was minimal, and much of what was produced was useless. In their cells, prisoners were permitted no hobbies, and only limited activity was allowed in the recreation rooms. Recreation itself consisted, as before, of carpet bowls, cards, chess, and table tennis. Those on the upper landing, because they were always locked up and exercised alone, had no hobbies or pastimes. Most of the time they refused to work.

Religious instruction consisted of periodic visits by a *padre*, who walked down the wing once or twice a week, and contact with the education officer appointed in early 1971 was similarly sporadic. Summing up general conditions in that year, Wickliffe wrote:

> Hobbies . . . just don't exist in D block . . . there are virtually no recreational activities at all for D block. Most of the inmates are locked up for 24 hours of the day, every day of the week for years on end. . . . Apart from the radio and newspapers, there's virtually nothing to occupy a man's time.[19]

The new regulations provided a spartan regimen in this section of the gaol, and a circumspect administration might have seen advantage in ensuring that promised amenities were provided. However, Buckley was beginning to lose control over the prison's rival factions, and his growing inability to cope was exacerbated by his lack of rapport with either. So instead of attempting to mediate with the intractables, or find concord between them and the officials, he aligned himself with the latter and tried to crush rebelliousness with force. Thus, Wickliffe reports:

> Buckley claimed that he had now been given the power to do with us what he saw fit, in order to preserve the harmony of the prison, even if that meant overriding the regulations. This in effect meant that we, the prisoners, had no rights whatsoever, while he had unlimited powers.[20]

Buckley had certainly received no such mandate, but inmates believed that he may have. In any case there was little prisoners could do about it. Although living conditions were worsened by the restrictions, therefore, their most significant impact was on the block's balance of power. Prisoners were now cowed by this situation, but it did add to their desperation. The wing's atmosphere worsened further. 'The minute you walk through the D. block sally port you feel it because there's no outlets for your tensions and frustrations, they build up and build up until they finally explode. We take it out on those things or people nearest to us, ourselves or the screws,' Wickliffe wrote.[21]

Although many of the men were now on medication, aggression thresholds remained low. Prisoners were ready to repel any slight, irrespective of consequences. Roberts remembered four occasions involving himself. Of a typical instance, he said:

> This screw comes up to me and he says, 'Oh, how's it like to be outnumbered?' And it's supposed to intimidate you, six to one. Immediately there was a challenge, you know. So I turned around and belted him. Regardless of who was there, it didn't really matter it was that sort of situation where they thought they had the numbers and had you subdued and under control. And it just made you react, you know.

Under these conditions a siege mentality became a permanent feature of D. block. The code of unconditional commitment to inmate interests, already strong, now matured to the extent that prisoners felt bound to support their peers, regardless of outcome and regardless too, of whether or not they felt justified. Roberts remembered a time when an inmate handed him a stick, asking him to look after it. When requested to surrender the stick he refused, on the principle that he had a personal commitment to take care of it. As a result Roberts was unlocked; felled in a baton attack, confined to solitary, and put on report. The stick was confiscated.

In a related incident the same day, Saifiti beat up a popular officer called Savage, simply because Savage had allegedly been late unlocking him to change his books in the block's book cupboard.[22] Although they felt Saifiti's action was excessive, Roberts and Rangi were firm on the axiom involved:

> S.R: Savage was a good guy, boy. He's a *fucking* good screw. But even though you know a screw's a good fucking screw, you couldn't sorta favour him.

> Because he represented the screws and *that* was *it*, man. After they started putting in these double doors and all that, [i.e., after the hostage incident] you know, you had to suppress your own thing and look at it from the guys' point of view and fuck your own. And so even though some of us thought Sai was being a bit out of hand breaking his jaw, you still had to back him up because of that.
> D.R: Oh yeah, you would. It wouldn't matter. But he was a bloody good screw. Just *any* screw, you know, you upset some of the guys, they're going to let you have it.

Wickliffe has made a similar comment:

> We had to stick together and we did so to the limit. If one man was provoked, persecuted, or unjustly treated, we would all back him up. Everyone was that man at some time or another. He could never be let down.[23]

A situation such as this — 'all for one, one for all' — made employment at Paremoremo unpleasant and risky. Staff assigned to D. block, in particular, could count on nothing but unmitigated hatred from their charges and knew they could be attacked at any time, on any pretext. The prisoners no longer seemed to care about release. Roberts, who had gone to prison in 1968 with eighteen months, finished up with nearly nine years. Saifiti, entering Paremoremo with a twenty-one month sentence, was now serving almost seven years, a result of accumulated assault charges.[24] Their behaviour had not changed at all, and the extra time they had gathered only meant that, now, staff would have to put up with them for that much longer. The prospect of remedy looked remote.

STAFF DISSATISFACTION

In the fifties and sixties, notes Thomas, it was becoming clear in Britain that staff were losing the power to direct the behaviour of inmates. 'Assaults on staff', he writes 'were constantly being discussed'.[25] Rising violence and general disorder were attributed to waning discipline, which began after the Second World War.

In New Zealand, we have seen that the same process began at exactly the same time, and progressed for more than two decades. By 1970, the new penology was firmly established. Although the real causes of friction were more complex, it was at Paremoremo that the perceived relationship between violence and liberalism reached its zenith. Here the reaction was sharpest. Paremoremo's record of dissatisfaction, which had begun in 1969, increased in the early seventies. In 1971 there were a further eighteen charges of assault against staff logged at the maximum security gaol,[26] and escalation in the trend nourished the officers' mood of anger and concern.

However, fear of assault was only one of the several sources of strain in the prison's early period. The staff village, which had been built nearby, remained poorly serviced[27] and it was some time before shops, a post office, a school, social amenities, and a regular system of public transport could be organized.[28] Employment conditions were less than satisfactory. Unlike other institutions, where staff could work outside part of the time and sometimes have a complete break from the gaol environment at lunch-time, officers at Paremoremo were locked in from the minute they clocked on to the time they quitted at the end of their shifts. Personnel often commented that while they were at work, they were as captive as the inmates.

The position of industrial employees in this regard was even worse, because in the workshops, instructors were not only locked up, they were often alone with hostile workers. A large portion of instructors' time was spent, not teaching work skills or supervising productivity, but trying to stop the sabotage which was constantly taking place. Attempts to discipline the men or increase their output were fruitless, resulting in wrecked machinery and frequent threats of violence. Buckley had given orders that security, not safety, was paramount. When these members of staff asked him what assistance they could expect in the event of an attack, the superintendent had replied, 'None whatsoever. Because in no circumstances are those grilles to be opened in case there's a man escape.'[29]

Buckley's aloofness from the institution contributed to its sense of alienation. Experienced old screws like Cavanagh and Tuyt had seen this when they first toured the place in 1968. Others felt it as soon as they began working there, and quickly handed in their notices. One former instructor who had remained there until 1975 told me that he would not work at Paremoremo again 'for two thousand dollars a week'. Because of these factors, staff turnover was high and bad publicity made finding replacements difficult. Volunteer workers were reluctant to assist, so from a staffing angle the prison remained chronically deficient. Although twenty extra personnel had been assigned in its first twenty months, this had been offset by a similar number of resignations. As a result, by November 1970 the gaol was still operating at sixty-five below complement, and two of the five blocks, — classification and C. blocks — as well as the medical section, could not be opened.

Existing personnel were overworked and could be rostered for duty on as many as thirty consecutive days. Children and wives in the village missed their fathers and husbands. In the prison's first twenty months, at least thirty had been assaulted, some seriously. Families too, therefore, feared for their men's safety when they ventured each day into the confines of Paremoremo's interior.

Thus the prison's problems were numerous, but the solution lay in more than merely dissolving the impasse in D. block. For there was endemic discord in the rest of the gaol, and like the intractables, standard block prisoners were becoming increasingly militant. Although less acute, unrest in the open section arose out of the same set of circumstances as those which had plagued the segregation unit. The major difference was that here, since opposition could not be overcome simply by locking up inmates, the keys to resolution were more obscure.

The Standard Blocks

Because of the concentration of publicity around the intractables' unit, little was heard of events in the standard blocks in 1970, nor for that matter, for most of 1971. From the point of view of the media, Paremoremo during this time might have consisted of the twenty-five or so men in the projection at the end of the prison, and not much else. For some time only three wings operated, but in 1970 the classification block began to work and for a time all adult males sentenced in the northern half of the North Island to terms of two years and over began their confinement at Paremoremo. Some of these were then posted to the standard blocks, others went to different institutions around the country.

In the standard blocks, the saga of troubles which had begun with the Christmas parcel strike in 1969 continued into 1970. Although commitment here was weaker than among the intractables, there was no shortage of spirit. The standard blocks had not responded to Wickliffe's request for action in sympathy with Saifiti, however, when they heard of what had occurred after the hostage incident there was restlessness, and a gradual hardening of resistance. In 1971 this cohesiveness was added to by a managerial initiative, providing for the establishment of inmate 'activities committees' within the standard sector.

Supported and sanctioned by Wellington, activities committees were created for the purpose of co-ordinating recreation among the relatively free. This role they did perform, but in *de facto* addition, committees became the focus of political activity as well. Although the Penal Institutions Act specifically forbids the formation of prisoners' unions,[30] the activities committees now gave a legitimate cover for attempts at self-determination.

In A. and B. blocks, three committee members were elected by the prisoners of either unit, and these men became its representatives. Apart from their regular activities, committees organized 'kangaroo courts' to deal with violators of inmate interests and also co-ordinated defiance of administration. Since negotiation was seldom fruitful at this time, confrontation became the principal means through which grievances could be expressed. As levels of dissatisfaction arose, therefore, committee-organized demonstrations emerged as a common feature of standard block life. One committee member told me:

> We used to go out on strike (about) every two weeks, in A. block. Over anything... So we were on the committee, organizing activities, but we were mainly, sort of, the union. Most of the activities we organized were these strikes and things. [Laugh]
> So we'd, say, strike over our meals, working conditions, or not being allowed to do something, and used to go out at the drop of a hat. In those days, if we called a block meeting, *no* cunt would go to work, eh. They'd *all* be at the meeting.

However, strikes were not the only means by which standard prisoners expressed discontent. There was much sabotage and many floods and fires as well. Gaoled student activist John Bower, who was one of the principal committee organizers, said that the reason for the protests was the pettiness of administration. In reaction to what they felt were pointless rules and conditions, for example, Bower described how he and the others in the bootshop conspired over a period of weeks to burn and smash almost every piece of equipment in the area. When destruction forced it to close down, the bootshop workers' example was followed by men in the other industries.

Although commitment to action was weaker here than it was among the intractables, a relentless campaign of pressure eventually forced minor adjustments. Slowly inmates gained some discretion over dress and hair style, eventually more outside groups were permitted, and restrictions over literature were relaxed. The superintendent still preserved a disciplinarian cover, but his implementation of it was softening. Buckley's zeal for stringency mellowed as his retirement approached. We shall see that it was not until he was finally replaced, however, that significant improvements appeared.

On 20 March 1971, there was another riot at Mt. Eden. Quelled, without the destruction of the riot six years before, the event was still serious enough to warrant a reappraisal of conditions. Mt. Eden gaol, which after 1965 had been partially rebuilt to accommodate a maximum of 210, had held 375 prisoners at the time of the riot. There were only fifty-four disciplinary staff to supervise these men, and twenty industrial instructors. Nevertheless, about 300 of the prisoners remained unemployed. The incident created headlines throughout the country and sparked a major debate in Parliament.[31] Embarrassed at the disclosures of neglect, a recommissioning of those parts of the prison which had been closed since 1965 was immediately ordered by Cabinet.

Dissatisfaction in prisons was widespread and small riots had occurred earlier that year at Mt. Crawford and Paparua as well. Then, less than two weeks after Mt. Eden went off, a concerted attempt was made by three prisoners to take over the classification section at Paremoremo. Although the bid was unsuccessful the classification incident, combined with the other troubles, added to the concern in Wellington. As the Government stumbled and wondered what to do, the Paremoremo subunion demanded to know why nothing had been done about security measures for D. block, which had been given urgency up to eight months before. Lighting in the wings had yet to be approved, slam locks had not been fitted to gates and the extra grilles, which had been proposed half-way down each landing, were still absent.

Record population rises between 1969 and 1971,[32] added to by vocal union complaints and a bad press, left Government with no choice but to act. Ceilings on the number of staff, fixed only two months earlier, were suddenly relaxed. At Paremoremo, twelve extra staff were appointed in as many weeks and C. block, unused since the gaol first opened, was able to receive its first inhabitants.[33] The daily average population of the prison, which had risen by only eighteen between 1969 and 1970, thus leapt by thirty-nine the following year, levelling out at 159 towards the end of 1971. From this point on, the maximum security prison was fully functional. With all of its units in operation, Buckley now instituted a progressive block system whereby prisoners, through good conduct and industry, could increase their welfare by promotion. Men would begin their terms in A. block — which had the least 'privs.' — then transfer to B. block, and finally go on to C. Accordingly, industry which was considered to hold the greatest degree of responsibility, such as the kitchen and laundry, was shifted from A. block to C. It was from the latter unit that prisoners were most likely to win transfer out of maximum security.

The graduated privilege system never worked at Paremoremo, and its failure was due to several factors. First of all, inmates settled within their own environments and most became socialized into one or other of the cell block cliques. The territorial feelings in the blocks and the fact that sports competitions usually took place along block lines meant that a 'team' spirit usually emerged within each wing. Prisoners developed a powerful sense of block identity and pride. At one point, for example, A. block members even began wearing little badges, proclaiming theirs as the best block in the gaol.

Secondly, transferring to another block meant leaving old friends, adopting a new identity, and being accepted by a new clique. Juggling with administration for a cell amongst friends and overcoming the problems of ingress upon a tight established community were other deterrents to transfer. Most prisoners felt that moving was not worth while.

Thirdly, the differences in privileges between A. and C. blocks were only slight. In C. block — unlike the others — the men had unlimited correspondence privileges. Their choice of employment was marginally better, and they were locked up an hour later at night. However, these were only small considerations and by and large the lifestyle of C. block was hardly different from elsewhere in the gaol.

The final reason the graduated block system failed was that inmates were ideologically opposed to competing against one another for privileges. To do so was felt to be personally demeaning, and prisoners knew that the effect of such competition could only damage their unity. In 1970 and 1971 there was constant friction with staff as the two groups jostled for power. The convicts knew that, lacking any statutory force, their only weapon was their ability to organize. The entire well-being of the community revolved around this essential quality, and it was this which, by the end of 1971, had become management's most formidable obstacle.

The Publicity Filter

If a policy of repression had been in any way responsible for entrenching inmate attitudes, there is no doubt that the condition was added to by the prison's treatment in the media. Up until the final months of 1971, the bulk of popular feeling about Paremoremo was pro-administrational. Because of the strict code of silence observed by loyal officials and because the voice of prisoners was gagged, the great proportion of knowledge reaching the outside came from formal publicity statements. The rest came mainly from union complaints. In 1970, when the acting Minister objected to 'sensationalist' media coverage of the troubles,[34] the press replied that it had been forced to rely on anonymous sources because of the Department's withholding of information.[35] Since nearly all of the information on the gaol came from staff or official sources, at least half of the explanation for its difficulties was obscured.

In many sectors a picture emerged of well-intentioned staff, beleaguered by amply fed, generously served thugs, who attacked without purpose, warning, or restraint. Such an image was exemplified and added to by the 'Con is King' article of 1969.[36] It was perpetuated by later reports[37] and by accounts in the daily press. Of the repression, the provocation, the retaliatory beatings, and the spartan conditions in D. block, little was known or understood. Since the debunking of I. M. F. McDonald, hardly any questioning of official accounts had taken place.

As far as acting Minister Thomson was concerned, D. block's men were dangerous recalcitrants, whose complaints had already been found to be without merit.[38] In spite of Saifiti's eventual pardon through want of conclusive evidence, Missen still believes that the case was used by inmates as an 'excuse' for rebellion. His easy dismissal of an event, which for inmates was a serious misfeasance, shows the size of the gulf between Wellington officials and the men that they administered. Throughout 1970, Missen and Riddiford had sought to combat the violence at Paremoremo by attending to its security, but both then shrugged off unrest as a result of overcrowding and a 'worldwide trend'.[39] It is doubtful whether the reasons underlying the other troubles — the assaults, the fires, the flooding, and the sabotage — were ever seriously entertained either.

So, although the invidious position of staff at the institution was understood, that of prisoners was not. It was the inability of the Department of Justice to comprehend, recognize, or accommodate the basic emotional needs of Paremoremo's men which fed the roots of trouble in its early years. The hardening of attitudes was considerable by the end of that period and the complex structure of acrimony which had grown, could not be brought down overnight.

NOTES

[1] The worst penalties that could be awarded under this Act were fifteen days solitary confinement with restricted diet, and forfeiture of up to three months remission (Penal Institutions Act s. 33(3)).
[2] The difficulties that prisoners encounter in seeking gratification through grievance procedures is well recognized. The problem is discussed, for example, in Bergesen (1972–1973); Campbell (1970–1971) 46–52; Cohen & Taylor (1976) 17–29, 37–48, 56–7; Hawkins (1974) 269–73; (1976) 148–57; Mitford (1977) 257–66; Tettenborn (1980) 74–89; Triggs (1976); Vogelman (1968) 393–5; and Zellick (1975) 4–12.
[3] *NZH* 14 September 1985.
[4] Wickliffe, reported in Bungay & Edwards (1983) 104.
[5] The practice of allowing D. block inmates to watch films in the gymnasium was a privilege they had only recently acquired.

6 Maximum sentence, three years imprisonment.
7 Crimes Act s. 189, maximum sentence, ten years' imprisonment.
8 NZPD vol. 367 (1970) 2339.
9 *NZH* 13 August 1970.
10 NZPD vol. 367 (1970) 2339; *NZH* 14 August 1970; *EP* 14 August 1970.
11 NZPD vol. 369 (1970) 4473.
12 Riddiford was apparently still out of touch: in the statement he said these conditions would relate to Paremoremo. Of course, such separation already existed in the prison and Riddiford can only have been referring to D. block, where special conditions were subsequently imposed.
13 *Report by Powles and Sinclair on the Saifiti Case* (1972) 10-11.
14 Ibid.
15 Although it was not questioned at the time, it is legally impossible for a person serving life imprisonment to receive a term cumulative on that sentence (*R* v *de Malmanche* [unreported] 28 March 1966; cited in *Wickliffe* v *Police* [unreported] 10 November 1966: M1265/76:2). See also *R* v *Foy* [1962] 2 All ER 246.
16 *EP* 13 October 1970.
17 *Report by Powles and Sinclair into Matters Pertaining to Paremoremo Prison* (1972) 5.
18 Wickliffe, reported in *Red* no. 2 Dec. (1971).
19 Wickliffe, interviewed in *Red* no. 2 Dec. (1971).
20 Wickliffe, in Bungay & Edwards (1983) 105.
21 Wickliffe, interviewed in *Red* no. 2 Dec. (1970).
22 This occurred in January 1971. Saifiti received an extra nine months for the incident.
23 Wickliffe in Bungay & Edwards (1983) 104.
24 Saifiti had previously shown little regard for release dates. In April 1968 he had had only one month to serve of a nine month term when he and a man called Chris Guest escaped from Hautu prison. On this occasion Saifiti had given himself up after his co-escapee had fallen into a boiling mud pool at Tokaanu and been killed. It was this escape which earned Saifiti his transfer to Paremoremo.
25 Thomas (1972) 189-90.
26 Greer (1983) 374.
27 This village consisted of eighty-nine houses by August 1969 (*NZH* 23 August 1969).
28 For grievances regarding village amenities see *PGM* 5 May, 21 July 1969, *NZH* 22 July, 23 August 1969; *Truth* 16 September 1969.
29 T. Hooper (pers. comm.).
30 Penal Institutions Act 1954 s. 32(2)d.
31 NZPD vol. 371 (1971) 632-58.
32 DAPs in 1969-1970 jumped from 2510 to 2752. 1971 figures showed the largest prison population rise ever recorded in New Zealand history. In this year alone, daily average populations increased by almost 500, or about 19 per cent.
33 Annual DAP figures for Paremoremo between the years 1969 and 1970 are 110, 128, and 159.
34 E.g., *AS* 5 December 1970.
35 *Truth* 24 November, 15 December 1970.
36 *Truth* 16 September 1969.
37 *Truth* 11 August, 24 November 1970, 7 December 1971.
38 NZPD vol. 376 (1971) 4739.
39 NZPD vol. 376 (1971) 4625; *App.J.* H-20 (1971) 12. There was in fact a surge of uprisings in overseas prisons during the early 1970s. Some of these are referred to in the next chapter.

16: THE BUBBLE BURSTS 1970–1972

When change finally did come to Paremoremo, it did not arrive in a vacuum. The reorganization of maximum security was only possible because the time was right for it, and although one man provided ignition, progress needed considerable fuel.

The Protest Movement and the 1960s

In this country, as in many Western countries, the 1960s was a period of cultural development and changing lifestyles. In areas of dress, attitudes, and values, adjustments took place which were in marked contrast to those of the previous decade. Educational levels increased and social awareness grew. Attendance at universities boomed, and in the United States a burgeoning movement for black civil rights began.

In New Zealand cultural and political trends followed those being set overseas, particularly in the United States. Enrolment at New Zealand universities almost doubled between 1959 and 1969, although for the first half of the decade university campuses remained quiet and conservative. The tremors of the civil rights movement which shook the United States after 1960 were initially hardly felt in New Zealand and it was not until after the Vietnam War emerged as a domestic issue in 1965[1] that organized political pressure and public demonstration became common. Students, whose numbers were now large enough to make them a significant social force, were at the forefront of these activities and joined the tens of thousands who marched against the war in 1970 and 1971. Accompanied increasingly by sectors from the general public, they also began to campaign on other foreign policy issues, as well as on matters such as sport with racist South Africa, the Treaty of Waitangi, homosexual law reform, women's liberation, and the British presence in Northern Ireland. Between January 1967 and November 1971, in fact, at least 339 different demonstrations were recorded around New Zealand. This represented an average of more than one a week.[2]

The quick escalation in political involvement brought many face-to-face with the law. At the demonstrations against the visit of Vietnamese premier Air Vice Marshal Ky in 1967, and that of United States Vice President Spiro Agnew in 1970, protesters were attacked and beaten by police, and many were arrested. In a rising spiral of commitment there were fourteen bombings of military and conservative establishments in New Zealand in 1969, which ended with the apprehension of student activist John Bower. Bower and five accomplices were eventually convicted and incarcerated for arson-related offences, with Bower, the ringleader, receiving five years. That event is significant because all were friends of radical dissenting groups in Auckland. Bower, who was sent to Paremoremo, was thus able to continue his defiance of authority not only inside the prison, but outside it as well.

The protest movement of course did not occur alone and was part of a complex cycle of political and social change which was sweeping the world. Another hallmark of the era was a jump in the presence of illicit drugs, which, like the political movements, had its origins in the United States. New Zealand once more followed the leader, and large amounts of cabbais, LSD, and later, opiates were either produced locally or imported from abroad. Reported drugs crimes grew from 107 in 1967 to 740 in 1971,[3] while the number of people imprisoned on drugs charges grew from seventeen in 1969 to seventy-three in 1972, exceeding 100 the following year. Numerous students were among those charged and they became the folk heroes of the student community. A mood of sympathy for the underdog and rejection of authority that emerged out of this intensified the pressure for change which eventually reached the penal system.

Prison Protest

In New Zealand, as elsewhere, the rise of prison activism followed behind the civil rights movement and the general expansion of political awareness in the 1960s. In the UK an era of prison protest began in October 1969 with one of the most serious riots in British penal history, at Parkhurst. From here the level of agitation gradually rose in England, reaching a climax in 1972. Between January and May 1972 some fifty collective inmate demonstrations were organized in Britain and on 11 May, the English prisoners' union PROP[4] was born. The formation of PROP coincided with a sudden surge in organized protests. The bulk of these sought vindication of grievances against staff and/or demanded a wide and varied range of extra liberties and privileges. Pursuit of these freedoms culminated in a 'national strike' of over thirty English gaols on 4 August 1972.[5] This was followed by a spate of disturbances at more than twenty British institutions, including Albany, Parkhurst, Peterhead, and Dartmoor.[6]

While this was happening, in the United States the political strife of the 1960s had permeated the penal system and culminated in a rash of prison disturbances in the early 1970s.[7] Racial tension soon reached a peak. In one week early in 1971, for instance, there were twelve stabbings (one fatal) reported in the State prison of California.[8] In 1970 and 1971, black activists Jon and George Jackson were gunned down while in custody, and their attorney, black civil rights leader Angela Davis, was arrested and charged with supplying weapons to Jon Jackson.[9] Conflict led to martyrdom and hatred, fanning the fires of further violence.

Open confrontations with correctional authorities now became popular. In November 1970, one of the longest stand-offs in American prison history took place: a three-week strike at Folsom, California. In the next two years there were no fewer than twenty-seven major prison riots across the United States including one of the bloodiest ever, at Attica in September 1971.[10] Much inmate agitation was politically motivated and the period saw the birth of organizations like the Attica Liberation Front in May 1971, followed by the Washington State Prisoners' Union, the California Prisoners' Union, the United Prisoners' Union, and numerous others.[11] By 1973, one of these, the San Francisco Prisoners' Union, boasted a membership of 3000, located in sixty-eight separate prisons and penitentiaries.[12]

Although the United States provided the vanguard of political movements at the time, prison pressure was not confined to the English speaking world. In 1972-1973, for example, the New Zealand press reported trouble in the prisons of a number of countries including Italy, France, and Hong Kong. In Sweden a series of strikes in August 1972 won support from 60 per cent of its inmate complement.[13]

It was upon this record of international tension that Missen[14] and Riddiford[15] based their assumption that the riots at Paremoremo were explicable in terms of a 'world-wide trend'. The idea also found favour with Robson.[16] It was of course inevitable that the men of D. block knew about this anti-authoritarian wave and it is understandable that as they did, they saw parallels in their own situation. Wickliffe writes,

> This was the era of Che Guevara and George Jackson and the Weathermen. All over the world street riots and bloody prison revolts like that at Attica were occurring. And we in D. block were having our own little war with our jailers.[17]

However, the explanation for unrest at Paremoremo lies far beyond some sort of international contagion. There was a general mood of scepticism abroad in the early 1970s, but this provided a catalyst to, rather than a cause of, the endemic troubles at the gaol. As we have seen, fundamental administrative difficulties had existed at Paremoremo almost from the day of its opening. It would be naive to think that had protest halted overseas disorder at Paremoremo would have ceased as well. What the Secretary for Justice was not yet ready to admit, but what he would eventually concede, was that the only means of smothering recurrent violence at the prison would be an extensive revision of managerial policy.

AWARENESS ABOUT PAREMOREMO

From the beginning, it was certain that the cover which obscured the operation of Paremoremo would eventually have to break. Reports of continuing disturbances could not be repeatedly denied: few prisoners could be held indefinitely; communication between inmates and the outside world could never be suppressed entirely. Outside the gaol the activity of pressure groups was intensifying and their interest in prison affairs was increasing. From 1968, awareness about the hidden world of maximum security had been growing and publicity was thus inevitable.

Fred Ellis had been an inmate at Paparua prison, Christchurch, before being unexpectedly transferred to the security wing at Waikeria, prior to the opening of Paremoremo. Ellis, a Māori, held a keen interest in ethnic issues, and having read of the anti-racist work of the Newnhams, decided in 1968 to write to Tom Newnham, then secretary of the Citizens' Association for Racial Equality (CARE). Tom and his wife, Kath, began corresponding regularly with Ellis, and when Paremoremo opened and Ellis was transferred to B. block, they continued writing and began to visit.

The Newnhams' contact with Ellis produced other acquaintances in the prison, and it was through the Newnhams and their affiliation with CARE that other active citizens began to take an interest in the condition of security prisoners. By the time Ellis was released in late 1969, a number of inmates were in contact, and a variety of organizations were concerned for their well-being.

After 1969, information obtained by interest groups about conditions inside the prisons came increasingly from convicted men. Unlike Ellis, many of these already had an appreciation of political issues before coming before the court. From about 1970, the surge of drug crimes and the beginnings of confrontation protest brought large numbers of young, often middle-class offenders, into the justice system. Charged with what were sometimes serious criminal offences, for the first time since the war, penal establishments began to fill with educated and/or politically active people. It became fashionable to be martyred through subjection to State justice, or if not, at least to know someone who had.

By late 1970, former student and popular activist Tim Shadbolt had already been locked up twice for refusing to pay his fines. That same year, John Bower was gaoled and Shadbolt mounted an energetic (but unsuccessful) campaign to have his friend released.[18] There was empathy too from the large numbers who were never caught. In 1971, when Colin Lum, expelled from Otago medical school after an LSD conviction, was refused admission to Auckland University, a vocal protest was lodged by the student fraternity.[19] Non-students received equal attention. In 1972, Arthur Thomas — convicted in 1971 of a double murder — and Rodney Davis — sentenced to ten years in 1968 for causing death by administering illegal narcotics — both had their claims of innocence given sympathetic airings in the Auckland University newspaper *Craccum*.[20] Davis' case was also taken up by the New Zealand Howard League and later the Community Action Committee, both of which petitioned the Government to release him. These attempts failed, but Thomas, over a relentless campaign, became a household name and was eventually pardoned in 1979.

Interest in, and knowledge about, the lock-up gaols was still restricted to isolated pockets at this time, and, in spite of growing pressure, many prisoners, isolated from their families and friends, received no visits whatsoever. The Newnhams were now visiting regularly, and overwhelmed with the work they felt needed to be done, began to request assistance from others. These calls were responded to, and through the Newnhams, the Bowers, and other interested parties, the network of politically active prison visitors began to grow. One of the earliest, and later more prominent of these, was Maynie Thompson.

Maynie Thompson was an old friend of Kath Newnham, the two having met years before when they worked together as school dental nurses. Thompson had been involved in the developing playcentre movement in the 1950s and like the Newnhams, Thompson was a member of CARE. When she heard of the Newnhams' predicament and the plight of some of the prisoners, Maynie Thompson felt prompted to go out to Paremoremo to visit one of them. Having done so, she rapidly became committed to assistance. Her initial feelings she described thus:

> It was such an emotional shock to walk into that prison. It was like something out of this world. I was so shocked to see Mike Birch. He was in D. block and I visited him in those little cells, [security booths] and to see this fellow with all these love-hate tattoos all over him. And he looked so pale and miserable and terrible and it suddenly came to me just then how absolutely *incredible* it is to keep men in cages.... And I just felt so overwhelmed at the sight of a man, a pathetic-looking individual, behind those jolly bars. And I was committed right then and there. I just knew I had to be involved.

It was this sort of emotional response to the prison's reality — the sallow appearance of inmates, the surly demeanour of staff, and the legends of tension, violence, and injustice — which aroused the compassion of so many of the early Paremoremo visitors. From that point, it was easy to commit them to the movement for reform.

To start with, attempts were made to improve conditions for the men and to question Buckley's policy. These efforts were met with obduracy. It was clear that no significant changes were favoured and that none, therefore, would be made. November 1970 saw the first assembly of sympathizers outside the prison fences. One Sunday afternoon about sixty 'hippies' gathered on the road running past the institution and for a few hours indulged inmates with a four-piece band and a fireworks display. Prisoners were addressed with a loud hailer, the object of the exercise being, a spokesman said, 'to give the prisoners a bit of entertainment'.[21] Concern about the incarcerated was beginning to get publicity, and with it, awareness of their condition increased.

Throughout 1970, Tim Shadbolt had been a vocal critic of New Zealand's penal system. In January 1971, newly released after a brief spell at Mt. Crawford, Shadbolt spoke at the Power, Justice, and Community Conference in Wellington. Now an accomplished orator, Shadbolt's attack on imprisonment succeeded in persuading the New Zealand Student Christian Movement (NZSCM) to make the nation's legal system one of its priority concerns. This group now joined in establishing contacts with ex-prisoners and developing a programme of regular visits. Other religious organizations then also became involved, and their concern was deepened in March 1971 when, during the course of a riot at Mt. Eden, a man was shot in the thigh.

In June, largely in response to this, Auckland church leaders organized a seminar, attended by senior Justice officials and a number of MPs, on penal policy and practice in New Zealand. Questioned during the seminar by a concerned citizen, MacKay admitted that he was having trouble finding work for D. block inmates. Although he denied that any prisoners in the block were in solitary confinement or that they were barred from church services,[22] it was later announced that controls would be eased once the new security measures were in place.[23]

The Minister was somewhat less forthcoming in penal matters and denied that even the Mt. Eden crisis was significant.[24] In truth, however, there was considerable disquiet over both of the Auckland establishments. The more sensitive of these was still Paremoremo, and rapid steps were now taken to upgrade it.

In October the position of full-time nursing sister was advertised, so that the gaol's expensive surgical and medical unit could finally come into use. A full-time teacher — Ken Travis — was appointed too, and one of his earliest moves was to approach community organizations in an attempt to attract volunteer workers. Although at this stage they could not be paid, these volunteers provided a valuable service and later became an important feature of prison life. One of the first such visitors was Paul Beachman, a local high school teacher. Beachman, persuaded by Travis, agreed with two others to visit and set up discussion groups within the open blocks. Another was Don McKinnon, of the Auckland Debating Association. McKinnon, as we shall see, was eventually able to establish a successful debating programme, which continued operating until inmate confrontations forced its closure in 1985.[25]

The effect of the private and public campaigns was considerable. In July, a full page spread in *Craccum* called for more volunteers to visit the prison,[26] and by September many sectors of the community were actively involved. Apart from the official workers, affiliates of the Church were writing and visiting, as were people from CARE and from University-based organizations. By the end of 1971, the protest movement and the political left were receiving regular details of events via inmate Bower, and Labour politicians were pressing for a re-examination of the Justice Department's most controversial facility.

The Pressure For Change

Soon after the Mt. Eden riot, Daniel Riddiford, a sick man no longer able to cope with the pressures of office, was replaced temporarily by the Minister of Defence and Police, David Thomson. Although Thomson was not a lawyer, he was, like Riddiford, an ex-prisoner of war and a hardliner on issues of justice and crime.

Since August 1970, the Department had been resisting pressure for an independent inquiry into Paremoremo. Recurrent problems at the prison had consistently been either played down or had been countered with assurances that the causes of difficulty were being looked at.

Thomson's approach to prisons would be little different from that of his predecessor. Acting for Riddiford when the latter fell ill after the June riot, for example, Thomson had claimed that the understaffed and seriously overcrowded Mt. Eden[27] was 'Virtually up to the grade of what was popularly known as a three star hotel.'[28] With Thomson now at the helm in Parliament, the prospect of improvement did not look bright.

In spite of Riddiford's growing incapacity, therefore, executive strategy did not change. The Government continued to insist that tighter restrictions had improved conditions in D. block[29] and claimed falsely that there had been a reduction in assaults on staff.[30] These myths were fragile ones, but in this manner[31] Paremoremo's governors kept a tight hold on the flow of information. Prisoners in D. block's top landing now had to take their visits in electronically monitored security booths, in which any communication was difficult. When a hunger strike, fires, and flooding took place in the block in August 1971, the results of a magisterial inquiry were withheld. The superintendent then suspended all D. block visits, except those from close relatives. Two visitors who were felt to be providing a bad influence were banned from the institution altogether. With an unsympathetic Minister in the House there was little which anybody could do about it.

Although their collective knowledge was considerable, the efforts of visitors to attract a following had so far been ineffective. There was no unity among them, and as yet the information they had was uncollated, sketchy, and often inaccurate. Because of its monopoly over official knowledge, the Department found little difficulty in parrying allegations of mismanagement. After September, the closing off of D. block threatened to contain information about that section of the gaol completely. The despondency which the isolation of D. block brought was considerable, but could not last for ever. For while communication with men inside could be strictly monitored, there was no way of silencing those who were discharged. It was inevitable, therefore, that sooner or later a release from Paremoremo would bring life to speculation about the prison's interior.

THE BIRTH OF PROJECT PAREMOREMO

During the year of 1971, after Shadbolt's description of prison conditions at the January conference, continuing interest had been shown by the New Zealand Student Christian Movement in the nation's judicial system. Together with the Newnhams and CARE, by mid-1971 this group already represented the beginnings of an organized committee on prison management. When Dean Wickliffe, one of D. block's most vocal members, finished a three-year term for armed robbery early in October it was to the NZSCM that he eventually turned for help.

Wickliffe is an intelligent man with strong feelings about justice, and freshly discharged from the toughest block in the land he was determined to continue the fight, which he had begun two years earlier. By nature an independent person, initially Wickliffe attempted to work alone and he travelled to Wellington to arrange an audience with the Secretary for Justice or the Minister.

Although (or perhaps because) he was already known to head office, the former prisoner obtained little satisfaction from his efforts. Refused permission to see either of the Department's top executives, he was granted an interview with the Director of Prisons, Mr R. O. Williams, who informed Wickliffe that he could offer no hope of redress.

Frustrated, despondent, out of work, and in need of money, Wickliffe returned to a more familiar activity. He committed a series of burglaries and within a few weeks was back in trouble with the law. Wickliffe decided it was time he sought outside help. Released on bail, he approached the reformist network and through it was able to make contact with Reverend I. D. Borrie, at that time General Secretary of the NZSCM.

For those concerned about the imbroglio at Paremoremo, Wickliffe's arrival was a timely blessing. Here at last was a source for the information they had pursued so vainly, and efforts to have the prisoners' grievances heard were renewed. On 26 October, a meeting was arranged with a group of opposition politicians[32] at which Wickliffe was able to describe D. block conditions personally. Surprised and disturbed by the revelations, a visit to the institution was planned and the politicians determined to take action. A seminar was projected for 20 November, and with it, Borrie became totally committed to the objective of reform at Paremoremo.

Details of the style and content of the Paremoremo seminar were complex, and Borrie conferred closely with Wickliffe and Tom Newnham to establish a workable format. In the meantime, Wickliffe continued his vain attempts to be granted an interview with Missen. Wickliffe was due for a depositions hearing on 15 November and fearful that bail would be revoked, Borrie joined in trying to arrange a meeting with senior Department officials. First of all, former Secretary, John Robson, was contacted and, although sympathetic, he was not prepared to become involved. Through Robson, however, an audience with Missen and Williams was arranged. The results of their meeting, on 12 November, were less than encouraging. Summarizing the interview, Borrie wrote, 'I was politely listened to and shown the door. As far as the Justice Department was concerned, Mr Wickliffe had had a spell of verbal diarrhoea since his release. He was an intelligent trouble-maker for whom nothing could be done.'[33]

The intransigence of the Department did little to deter Borrie, and other organizations soon rallied in his support. Important among these was the Race Relations Council, which also approached Missen and Williams to arrange a meeting. Once again the officials were reluctant. The Secretary for Justice accepted the Councillors' request, but permission for Wickliffe to attend was still refused. When the meeting eventually took place, the delegation was told that although improvements were being made to the institution's workshops, upper-landing D. block men were an insoluble problem. As far as Mr Williams was concerned, the visiting privileges the men enjoyed were too liberal already. Resistance against improving D. block visits was also backed by the Minister, and it became apparent that the Department and the Government stood united in their opposition to any significant relief in the intractables' condition.

The pressure to which Justice administration was being subjected had mounted rapidly since Wickliffe's release, and his version of the situation at Paremoremo was producing increasing discomfort. Wickliffe was still awaiting depositions, and Borrie had good reason to believe either that Wickliffe would abscond before his lower court hearing took place, or that the Justice Department would have his bail revoked when he appeared in court on the fifteenth. If this happened, the main thrust of the reform movement would be lost. On this account, Borrie's fears were compounded by the negative reception he had received on the twelfth and so, two days before the hearing was due, the NZSCM arranged for Wickliffe to make a 2000 word tape, outlining his main areas of complaint.

Typed out, cyclostyled, and distributed among interested bodies, Wickliffe's statement to Borrie had a sharp impact on those who read it. Unfortunately, because of the gravity of its allegations, the document's credibility was considered doubtful and none of the media organizations approached would publish it. It was not until some of its points were raised in the House of Representatives[34] that the matter was given any serious attention. The following month the text of the document appeared fully in the small socialist publication *Red*.[35]

It was the material in the Wickliffe statement which finally prompted the independent pressure groups to take co-ordinated action. Accordingly, on 20 November 1971, a seminar convened by Borrie, Wickliffe, Newnham, and others took place in Auckland. Organized — at opposition MP Mrs Whetu Tirikatene-Sullivan's suggestion — in conjunction with the Māori Women's Welfare League in Auckland, the closed meeting was attended by fifty invited guests. These included several opposition Members, ex-inmates, and representatives from various voluntary organizations.[36] This body now formed a committee, under the auspices of the Howard League. To be known as 'Project Paremoremo', the functions of the committee would be 'to expose to the New Zealand public the details of the inhumane aspects prevailing in "D" Block and the prison'.[37] For this purpose two subcommittees were set up, the first to circulate a petition demanding an improvement in prison conditions, the second to obtain and collate further statements and information on the situation in the intractables' block.

From the beginning, the liberation of D. block was at the heart of Project Paremoremo's goals. Besides this, however, from the time he was released, Wickliffe had been barracking strongly for a re-trial of the wrongly convicted Saifiti. Many had found Wickliffe's arguments credible and sincere, but the attitude of head office made it clear that an internal review would be useless.

To the suggestion that Saifiti was innocent came the retort that his plea had already been rejected by the Court of Appeal.[38] In the meantime, complaints about poor facilities were still being met with replies indicating that the inmates had brought their conditions upon themselves, that conditions would be improving, or that conditions were not as bad as the violent and rebellious recalcitrants had claimed them to be.[39] To the members of Project Paremoremo, committed to positive action, the only solution now seemed to be an inquiry, organized by an independent authority.

THE FIRST POWLES–SINCLAIR REPORT (1972)

The office of the Ombudsman had been created in this country in 1962 for the purpose of investigating complaints about the decisions or activities of Government and Government-related agencies. Since the office came into being, the Ombudsman for New Zealand had been Sir Guy Powles, formerly High Commissioner in India and Ceylon, and it was to him that Project Paremoremo turned. On 24 November, only four days after the committee had been created, Dean Wickliffe and other members of Project Paremoremo met with Sir Guy in Wellington. Over a period of two hours, the group's principal areas of concern were mapped out, and an investigation was requested.[40]

From the time Paremoremo had opened, the number of complaints received from prisoners and others about gaol conditions had risen hugely. In 1969, there had been seven complaints, the following year, twelve. In 1971, the year that Paremoremo became a major issue, the number of objections increased to forty-nine, which was five times more than the number received from any other institution, and more than twice the total of complaints received from all other institutions combined. When pressure group activity was rationalized at the end of 1971, the level of outside agitation increased as well. After the formation of Project Paremoremo, the rate at which complaints about the prison were received jumped by 900 per cent.

Against this pressure, there was a counter-reaction from staff. On 30 November, the subgroup at Paremoremo expressed concern about news releases by Wickliffe and Labour MP Eddie Isbey, claiming that they had endangered officers' safety and unfairly reflected on staff integrity.[41] Isbey was unrepentant and replied, *inter alia*, that he would not be gagged by the subgroup in the performance of what he saw as his parliamentary duty.[42]

Upset by the response, and keen to present their own case, Paremoremo staff held a meeting with MPs several days later, having first escorted them around the prison. Impressed, National MP Frank Gill then arranged for a repeat tour by Government Members in January. Staff now saw their position gaining credibility, but needed a catalyst for action. This was provided by an assault on a D. block officer on 12 December. The following day, prison officers announced their intention to work to rule from 20 December, in support of a salary and leave claim, being considered by the State Services Commission.[43]

In the meantime industrial threats were spreading. On 20 December, Mt. Eden staff joined those at Paremoremo in support of the latter's appeal. Sympathetic action was also pledged by Waikeria and Waikune, but a rushed meeting between Public Service Association representatives and the State Services Commission soon cooled the conflict by giving in to most of the officers' demands.

On 13 December, the same day that the subgroup announced its intention to take direct action, the acting Minister of Justice announced an independent inquiry, to commence within a few days. Known loosely as the 1972 Powles–

Sinclair report, this first major investigation into Paremoremo prison was conducted during the months of December 1971 and January 1972.

At the same time as the officers were agitating in support of their work claims and the investigators were visiting the prison to collect data for their report, the situation within it was deteriorating. Possibly as a result of pressure from the prison officers' subgroup and possibly to impress the two investigators, in the latter half of December, the Paremoremo administration laid down some 'firm rules' for the conduct and routine of inmates.[44] As had frequently occurred in the past, the imposition of more stringent measures had a reactive effect. Instead of suppressing discontent, the new measures aggravated it. Thus it was that in the three weeks after Christmas 1971, levels of disorder increased markedly.

Not long after Christmas, there was an apparent escape attempt from the gymnasium. Soon afterward the A. block recreation area was set on fire and flooded. In the workshops destruction was continuous. Machinery was sabotaged, telephone cables were cut, and there were many fires. A. block was the seat of the trouble and it was from here that protester-bomber John Bower organized many acts of sabotage, including one (failed) attempt to blow up the bootmaking plant. By 11 January, four of the six workshops had been put out of action completely. Inmates were conducting a go-slow campaign and were refusing to be locked up at night. It was reported that so many staff had been assaulted lately that violence was now considered normal.[45]

The situation at the prison was more dangerous than it had ever been, and Buckley's authority, always fragile, was now beginning to crack. Confronted with allegations that things were close to anarchy, both Buckley and Powles claimed the reports were exaggerated. Buckley conceded that there had been a few fires, but he said little or no damage had been done to the institution.[46] The following day there were five more fires.

While Buckley insisted that he had no idea why the trouble had started, the *The New Zealand Herald* reported that the reason for the protests was the tightening up of discipline.[47] Confronted with the suggestion, Deputy Secretary Cameron and Director of Prisons Williams both described claims of destruction as 'highly exaggerated'.[48] Both denied knowledge of what was causing the trouble and Cameron said there was no need for action. None the less, the threat was considered serious enough to warrant a doubling of the guard at the institution. It was reported that this was to defend against the possibility of a full scale riot.[49]

On 13 January, Powles again claimed that latest news reports — which this time had come from a prison officer — were exaggerated and he criticized the press for inflaming the situation. Missen too said the reports were wrong, but because the information had come from unofficial sources, he refused to discuss the details of any inaccuracies. On 14 January, after a visit to the gaol by the Inspector of Prisons, Missen finally conceded that 'minor' fires and sabotage had occurred. He too, however, denied knowledge of their causes.[50]

The co-ordinated campaign of destruction at Paremoremo was incidental to, but coincidental with, the visits of Powles and Sinclair. To these men, intelligent, educated, and not unfamiliar with the nature of penal administration, the situation was disquieting. During the course of their inquiry the two not only familiarized themselves with the layout of the institution and observed at first hand the damage done, they conducted many interviews with staff, specialist personnel, and inmates as well. They also received a comprehensive list of complaints from D. block.

On 21 January, Powles and Sinclair's sixteen-page report on Paremoremo was presented to the Government, and was released publicly five days later. The 1972 Powles-Sinclair report covered matters such as procedure for admission to, and transfer from, Paremoremo, the system of internal classification and designation to D. block, and the enforcement of formal and informal discipline. The report also considered matters pertinent to the recruitment, training, and management of staff.

Taking into account the haste with which it had to be prepared — the investigation had been conducted and its results presented inside a month — the report was quite extensive and detailed. On the positive side, continued existence of D. block was defended and none of the fourteen men resident in it were judged to be there without justification. Allegations of staff brutality were groundless, the food was good, and religious ministration and medical care were adequate, the report said. Although some of these findings were contentious, the report did contain numerous recommendations for change. In particular, it was felt that improvement could be made in the structure of visiting booths; that visiting, as a rehabilitative aid, should be considered a right rather than a privilege; that more welfare, social, and recreational activities should be provided; and that a set of guiding principles should be drafted, governing the policy of sending people to maximum security.

In the opinion of the investigators, too many medium security prisoners were held at the institution and perhaps only forty of the men there were really in need of such confinement.[51] At the root of the trouble, the investigators found, was a shortage of qualified personnel. This resulted in a serious two-way problem: there was tension between staff and inmates, as well as between staff and the administration. Improved recruitment and training procedures were advised. In the end, many of the report's proposals were ignored and there was little change in the prison's organization. One important suggestion, however, the appointment of extra superintendent's assistants, was adopted. These assistants, who became known as divisional officers, were established a month later and, as we shall see, had an important influence on formal relations.

Another issue was that of Atenai Saifiti. In relation to this, the investigators wrote that they had been able to satisfy themselves that 'there is a reasonable ground to suspect that a miscarriage of justice may have occurred.'[52] A separate submission on Saifiti was subsequently prepared. This paper, released some months later, examined details of the case and recommended a free pardon for Saifiti under section 407 of the Crimes Act.[53] The recommendation was acted upon by the Governor-General, and in May Saifiti's conviction for the Smith assault was expunged.

Conspicuously missing from the document was any reference to the superintendent, other than to say that although there was tension between staff and administration, '[administration] keeps the prison running and has the confidence of inmates.'[54] The reason the absence of further attention is noteworthy is that dissatisfaction among staff and prisoners with Buckley's regime was, as suggested, one of the major foundations of dissent. Even some of the visitors had got the clear impression that Buckley lacked the ability to administrate. Maynie Thompson, for example, who met the man only on a few occasions, said of him:

> Well we once had some kind of an interview with Buckley right at the beginning and he was pathetic. He should *never* have been in charge of a big thing like the prison [at Paremoremo]. He was right out of his depth. Intellectually and in every other way.

Paul Beachman's first and lasting impression of Buckley was similar:

> ... we were impressed by the fact that Buckley seemed so scared. That he was actually fearful about running the institution. And rather than getting the impression that he was this almighty sort of a person, you had the impression of someone who was totally terrified about the way he was going to run it and how he was going to contain all these guys. He was very much out of his depth and you gained that impression straight off. It stood out a mile.

As we shall see, the notion that Buckley was no longer competent was prevalent at head office also, and if Powles and Sinclair had not discerned this during the course of their own research, it would be surprising. Whatever the case, the investigators' failure to comment on managerial capacity of the superintendent had little impact on his future.

REACTION TO THE REPORT

J. K. Hunn has written, 'In the Public Service it is traditional for the authorities to prefer conservative rules liberally interpreted, rather than the other way round.'[55] This is as true of the Justice Department as of any other Government agency. In prisons, the very broad interpretation of administration as laid down in the Penal Institutions Act, the Regulations, and various procedures manuals leaves a good deal of discretion to the superintendent. His philosophy, his orders, the extent to which he delegates or monopolizes authority have a direct bearing on the attitudes, demeanour, and morale of officers, and through them, the behaviour of prisoners. It is the superintendent, not head office, who decides the day-to-day course an institution will steer, and like a ship's captain, it is the superintendent's responsibility to navigate hazardous waters.[56]

As would be expected, the release of the Ombudsman's paper and discussion of it in the media caused interest in Paremoremo to grow. On 27 January, a dozen National MPs visited the gaol in order to gain a better understanding of how it was run. The Powles–Sinclair report had been welcomed by the public service and Government chiefs but the independent organizations, whose activity had precipitated it, were unhappy with it. As far as they were concerned, the worst aspects of the institution had been ignored. The neglected issue of management was what was most unsatisfactory, for it was here that the main difficulty was seen to lie. Less than a week after the report was presented, therefore, the Paremoremo Prison Officers' Association did what the Ombudsman had failed to do, and delivered a broadside at the gaol's administration. When the MPs visited on 27 January, a delegation of five members of staff met with them to air its grievances.

Staff insisted that the weak link was high-level directorship: the superintendent, his deputy, and the senior chief officer. A wide gap existed between this élite and those beneath it. There was little communion between administration and the rank and file, and officers felt that they lacked support in their duties. There was deep dissatisfaction with the way the prison was being run, and the majority of staff, the delegation said, had no confidence at all in the superintendent. Moreover, it was declared that newspaper reports of sabotage and rebellion, which earlier that month head office had described as 'highly exaggerated', did in fact present an accurate picture of what was happening. This state of affairs was a direct result of weakness in the regime of Buckley and his lieutenants.[57]

To this point, no progress on promised improvements in pay and holiday allowances had yet been made. The Department knew industrial tension was peaking, and when further action was threatened the acting Minister quickly moved towards appeasement. He also hinted that after the official inquiry, there might be a change in policy. Exactly how great the change might be and what it might entail, the acting Minister declined to say.[58] It seems, however, that a major decision about the prison's future had by this time been made. What was delaying it was deciding on the wrapping which should be used to present the resolution in public.

THE DEPARTURE OF BUCKLEY

Towards the end of January 1972, Buckley had a car accident and as a result of a back injury and 'other illnesses', was forced to take some weeks off work. Missen reported that the superintendent was 'seriously indisposed'.[59]

Criticism by the prison officers' subunion had been directed not only at the superintendent, but at his deputy Mr Ward. While Buckley was in convalescence, Thomson toured the institution and in order to restore confidence in second-line administration, commented that he was much impressed with the competence of Ward and other senior personnel.[60] No word was uttered about the ailing Mr Buckley. It is possible that this omission was due to Buckley's absence, but one wonders whether Thomson's statement was not an indication of what was about to happen. For while Paremoremo's first superintendent lay confined in his sick-bed, the stage was being set for his removal.

Misgivings about Buckley had been growing since the previous decade, when, not two and a half years after taking over Mt. Eden, that institution had been wrecked in a riot. 'Not the ideal choice' when he was given Paremoremo, it too had soon become wracked with strikes, assaults, rebellion, floods, fires, and sabotage. However, not only was head office worried about Buckley's performance, he was also losing confidence in himself. Eric Missen, the Justice Secretary, told me that after the 1965 debacle and further trouble at Paremoremo, Buckley lost the ability to deal with matters on the spot. 'He was almost *daily* on the blower to head office,' Missen said, 'over things that you'd expect a superintendent to straighten out himself . . . it was quite clear that Buckley had lost his nerve'. It was Missen's opinion that the 'extreme situations' which developed at Paremoremo could have been avoided if Buckley's handling of them had been more effective.

On 5 February, it was announced that Jack Hobson of Mt. Eden was to take over Paremoremo as acting superintendent while Buckley was sick. The announcement is puzzling, as Hobson is emphatic that his appointment was permanent and that at no time was he at Paremoremo in a temporary capacity. Not surprisingly, therefore, six days later, Hobson was named as the new superintendent of the prison.[61] Although it was reported that Hobson had been filling in for five weeks,[62] he had not in fact entered Paremoremo until 7 February, and he had taken full command from that date.[63]

In the meantime, Buckley had ostensibly been promoted. He now became regional co-ordinator for Auckland prisons, with the official title: 'Superintendent (advisory and special duties)'. When Hobson arrived, he moved into an office right at the heart of the prison. Buckley was allowed to remain in his old office in the administration block, well away from the institution proper. Nobody seemed to know exactly what Buckley's new job actually entailed, but it was widely rumoured that the appointment was a sinecure, created to avoid any embarrassment. This was confirmed by Missen, 'Well, finally of course we had to take Buckley out . . . We took Buckley out and we sidelined him. In fact we gave him a special job and put Jack Hobson in.' It was noted by *The New Zealand Herald* that the position had not existed hitherto[64] and when Buckley retired two years later, that a new co-ordinator was not likely to be named in the near future.[65] The job remains unfilled to this day.

NOTES

[1] In May 1965 the New Zealand Government announced its intention to send combatant troops to assist United States forces in Vietnam.
[2] Jackson (1973) 164.
[3] Cameron (1972) 311.
[4] Preservation of the Rights of Prisoners.

5 Fitzgerald (1977) 136-58.
6 Ibid., 161-79.
7 Cressey (1973) 145-7. Those interested in this subject should not overlook Jackson's published letters to Davis, composed in prison before he died (Jackson 1970).
8 *NZH* 16 March 1971. The prison referred to is San Quentin.
9 Jonathan Jackson and two other San Quentin inmates were shot to death in the parking lot of Marin County Courthouse after taking their trial judge hostage on 7 August 1970. Davis was arrested on 13 October. 'Soledad Brother' George Jackson, serving one year to life for robbery, and recently convicted of murdering a prison guard, was shot dead while allegedly trying to escape from San Quentin on 23 August 1971.
10 Select Committee on Crime (1973).
11 Fitzgerald (1977) 210-52.
12 Mitford (1977) 291-2.
13 Fitzgerald (1977) 157-8.
14 *App.J.* H-20 (1971), 12.
15 *NZH* 8 February 1971.
16 Robson (1974) 37-46.
17 Quoted in Bungay & Edwards (1983) 127.
18 *Craccum* 19 March, 30 July 1970.
19 *Craccum* 8 April 1971.
20 *Craccum* 8 June, 7 September 1972.
21 *NZH* 9 November 1970; *EP* 10 November 1970; *Truth* 24 November 1970.
22 *NZH* 7 June 1971. The Department's usual rationalization for denials of this type was that top-landing prisoners, though locked up and exercised alone, could communicate at all times with others in their wing. And while religious services did not exist for these men, they were visited in their cells from time to time by ministers of the Church.
23 NZPD vol. 376 (1971) 4739.
24 *NZH* 22, 23 March 1971.
25 In 1978, McKinnon entered politics as National MP for Albany. After serving for seven years as party whip he became deputy leader in September 1987.
26 *Craccum* 22 July 1971.
27 Concern about crowding had been publicly voiced by the Department as recently as January (*NZH* 28 January 1971).
28 NZPD vol. 376 (1971) 4630. The report that Thomson used to back this claim had been researched by Government officials some weeks after the cleaning up of the prison had begun. It dealt solely with the provision of sanitation and related health matters.
29 NZPD vol. 374 (1971) 2703; *NZH* 27 August 1971.
30 On 26 August 1971, Riddiford said that there had been just two (minor) attacks on staff so far that year (*NZH* 27 August 1971). In fact he was only referring to D. block. Figures show that in 1971, there were eighteen charges of assault on an officer laid at Paremoremo (Greer 1983, 374). This indicates an assault rate that was substantially unaltered from previous years.
31 See Agger (1956); Etzioni (1968) 243-4; Kuhn (1963) 545, 721-7; Lasswell & Kaplan (1950) 88, 101-2; Loewenstein (1957) 55-61. For some specific examples of the control of information in totalitarian systems see Inkeles (1958); McCleery (1957) 375, (1975); Shirer (1977) 244-8.
32 The politicians involved were Phil Amos, Mat Rata, Whetu Tirikatene-Sullivan, Martyn Finlay and Eddie Isbey.
33 *NZSCM Newsletter* 25 November 1971.
34 NZPD 376 (1971) 4738; *NZSCM Newsletter* 25 November 1971.
35 *Red* 2 (Dec. 1971).
36 Organizations represented included NZSC, the Auckland University Students' Association, the Howard League, the Campaign Against Repressive Legislation,

CARE, Nga Tamatoa, the Women's Liberation League, and the Auckland Humanist Society.
37 *8 O'Clock* 20 November 1971.
38 For example, NZPD vol. 376 (1971) 4740. The reaction was also expressed in a letter from Thomson to T. Newnham, 19 December 1971, in reply to a letter from T. Newnham to Thomson 23 November 1971.
39 See for example NZPD vol. 376 (1971) 4738-40; *EP* 19 November 1971; *NZH* 19 November 1971.
40 Identical letters from T. Newnham to Thomson, Finlay, and Powles are dated the day before the meeting, 23 November. Further correspondence with the Ombudsman dated 28 November 1971 is referred to by Powles in a letter to T. Newnham 2 December 1971.
41 *NZH* 1 December 1971; *EP* 1 December 1971.
42 *EP* 2 December 1971; *NZH* 3 December 1971.
43 *NZH* 14, 16 December 1971.
44 *NZH* 12 January 1972.
45 Ibid.
46 Ibid.
47 *NZH* 13 January 1972.
48 Ibid.
49 Ibid.
50 *NZH* 15 January 1972; *Sunday Times* 16 January 1972.
51 *Report by Powles and Sinclair into Matters Pertaining to Paremoremo Prison* (1972) 14.
52 Ibid., 16.
53 *App.J.* A-6 (a) (1972).
54 *Report by Powles and Sinclair into Matters Pertaining to Paremoremo Prison* (1972) 11.
55 Hunn (1959) 19.
56 For a good literary treatment of the effect of a naval commander on the morale of his crew see Wouk (1951).
57 *NZH* 28 January 1972.
58 Ibid.
59 *NZH* 5 February 1972.
60 *NZH* 29 January 1972.
61 *NZH* 11 March 1972.
62 Ibid.
63 Hobson (pers. comm.).
64 *NZH* 11 February 1972.
65 *NZH* 6 June 1972.

17: A NEW REGIME 1972

Sir R. E. Jack, Minister of Justice

At the time that Paremoremo's new superintendent was appointed, the second National Government, then in its twelfth year, was endeavouring to reorganize itself for a general election. Early in 1972 the Prime Minister, Sir Keith Holyoake, had retired, and in February his position was taken by Deputy Leader and former Justice Minister, John (later Sir John) Marshall.

The Department of Justice had been without an effective political head for most of 1971. Since the Mt. Eden riot in March, Riddiford had been ill, and for almost a year his position was filled by David Thomson. As that year progressed, it became increasingly likely that Riddiford would not be able to return to his portfolio, and one of the first duties of the new Prime Minister was thus to select a new Minister of Justice. The Minister chosen was Sir Roy Jack.

Roy Emile Jack was born the son of a prominent Wanganui jurist in 1914. Educated at Wanganui Collegiate and Victoria University, he took an LL B and served as a Judge's Associate between 1935 and 1938, before joining the RNZAF at the outbreak of war. By the time of his demobilization in 1945, Jack had attained the rank of Flight Lieutenant. At the end of the war, Jack entered the family law firm, and in 1946 he was elected to the Wanganui City Council. The following year, he became the city's Deputy Mayor, where he remained until 1955. In 1954 he entered the arena of national politics, successfully contesting the Taranaki seat of Patea. For the rest of his life, Roy Jack remained a Member of Parliament. Although a keen debater, in 1967 he became Speaker of the House. Jack relished this position, and retained it until February 1972 when the new Prime Minister asked him to relinquish it and take over Justice.

Sir Roy Jack had not wanted to become a Minister of the Crown. The role of Speaker suited his self-image and temperament, but Marshall, facing a dearth of qualified lawyers in his caucus, decided Jack was the only realistic choice. It was only after some persuasion from the Prime Minister that Jack reluctantly agreed to give up the chair.

Marshall's selection was not popular with departmental staff. A conservative man, who believed that women should be excluded from certain male domains out of 'gallantry',[1] Jack once said in an interview, 'I like a sense of decorum, reasonable dignity, reasonable dress, a sense of tradition. I like a sense of humour and an avoidance of rudeness, acrimony, and vulgarity.'[2]

A more unlikely candidate for dealing with the coarse realities at Paremoremo would have been difficult to find. Unhappy as he was in work which he found distasteful, Jack's brief reign over Justice was characterized by inaction and a general lack of enthusiasm. 'He really wasn't interested in getting things done unless they were fairly easily done,' Missen told me. 'He just wouldn't fight for things at all . . . when Roy Jack was there you just had to battle along as best you could.'

Sir Roy Jack did not remain in his new office for long, because in November 1972 the National Government was ousted in a general election. However, in the preceding months, when the situation at Paremoremo again became critical, it was fortunate that despite a reluctant Minister, the prison had at its helm a man as capable as Jack Hobson.

J. HOBSON, SUPERINTENDENT

In the same month that Marshall became Prime Minister and Roy Jack took over Justice, Jack Hobson was made superintendent at Paremoremo. Aged forty-seven at the time of his appointment, Jack Hobson had arrived in New Zealand in 1946 after the close of the Second World War. The son of a Lancashire policeman, Hobson was born in 1925 and joined the Royal Navy in 1941. Posted to a troopship where he served as a gunner, the sixteen year old awaited the end of conflict then left the navy and emigrated to New Zealand.

In Wellington, Hobson enlisted with the army, again as a gunner, but having reached the rank of sergeant, became bored with the life of the peacetime soldier and resigned in 1948. A service occupation still appealed to the young migrant, so he travelled to Wellington, where he was interviewed and accepted into both the Prisons and Police Departments. Unable to decide between the two, he tossed a coin which came down heads, and set off to join the prisons service. To the end of his career, Hobson was to wonder whether his decision had been the right one.

Hobson reported for duty at Mt. Crawford prison where he passed his junior and principal officers exams before transferring to Waikune as deputy principal officer in 1952. Here he met superintendent Eddie Buckley, and Hobson served three years under Buckley until the latter was transferred to Mt. Eden. It was while he was at Waikune that Hobson's high promise as a prison officer was first recognized. For much of the time he ran the quarry, joining the inmates in work which was hard, hot, and demanding. The gaol was badly understaffed and living conditions were primitive. Hobson soon won a reputation for resourcefulness and an ability to get on with prisoners. The men liked Hobson. He was tall, lean, and strong, and they respected his open manner and easy confidence. A reputation for straightforwardness remained Hobson's hallmark throughout his career. It became known that he would not back down from a challenge and as a young officer he was reputed to have sometimes invited inmates to 'go around the back' with him, as an alternative to being put on report. Hobson himself describes these claims as exaggerated, but does admit to having fought with prisoners in self-defence, or after a personal affront.

Whatever the case, Hobson's competence and the respect he gained grew with his professional experience. He rose from the rank of deputy principal to principal officer, and by the time he reached the grade of chief, he had also become the first prison officer in New Zealand to have been granted a quarry manager's certificate. In 1961, Hobson's merits were endorsed when Robson, who had just succeeded to the Secretary's office, had him promoted superintendent of Waikune. As if to underline the event, the Justice Department then pledged to spend 100,000 pounds rebuilding the forty year old prison.

As superintendent, Hobson operated with efficiency and independence, without allowing bureaucracy to stand in the way of his initiative. Capitalizing on the adventurous mood then alive in Wellington, Hobson was able to establish a regime at Waikune which was notably innovative. Inmates gained a degree of freedom which would be considered extreme, even by today's standards. His methods soon became legendary, and by the time he was transferred, his reputation as an experimenter, humanitarian, and administrator was firmly established.

From 1 March 1964, a pilot first offenders' classification scheme was established at Wi Tako. The object of the scheme was to inaugurate a prototypical classification system and to analyse its effectiveness, by use of a staff of highly skilled experts. The existing superintendent was due to retire, and when his place was vacated early the following year, Hobson was made superintendent and chairman of the classification programme. The first annual report of the Wi Tako classification committee was presented to the penal group by Hobson on 20 May 1965. The whole experiment was considered a success, and in 1967, like Waikune before it, Hobson's prison at Wi Tako was completely rebuilt.

Hobson enjoyed working at Wi Tako. After 1967, it was a new institution and the types of inmate sent there were less problematical than elsewhere. Moreover, as an experimental, 'show' prison, Wi Tako had great scope for new ideas. This interested and satisfied the young superintendent. When the position at Mt. Eden was advertised late in 1968, Hobson made no move to apply for it. Not until he had been summoned to head office and cajoled by MacKay and Robson did Hobson agree to take control of the old prison.

With reluctance, Mt. Eden's new superintendent arrived for duty early in February 1969. Compared with the openness of Wi Tako, and in spite of the summer's warmth, his introduction was depressing. The gaol was old, grey, dirty, and overcrowded, and morale there was low. Parts of it had remained untouched since the fire of 1965, and in those sections which were still inhabited, security and discipline were stringent.

Hobson was reserved about Buckley's policies, and as soon as he arrived he set about the task of reorganization. The move was monitored at head office. At a penal group meeting some months later, it was observed with satisfaction that security at the institution had been relaxed, and the evening recreation programme extended.

Unlike Waikune and Wi Tako, the scope for improvement at Mt. Eden was limited. The gaol was archaic and solid, and its architecture prevented anything more than superficial change. More importantly, populations were rising again in 1969, and this put a stop to further progress. Overcrowding and understaffing again became endemic. Reopened in 1965 for no more than 210, Mt. Eden's population had risen to 289 in 1966, to 320 in 1969, and by March 1971 the gaol held around 375 inmates. Despite being overshadowed by interest in Paremoremo, recognition of the problem of overcrowding at Mt. Eden had been growing since at least August 1966. Alarm about it had been recorded as recently as December 1970,[3] January 1971,[4] and finally, on 15 March 1971.[5]

The 1971 riot took place five days later. On the afternoon of 20 March 1971, eighty-three prisoners at Mt. Eden refused to come in from the yard. After presenting a list of demands they began bombarding staff with missiles and lit fires with diesel. Hobson adopted a policy of containment, but when the rioters attempted to break out of the yard and into the prison buildings, he issued sidearms to his staff and ordered them to open fire, aiming low or over the men's heads. One inmate was grazed by a bullet, but when a second received a .38 slug in the thigh, the riot came abruptly to a halt. About eighteen rounds of ammunition had been expended and the rising had lasted less than four hours. Six officers had been injured, but damage to the institution was minimal.

At head office, reaction to Hobson's action was effusive. The humiliation of a 1965-style onslaught had been averted and even better, this time the whole incident had been conducted by the prison authorities. Although squads of armed police had stood by, Hobson had been in command throughout.

In the House of Representatives, Riddiford commended Hobson for exemplary conduct,[6] and noted that he and his officers had prevented a repetition of 1965, without the need of police assistance.[7] Missen recalled 1965 too, 'Certainly [the 1965 riot] was a different situation than what happened in that later episode when Jack Hobson was there . . . Jack Hobson got guns going and very quickly brought it to a halt.'

In recognition of his service during the riot, Hobson was awarded an MBE, and while he was already being groomed for the top job, it was his handling of the Mt. Eden riot which hastened the decision to send him on to Paremoremo. Hobson's decisiveness at Mt. Eden commended that brand of alchemy as right for Paremoremo as well. It was less than twelve months after the 1971 riot that Buckley was relieved and Hobson was installed in his place. The unusual means by which the vacancy was filled — contrary to official Public Service procedure — was similar to that employed when the Mt. Eden job had been advertised. Hobson was contacted by telephone and asked if he would take over Paremoremo. When he expressed reluctance to do so, the superintendent was summoned to Wellington and pressed into applying for the job. With this formality over, the responsibility of Paremoremo was his.

Hobson's First Weeks

Jack Hobson took over the afflicted institution at Albany on 7 February 1972, and almost from the day he arrived, the difference in his method was apparent. Unlike Buckley, who had always remained aloof, Hobson operated from its centre. Buckley's infrequent tours had been like State visits, with the superintendent accompanied wherever he went by a heavy retinue of subordinates. Hobson, on the other hand, made a habit of walking about the gaol on his own, talking casually to prisoners as he went.

The attitude of inmates to authority was openly antagonistic and their reaction to Hobson was hostile. A good example of how he approached the situation occurred soon after he had arrived, at a concert held in the gymnasium. This was attended by most of the inmates and a number of visitors as well. When the superintendent was introduced partway through the performance, prison visitor Paul Beachman described the men's reaction:

> There was the greatest uproar — whistling, cat-calls, boos, God knows what. From sitting in what had been a fairly warm sort of audience, suddenly there was this really hostile transformation. And I must admit I felt pretty scared. And Henare Dewes [the inmate master of ceremonies] looked worried too. And Hobson got up on stage and I thought, 'Oh, Christ! There's no way a guy can handle this situation!' I really had felt that the screws might step in and shut the concert down. But Hobson stood there and didn't bat an eyelid. Eventually he took the microphone and told a couple of funny stories which he told so well the guys actually laughed. Then he wandered off the stage and the whole thing was defused just like that.

Such was the easy manner with which Hobson deflected hostility. Aware that the embittered prisoners would do their best to unnerve him, Hobson had determined from the outset to meet their challenge head on. It was his power to deal with rejection which was the key to his later success.

Of all the men in Paremoremo, those in D. block were the toughest and most implacable. Buckley had been terrified of D. block and seldom ventured near it. Entering D. block was a daunting prospect for anybody, so Hobson thought hard about his approach. He knew that his first impression would be crucial. If fear or hesitancy were shown, his ability to confront the men in the future could be ruined. What was called for, he knew, was an air of casual yet confident authority, and he hoped that the same command which had silenced the uproar in the gymnasium could be summoned up again.

Hobson ordered the cells to be opened and he entered the deserted corridors of D. block, unaccompanied and alone. A stunned silence fell over the inhabitants as they watched the tall figure of the superintendent ambling down the wing towards them. Roberts and Rangi described their feelings in this way:

> S.R: Oh yeah! Oh fuck!
> D.R: Shit!
> S.R: What a surprise that man was!
> D.R: It sorta shocked us.
> S.R: Yeah, that was a shock.
> D.R: Walked in by himself.
> S.R: *That* was a shock!
> D.R: Opened all the grilles.
> S.R: On his own, boy, on his own.

D.R: Yeah, we had those grilles dividing us.
S.R: Me and Bernie was on one side, you fellahs were on the other side and ah — fucking what!
D.R: Hobson come straight through, we didn't — well I didn't know him — I don't know whether Sai recognized him or not. I think he didn't 'cause he said, 'Who the hell are you?'
S.R: Jesus, what!
D.R: He actually opened all the grilles.
S.R: Came in on his own, boy, this is no fucking crap.
D.R: He unlocked the grilles, me and Sai were together, just opened it, didn't lock it behind him, and he spoke to both of us, you know.
S.R: And that's the first fucking time anything like that had ever happened.
D.R: And it bloody stunned us, eh!
S.R: Oh, fucking what! Brought immediate shock, and then a sort of reluctant respect, you know. Just come in eh, and says, 'Oh yeah, how ya goin' Rangi?' I remember him saying that to me, boy, fuckin, 'How ya goin' Rangi.' . . . Come into my cell, man. Just come in on his own, eh. FUCK! But fuckin' no screws with him boy.
D.R: Psychological impact eh! Shattered everyone.

Having negotiated the formidable hurdle, Hobson wasted no time in liberalizing the institution which had been kept so rigid for so long.

Hobson ran Paremoremo according to the same philosophy as he had at Mt. Eden, and the show prison at Wi Tako before it. Most superintendents profess an 'open door' policy with regard to inmates. Few of them really have one. Hobson was an exception. He believed that the traditional, formal interview situation — which is still used in most institutions — was wrong, and he promptly got rid of it. He dispensed with interview application slips, and there was no 'mat' in front of his desk. There was never any saluting; no standing to attention. His door was always available for prisoners to knock on without prior appointment, to discuss any situation or problem. Usually these interviews took place without the presence of other staff. Prisoners were invited to take a seat and made to feel at ease when addressing the superintendent. The D. block men, who had no freedom of movement, were promised direct access to Hobson, his deputy, or a chief officer to discuss problems when and as they arose.

On Saturday mornings, Hobson continued his long-established practice of touring his gaol. Dressed casually, reading glasses on top of his head and a piece of paper in his pocket, Hobson would walk about the prison unattended, talking informally or passing the time of day with as many inmates as he could. In the first few months of 1972, a multitude of changes were put into effect. The divisional officer (DO) system, recommended by Powles and Sinclair, was installed. Now, every cell block had a permanent senior superior, who regularly consulted with the superintendent. These men, dressed in a civilian style uniform, made administration more accessible than at any previous time.

Officers no longer rotated their duties between blocks and were assigned to specific areas on a semi-permanent basis. In this way, staff and inmates became more closely acquainted and learned to understand one another's strengths, weaknesses, and idiosyncrasies. Relationships were thus able to stabilize to a degree not possible under the rotating roster. Use of forenames between staff and prisoners, which began as a result of this greater intimacy, was also permitted by the prison's new head.

In order to consolidate his power, by November, recognition of inmates' committees had ceased. Hobson felt they were now unnecessary, but more importantly, he saw them as a potential threat to discipline. The prison was still volatile, and he knew that inmates were more easily controlled individually than as a unified body. One of the keys to the new system was that the superintendent would have supreme command.

The block progression policy, wherein prisoners improved their privilege status by promotion towards C. block, was abandoned. From soon after Hobson's arrival, all blocks shared the same privileges except C. block, which was locked up an hour later at night to allow the kitchen crew their recreation time. In addition to this, the closed-block policy was relaxed, and in the evenings, groups of inmates could now visit other wings for the purposes of recreation and sport.

General conditions were eased. Correspondence restrictions were lifted, and inmates now could write and receive as much mail as they wanted. Postage within New Zealand was all paid for by the Government. Visiting was liberalized. The half-hour per week regulation was extended to one hour and the smoking ban was lifted from the visiting room. Permission was given for prisoners to wear personal items such as watches and rings. Reading restrictions were lifted, and generally any literature legally available on the street would be allowed into the gaol. This included 'girlie' magazines and law books, which hitherto had been prohibited.

Involvement in hobbies was encouraged. Special rooms in the basements of cell blocks were set aside for activities such as wood carving and basket weaving, and purchase of equipment and raw materials for them was made easy through two, full-time activities officers. A Māori culture club was formed, weight-lifting began in the gymnasium, and the debating programme which Beachman and McKinnon had started was extended so that outside teams could visit more often.

There was a deliberate policy of community participation. Sports teams began coming in regularly. Education courses started. An Alcoholics Anonymous group was established, and the Jaycees, a voluntary service association, became involved. Concert parties entered the gaol and the Prisoners Aid and Rehabilitation Society (PARS) was permitted to take a more active role in contacting inmates. In the block yards, volleyball and basketball courts were marked out, nets and other equipment were provided, and musical instruments were allowed into the wings. Later in the year an inmate-produced newspaper was authorized. With the assistance of a typewriter and a duplicator donated from the outside, the first edition of *The Paremoremo News* (later *The Parry! News*) was issued in October 1972.

In essence, then, there was a complete metamorphosis in administrative philosophy at the prison after January 1972, and with it the atmosphere of the gaol brightened encouragingly. However, while the improved routine in the standard blocks was marked, the deepest impact was felt in D. block.

Under Buckley, it will be recalled, the number of men classified as 'intractable' was high and steady. From the time the prison opened there had been a constant population of about twenty in D. block, which was suddenly reduced to fourteen when the announcement was made of a pending investigation. Tight conditions had been the cause of considerable friction.

Hobson was against the idea of a segregation block, and wanted the unit emptied. By 6 March 1972, therefore, within a month of his taking over, Hobson had had the number of men classified to the top security wing reduced from fourteen to six. By June, it had been cut to three. Although he felt that eventually these too would be moved, Hobson believed that to release all the men together would be a threat to discipline. Instead, he decided to discharge them gradually, over a period of several months. However, because their sense of group commitment was so strong most of the men refused to move until the block was ready to be emptied completely. Thus, in June 1972, there were still eight prisoners in D. block.

In the meantime conditions were liberalized. In fact, the change in administrative attitude towards D. block was even greater than elsewhere. One of the first initiatives Hobson took was to allow free movement and association. Although staff insisted that the three to one staff:inmate ratio be preserved whenever segregation men were being moved, solitary confinement ceased, and there was mixing between all D. block members in work and recreation. Activities in the block were broadened. Guitars and tape recorders were permitted and Hobson arranged for the men to have use of the gymnasium from 6.00 to 8.00 on Wednesday evenings. D. block was also allowed to use the library once a week, and all films shown to the larger population were now re-screened especially for the intractables. Attendance at concerts was allowed and in the first month after Hobson took over, there were two concerts given at the institution, both attended by D. block.

The degree of isolation of the intractables was reduced. Discussion groups and education classes were held and the education officer began devoting half of his time to these men alone. PARS officials were now given entry to the unit and teams from C. block were sometimes also allowed in for recreational purposes. Visiting privileges were broadened. Instead of receiving their visits discretely from the rest of the institution and/or in security booths, the D. block men were now permitted to see their visitors on Saturdays in the same open visiting room as other inmates.

The manner in which Hobson so quickly relaxed conditions at the prison was applauded in many quarters. The Department of Justice surveyed developments with satisfaction. The watchful Project Paremoremo was pleased with events in these early weeks, and if no news is good news, it was auspicious that press releases were few in the autumn of 1972.

To begin with, the prison was quiet and the new routine ran smoothly. It may have seemed that the simple solution to the long-standing problems at Paremoremo had been the humane treatment of its inhabitants. However, the period of calm — the honeymoon period of 1972 — was short-lived, and by May, problematical elements at the institution had again began to stir.

Early Difficulties

It might have been thought unlikely that Hobson, after creating a programme more liberal than any known in New Zealand's maximum security since 1960, would encounter serious opposition from his charges. Despotism had spelt the downfall of his predecessor. Under the relaxed conditions of 1972, certain problems may have been expected, certainly, but organized opposition from inmates was probably not one of them.

Among students of revolutionary change, it has been observed that likelihood of insurgency frequently bears no relationship to deprivation. Leites and Wolf point out a number of cases — anti-colonial resistance in Vietnam, the Cuban revolution, and the riots of Watts and Detroit in the mid-1960s, for example, — where upheaval was sparked in localities which were relatively better off than those of inactive areas surrounding them.[8] Brinton[9] and Hoffer[10] give further instances of risings which took place not during times of recession, but of economic upturn and increasing hope. In the 1950s, the Mau Mau uprising in Kenya occurred at a time of prosperity and contentment.[11] It is the case that liberalizing regimes are frequently less stable than the oppressive ones they replace. A classic example of this is the French Revolution, which took place at a time of fiscal recovery and politico-social reform.[12]

At Oahu prison, Hawaii, an innovative era of post-war revision led to revolt and anarchy.[13] Chaos in the State prisons of Massachusetts in 1973 was blamed on a similar process[14] and adminstrative change — albeit beneficial — has been cited as a principal cause of the riots which swept the United States in the early fifties.[15] This was approximately the situation that Hobson faced at Paremoremo in 1972. With expectations lowered by years of deprivation under Buckley, inmate reaction had initially amounted to an angry kicking-back at antagonism. After conditions were relaxed, the face of disobedience changed, and now arose principally from abuse of, or attempts to gain extensions to, newly won privileges.

There are two aspects to the problems encountered by Hobson at this time. The first of these involved the inmates, the second had to do with staff. Hobson took over an institution which had for nearly four years struggled under the yokes of asceticism and disciplinarianism. Resentful and suspicious of authority, inmates were in no hurry to recognize a structure whose job, after all, was their continued confinement. The bulk of men at Paremoremo were totally unfamiliar with the concept of self-determination. In the past, custodial authorities had monitored their lives closely and conservatively. Having had little opportunity for personal choice, prisoners had become used to taking what they could get.

In a few short weeks of 1972, this situation changed completely. Inmates welcomed Hobson's many improvements, but they were not long satisfied with their newly gained liberties. Having quickly accommodated their improved situation, their levels of expectation increased. The pattern of what happened at Paremoremo is similar to that of Oahu:

> [The] extension of privileges . . . produced an atmosphere of good feeling and

relaxation in the community and the lowest rate of internal disorder in its recent history. That condition was shortlived, however, for it led to a stage, familiar in the government of any suppressed population, in which the granting of one privilege leads to redoubled demands.[16]

In both cases, destruction of old standards necessitated the setting of a new balance, and it was in the re-establishment of this balance that the first conflicts arose.

The second problem that Hobson faced in this period involved his staff. Like the prisoners, staff at Paremoremo had been used to strict orthodoxy. They had fixed opinions about the way a maximum security prison should be run, and many had ingrained feelings of antipathy towards criminals. Such sentiments had been aggravated by prisoners, with whom there had been constant conflict. By 1972, a great many of Paremoremo's officers harboured private but impelling desires for revenge and hoped that with Buckley gone, their opportunity for retribution would arrive. The officers were to be disappointed. For as we have seen, instead of increasing disciplinary powers, Hobson relaxed them. Rather than adding to the authority of the rank and file, Hobson centralized executive procedure in the hands of senior officials.

The more militant elements of the Paremoremo prison officers' subgroup soon became disaffected with the new superintendent. Apart from disagreeing with what they felt was a 'soft' policy on criminals, many were put out that their opportunity for vindication had been foiled. Conservative attitudes towards the treatment of convicts and opposition to reformative change are common problems faced by liberalizing penal regimes.[17] As a rule, penal employees lean politically to the right, and at Paremoremo, staff and their wives have tended to poll against parties and candidates professing an interest in penal reform.[18] Furthermore, in the face of progressive change, custodial officials are often associated with activity that undermines the credibility of adminstrations with which they disagree.[19]

This is what happened at Paremoremo. Here, the more entrenched members of the prison officers' subgroup embarked on a tacit but deliberate campaign to subvert the administration of Jack Hobson. Although Hobson, himself, is noncommital on this point, the fact was commented upon by Powles and Sinclair,[20] and is generally recognized by inmates and staff. Thus were difficulties at the institution in mid-1972 compounded. Serious rule infractions were ignored, managerial directives were deliberately misinterpreted or misapplied, and there was purposeful aggravation of prisoners in order to encourage violent responses. If problems with staff were not enough, it was at that very crucial point that Dean Wickliffe was returned to prison.

Wickliffe, whose contribution to pressure for an inquiry had been so great in October-November 1971, had in December moved out of the commune in which he had been staying.[21] From there he had moved to a brothel with his girlfriend, and having soon after this stabbed a man in a night-club, he bought a revolver for his own protection. He now recommenced doing burglaries. Wickliffe's involvement in crime and violence increased in the early months of 1972. He committed a number of thefts, but as before, his luck was short-lived. In the middle of March, during the course of a robbery which went awry, Wickliffe shot and killed the son of a jewellery shop proprietor. Arrested at the scene, he was tried for murder and sentenced to life imprisonment less than two months later.

It was in May 1972, coincident with the return of Wickliffe to Paremoremo, that 'a certain deterioration in attitude and conduct' became apparent.[22] B. and C. blocks were comparatively quiet, but as before it was in A. block that the principal trouble emerged. In June, discipline slipped even further. Prisoners now began to refuse to strip-search before seeing their visitors, drug-taking began, and a concoction of alcoholic beverages started to become more evident.

Although Hobson blamed the genesis of A. block's problems on the presence of Wickliffe, in D. block trouble was also brewing, and it had begun at the same time as that of the standard section. For it was also in May 1972 that D. block's positive response to the reforms stopped, and a noticeable change in behaviour took place. Threats of arson began in order to gain even more privileges, and existing concessions were misused. In May 1972, two D. block men were disciplined for using obscene language and for minor assaults on staff during visiting time. Other D. block prisoners reacted to these incidents by refusing to be locked up. As in the standard blocks, the level of disorder increased in the first weeks of June. Staff were abused, more fires were lit, property was damaged. Threats and general disobedience continued.

It was at this time that five of the remaining eight D. block men, who had been reclassified to the standard blocks, refused to move unless the whole unit was emptied. A concerted effort now began among them to ostracize all officers who worked in the unit. They refused to speak to screws about any matter at all and would only communicate with their DO or officers holding the rank of chief and above. As the situation grew worse and showed no sign of improvement, a rapid solution was sought. Without it, it seemed, there would be no end to crisis.

THE 1972 BATON CHARGE

Hobson had not been in charge of Paremoremo for five months before his first major confrontation with prisoners took place. Although it began in D. block, throughout the episode the rest of the institution was involved. After May 1972, in spite of the fact that disobedience was spreading, participation was never more than fractional. There was no doubt that continued disruption would jeopardize newly won freedoms and in view of this, a number of prisoners became concerned about the direction of events. Thus it was that early in June 1972 one of the leading members of B. block — a man whose nickname was 'Rabbit' — posted a notice offering a five dollar reward to anybody who would tell him the identity of a person who had lit a fire in the block the previous day.

Although perhaps true that this gesture was made in the inmates' wider interests, there is no doubt either that it constituted a gross departure from the most sacred tenet of the inmate social code: that against collaboration. Unqualified commitment to convict interests was paramount at Paremoremo, and Rabbit's gesture was taken as one of supreme treachery. Numerous inmates were offended by it, which was worsened by the fact that Rabbit and a supporter called R.A. were also active in opposing Wickliffe's efforts to establish a prisoner's union. None were more incensed by this than the most deprived and highly committed of all, the remaining eight prisoners in D. block. Seeing themselves as the vanguard of prison morality, they felt compelled to take some action.

D. block was still largely separated from the rest of the community, and the only opportunity for vindication lay in the visiting room. Thus it happened that on Saturday 24 June, about two weeks after the note had been posted, Saifiti and Roberts, accompanied by four more of the D. block members, were escorted out of the unit for a visit. Although Roberts and Saifiti were sober, at least one of the other four was drunk,[23] and it was only after some discussion with senior staff that the visit was allowed to proceed. The next hour or so was uneventful, but towards 4.00 p.m., as men began filing out of the room, Roberts and Saifiti followed Rabbit and R.A. into the strip area next door and confronted them. Roberts' understanding was that the two would be attacked immediately and without debate. However, when Saifiti began to argue with them, the offenders apologized and recanted, and the matter was peacefully resolved. The men were then escorted back to their blocks without incident.[24]

The following day, 25 June, a concert was scheduled in the prison gymnasium. Outside guests were billed, including Bunny Walters, a well-known New Zealand entertainer. This was an event of major significance to inmates, and as noted, on two occasions since February, D. block had attended concerts. In this instance they had no reason to assume that the custom would not continue. On the morning of 25 June the men in D. block were again drinking. Mysteriously, duty staff chose to ignore it. However, when Hobson and the chief suddenly appeared in the block on a brief tour of inspection, it was obvious that some of the men were drunk. Others were drugged as well.[25] The two officials confiscated the beverage immediately and poured it down a toilet.

Hobson and the prisoners are divided on what happened next. Hobson insists that at no stage was any indication given that D. block would be attending the 25 June concert. Inmates just assumed they would go because they had been allowed to before, he says. Roberts says that the men had been told they could go, but that this was later changed — possibly after the home brew was discovered. Whatever the case, and whatever the reason for it, the D. block inmates believed they were going to the concert, and when informed that they would be excluded, no reason for the decision was given. The men all swear that constant attempts were made during that day to see the superintendent. He refused to answer their calls until late in the day, after the concert was over.[26] There are three reasons why Hobson, on this occasion, might have chosen to depart from precedent.

First of all, the men were drunk, and it is possible that Hobson felt, especially after the incident of the day before, that the presence of D. block would be a threat to security and discipline. Secondly, although it is not widely known, there was a plot afoot in D. block at the time for one of the men to settle a score with a standard block prisoner, with whom he had had a longstanding feud. It is possible that the administration somehow got wind of this and decided to keep D. block locked up for fear of violence at the concert.

Hobson himself refuses to be drawn on either of these points. His explanation is the third possibility, that he had decided arbitrarily not to admit D. block on this occasion. Although he is vague as to why, he commented to me that at previous concerts the behaviour of D. block had been unsatisfactory. In the event, realizing that they would be excluded and still seeking a clear explanation for it, six of the eight D. block prisoners declined to be locked up that lunch-time. Lifers Gillies and Rangi remained uninvolved and were later transferred to the standard blocks. Meanwhile the six recalcitrants — four on the east side of the lower landing and two on the west — refused to move. Four of them armed themselves with broom handles and broken chair legs, to defend against any attempt at forced entry.

At 4.30 p.m., when most of the institution was locked up for the night, Hobson spoke to the prisoners and asked them to give in. They refused. Hobson then told them that unless they went back to their cells they would be removed by force. Having received no positive response, he then made plans for an assault. At 6.00 p.m. at least eight prison officers were briefed by the superintendent.[27] Issued with batons, shields, and protective helmets, and led by staff training officer Bob Severne,[28] they were ordered to clear prisoners from the D. block corridor.

Paul Morrison, who worked in the kitchen, was unaware of the situation at the time, and he described what he saw when he traversed the 'airstrip' to C. block (opposite D. block) soon after 6.00:

> As I was going through to cross over to C. block, a whole herd of screws came belting down the airstrip, pushing each other out of the bloody way, fighting to be first into the sally port to get in there. There was heaps of them and I was just so shocked. And I remember walking zombie-like into the block . . . my whole body was vibrating from seeing, like a herd of elephants on a stampede, throwing each other out of the way, fighting to get into the sally port. You know, I just couldn't believe it.

Hobson insists that Severne's squad was disciplined and controlled and that he personally supervised its activities. Evidence of what happened in D. block that evening suggests differently.

To begin with, the squad entered the west side, occupied by inmates Saifiti, Yorker, Barbarich, and Balsom. Three of these made an attempt to defend themselves, but were rapidly overcome. While this went on, Roberts and Rosel listened apprehensively from the other side. 'You could actually hear them pounding away', Roberts said. 'And I tell you it was a situation where it would have broken you, you know, if you weren't psyched up to it. It would have bloody shattered you . . . just to hear those bloody sticks going, laying into them.' Unnerved at the sound, Rosel began to have doubts and wanted to give in, but when Roberts threatened him, Rosel decided he had no choice, and remained sitting where he was.

With the first four prisoners despatched, the riot squad entered the east side. As the squad closed, Roberts tried to put up a fight. The effort was hopeless. In a few seconds he was felled, then he regained his feet only to be felled again. Kicked and beaten where he lay, he was finally thrown into an empty cell, still sensible. Rosel was less fortunate and received the most serious injuries of all. Although he was unarmed and did not attempt to defend himself, Rosel was beaten severely before being dragged into a cell, deeply unconscious and soaked in blood.

With the six men thus despatched, one by one they were either carried or escorted to the punishment unit, where attacks on some of them continued. All were injured and three staff received minor cuts and bruises during the fracas. It was not until the following day that the full significance of the evening's events appeared to the population of the standard blocks. A. block was already aware of the situation and Morrison's account had alerted C. block. In A. block that Sunday afternoon, a meeting had been organized to discuss what action should be taken about D. block's exclusion from the concert. A joint representation to the administration had been planned for the following day, but this became redundant when the details of the attack were known.

Rangi, who had surrendered partly so that the standard blocks could be informed of D. block's action, was especially distressed by the news. Plagued with self-recrimination at having deserted his friends, he committed himself to seeking redress on their behalf. Individual meetings were organized in each block to discuss possible courses of action and the inmate committees were revived. It was concluded that the open sections would go to work as usual, but that during the morning, representatives from the three blocks would discuss the matter with the superintendent. Hobson agreed and a meeting with delegates was held in his office in the morning of 26 June.

Hobson attempted to explain his actions and was questioned intensively by the representatives. Inmates found him defensive and at the conclusion of the meeting, decided that a number of queries had not been answered to their satisfaction. The most important of these were:

1. Why had no explanation been given as to the reason for D. block's exclusion from the concert?
2. Why had force been necessary when the strikers were already in a secure part of the gaol?
3. To what extent had batons been used in the forcible removal of inmates?
4. How serious were the injuries sustained by prisoners during the course of this removal?[29]

The most aggressive of the committees was in A. block. This unit had a tradition of militancy and Bower was still a member, as were Wickliffe, Rangi, and Waha Saifiti, Atenai's younger brother. The disruptive capacity of this group was considerable and it was largely in response to it that Hobson agreed to allow a meeting of all standard block inmates in the gymnasium that lunchtime. It was also agreed that a number of inmate-selected conciliators would be invited to investigate. As a result, criminal lawyers, Peter Williams, Kevin Ryan, and Barry Littlewood; Māori leader Dr Pat Hohepa; and Catholic priest Leo Downey were requested to visit the institution that afternoon. They were joined by priest John Kebble who later arrived unexpectedly.

During the early afternoon of 26 June, as planned, a meeting of prisoners was held in the gymnasium. When this was over, the men returned to their blocks while the mediators conferred with the superintendent. Tension in A. block was running high, and a plot was hatched between three inmates to stab an officer to death in reprisal. John Bower was an unwilling conspirator in the plan and, hoping to redirect the others' anger, began smashing the basement windows with billiard balls. The remaining prisoners then joined in, burning and wrecking whatever Government property they could find. By the time their energies were spent, the murder plot had been forgotten.

At 5.30 p.m. the mediators returned and tension rose once more. After reporting to the superintendent's office, the six visitors divided into three groups, each taking a separate cell block. Considerable force on behalf of staff did appear to have been exercised they said, and all six of the prisoners had been injured. Four of them had sustained serious scalp lacerations requiring sutures, and Rosel, who needed eighteen stitches to the back of his head, had been removed to the prison hospital. At this news, destruction in A. block began afresh. Floors were flooded, locks were broken, and inmates refused to go to their cells. The activity continued for an hour or so; however, enthusiasm rapidly dissipated, and after some judicious talking from the mediators, prisoners eventually submitted at about 8.30 p.m.

Tension in the gaol was still at a dangerous level. Meetings between the blocks were planned for the following day to decide whether further action should be taken, and it was rumoured, in the press at least, that hostages would be seized. There were also indications that further fires and another sabotage campaign were planned in the workshops when routine returned to normal.

A conference of senior prison officers was called and Hobson ordered a general lock-up. During the next two days a systematic search of the institution took place, revealing an assortment of prohibited implements, including eight knives, a club, and a length of plaited rope. Although recovery of such an array would be unremarkable in any gaol, in June 1972 the finds were used to show that the situation was serious, and to support the measures Hobson had taken. During the search, which involved stripping every cell and every prisoner, fifteen men were taken from A. and B. blocks and isolated in the classification unit. This section, which had closed down several months before, was now re-opened.

Wickliffe was singled out as the principal trouble maker. Beaten, abused, and taunted in an apparent bid to make him retaliate, he was returned to segregation, where he was to remain until his release in July 1987.

With this, stability returned, but the gaol was never the same again. C. block was unlocked on 28 June and on 29 June A. and B. blocks returned to normal. On 28 June there was trouble in classification when fire broke out, but this was only minor, and by September, all of the classification men had been replaced among the main population. In D. block, conditions reverted to what they had been before Buckley retired. Still smarting from the June experiences, Hobson had the liberalization experiment stopped. He changed his mind over D. block policy, and now believed segregation to be an indispensable component of maximum security administration. Moreover, he saw dangers in allowing D. block prisoners to exist in anything but complete isolation from the rest of the prison community. From that point on, no standard block prisoner would be allowed to enter the unit and D. block men would be kept apart from other inmates at all times. Visits would be held separately, there would be no attendance at concerts, nor would D. block have access to the library or gymnasium. When being moved about the prison — such as for medical purposes or visits — D. block men would still be escorted by three officers and they would not be permitted to converse with any other inmates. They would be locked up for twenty-three hours a day.

Although routine was soon relaxed a little, (lock-up, for example, was later reduced to eighteen to twenty hours), conditions have remained essentially unchanged. Today, only limited work is provided, few hobbies are permitted, members of the unit associate for limited periods of time and only in very small groups. Prisoners have little incentive to work or maintain institutional grooming standards. Life is characterized principally by its high security, long hours of lock-up, solitude, and inactivity.

In the latter half of 1972 there were between seven and eight prisoners in D. block. Today the figure fluctuates, but tends to exceed around twenty-five. Length of residence ranges from a few weeks or months, to many years. D. block remains as one of the most regressive and problematical aspects of an otherwise fairly liberal penal programme.

Significance of the Baton Charge

The 1972 batoning incident had a number of wider implications for Paremoremo. It was Hobson's *auto-da-fé* and in spite of inevitable pockets of criticism, he emerged from it well. The worse aspects of the incident were glossed over by the media, and the prisoners' side was never fully told. The second Ombudsman's report and an LL B (Hons) research essay followed a similar line and appeared to exculpate the superintendent entirely.[30] A 50,000 dollar lawsuit brought by prisoners in October 1973 failed. However Hobson, more than anybody, knew that there was more to the incident than had met the public eye, and it taught him some valuable lessons.

First of all, Hobson learned that liberalizing the conditions of men locked in maximum custody had pronounced practical limits. From then on, no relaxation of security would take place before all possible implications had been weighed. Secondly, it gave him good cause to question the judiciousness and loyalty of some of his staff. D. block personnel had ignored the fact that inmates were drinking on the morning of 25 June, and had the matter been attended to earlier it is possible that trouble could have been avoided. Moreover, although Hobson has refused to acknowledge it, it is apparent that sticks were deployed over-zealously and that some officers used the incident to carry out personal reprisals against prisoners. Whether the worst injuries were sustained during the initial charge or, as Hobson maintains, later in the punishment block, it was evident that the force used was considerably stronger than it needed to be.

Reflecting on the matter, Hobson must have realized that his overall approach to the problem contained mistakes. When he had arrived at Paremoremo, Hobson had promised D. block direct access to senior staff in times of trouble. Yet on Sunday 25 June, after confiscating the home brew in the morning, the superintendent ignored repeated calls and did not see the prisoners again until 4.30 that afternoon. Had he appeared earlier and explained his action, it is possible (though in my view, not likely) that a crisis would have been avoided.

Finally, Hobson knows that his decision to remove the strikers by force was unnecessary. The men were locked in their wings, in a small, bare, and totally secure area. A great deal of trouble could have been avoided simply by leaving them where they were, and waiting for a voluntary surrender. On subsequent occasions, when similar situations arose, he told me, this is what he did.

The inmates' opinion of the incident is varied. Although their public response was recriminatory, Hobson claims that Roberts said to him afterward that it had been a 'damn good fight'. Yorker relates the incident proudly, and told me how, before being felled, he had managed to stab a screw in the face with a trowel. 'Those were the days, all right,' he said, wistfully. Lawyer Peter Williams, who had attended A. block as a mediator, indicated in a newspaper column he wrote that the injured men had expressed little resentment against administration, 'and in several cases stated quite frankly that they had received the violence they expected and deserved'.[31] Other mediators agreed and it was reported that inmates had the highest regard for the superintendent, that he was a man of great mana among them, and that the majority of prisoners were pleased with progress at the institution.[32] Roberts told me that as far as he was concerned, it was the inmates themselves who had been responsible for the loss of privileges in D. block. Although Rangi disagrees, both continue to hold Hobson in the highest regard. Nearly all of the ex-prisoners I have spoken to feel the same way. Danny Yorker, one of the more seriously injured victims of the June baton charge, said of the man who ordered it, 'But I'll tell you what . . . the best super I ever met was fucking Hobson. He's a cunt, and yet I know I'd talk to him.'

Long after the incident had ended, the matter was still not forgotten. The writ against Hobson for damages caused by the batoning was pending, and a flush of complaints from Project Paremoremo appeared in the columns of Auckland newspapers. There was an official report on the matter and Sir

Jack defended Hobson's actions, insisting that what had occurred was no fault of administration. In the budget debate a few weeks later, the Minister of Justice devoted nearly all of his speech to praising Paremoremo in the context of progressive penal policy,[33] and in his annual report for 1973, Missen pointed out that the work of the superintendent had been endorsed. Loss of privileges in D. block, he said, had been found to have been vitiated by the prisoners themselves.[34]

So in the end, Hobson's credibility and his hold on the institution was strengthened rather than strained by the 1972 attack. Staff morale had been raised, executive confidence had been restored, and the respect of inmates retained. Although a liberal man, Hobson had shown himself to be a firm one. As at Mt. Eden, he had proven capable of taking decisive measures to secure his control and command.

NOTES

[1] *New Zealand Women's Weekly* 24 June 1968.
[2] *EP* 27 December 1977.
[3] *NZH* 30 December 1970.
[4] *NZH* 28 January 1971.
[5] *EP* 16 March 1971.
[6] NZPD vol. 371 (1971) 635.
[7] *NZH* 8 April 1971.
[8] Leites & Wolf (1970) 17–18.
[9] Brinton (1952) 31–3.
[10] Hoffer (1951) 27–8.
[11] Majdalany (1962) 56.
[12] de Toqueville (1955) 158–69, 177, 189–93.
[13] McCleery (1969) 127–8.
[14] *NZH* 23 June 1973.
[15] Schrag (1960) 138.
[16] McCleery (1961b) 171.
[17] A brief discussion of literature in this area appears in Boyle (1977) 245–6, 259–60; Cressey (1975) 109; Hawkins (1976) 167–73; McCleery (1969) 127–8, (1961b) 155, 172, 187, (1975) 69, 75–6; Morris & Morris (1963) 95–7; Thomas (1972) 189–90, 220.
[18] Election results for this period can be found in *App.J.* H-33 (1970), E-9 (1973), E-9 (1976), E-9 (1979), and E-9 (1982). See also Levine & Robinson (1976) 191–2 and Levine (1979) 191–8. Don McKinnon, MP, told me that although Paremoremo's officers traditionally voted National, after he became National MP for the area in 1978, National's returns there dropped noticeably. This he attributed to his long-standing association with the security inmates' debating programme. In chapter 9, we shall see that Finlay's attempts at penal reform made him unpopular with prison staff also.
[19] American examples are furnished by Cressey (1973) 142 and McCleery (1969) 128, 138; (1975) 69–70. At Carrington psychiatric hospital in Auckland a similar problem arose when a progressive medical superintendent took over in 1973 (Dr F. McDonald, pers. comm.). Similar problems have been reported at Oakley (forensic) psychiatric hospital after a change in administrative direction there in 1983 (*AS* 22 May 1984).
[20] *Report on Auckland Prison* (1973) 21.
[21] Wickliffe had been staying with philosopher-poet James K. Baxter. J. K. Baxter

was the son of conscientious objector and author Archibald Baxter, whose work I have referred to earlier. J. K. Baxter wrote a poem about the hanging of Eddie Te Whiu, *A Rope for Harry Fat* (1960).
22 *Report on Auckland Prison* (1973) 34.
23 The most noticeably affected was Danny Yorker.
24 According to the *Report on Auckland Prison* (1973) 36, all of the men had been drinking and had refused to be locked up at lunch-time. The inmates I spoke to denied this. Roberts was adamant that he and Saifiti had deliberately refrained from drinking that day, because of the confrontation that they knew was impending.
25 Some of the men had earlier saved their medication of tranquillizers and sleeping draft for the concert (D. Roberts and S. Rangi pers. comm.).
26 *Report on Auckland Prison* (1973) 36.
27 Ibid., 37. *The New Zealand Herald* (26 June 1972) reported that there were ten officers involved, and inmate estimates are considerably higher. However, in view of the narrowness of the D. block corridor, it is unlikely that more than ten officers in riot gear would have been of any use.
28 When I contacted him, Severne refused to be interviewed.
29 Paremoremo Inmates' Association Press Release (1972).
30 *Report on Auckland Prison* (1973) 34-7; Weiss (1973) 80-2.
31 *Sunday News* 2 July 1972.
32 *NZH* 28 June 1972.
33 NZPD vol. 379 (1972) 1102-3; *NZH* 15 July 1972.
34 *App.J.* E-5 (1973) 9.

18: CONFLICT AND CHANGE 1972

It was not until June 1972 that the advice of the Ombudsman was taken, and Atenai Saifiti was pardoned for the assault on prison officer Smith, two years before. However, while it was acknowledged that the conviction had been wrong, the Government would still not concede that the behaviour of those who had supported Saifiti could be excused. Questioned by Dr Finlay about it, Jack maintained that irrespective of whether or not the grievance had been of substance, the men's actions had been illegal, the convictions were correct, and there was no reason to adjust their sentences.[1]

TROUBLE RETURNS

After the disturbances in June, the population of D. block was eight, then seven men.[2] Although it was relaxed slightly, the spartan life-style in the unit remained, and continuous pressure from outside to have Hobson review the situation failed.

On 1 September a hunger strike took place in D. block, with all seven prisoners attempting to force a restoration of privileges. The strike was accompanied by external pressure and wide media coverage, but the attempt had no hope of success. Certain of executive support, Hobson refused to compromise, and less than two weeks after it had started, the strike was called off. The only significant effect of the action was a deterioration in the relationship between the superintendent and the organizers of Project Paremoremo. One Auckland reporter had been tipped off about the strike two days before it had even begun,[3] and Hobson openly speculated that the whole affair had been engineered from outside.[4]

In the meantime the situation in the standard blocks, which had improved slightly after the June lock-up, again began to worsen. By early September, the last of the men segregated in classification were being returned to the open division, and it was from this time that inmate demands and general disobedience increased. On 13 September a petition signed by almost every prisoner was presented to the superintendent. The men wanted permission to grow beards and moustaches, and some were refusing to cut their hair at all. Hobson deferred his decision until after the superintendents' conference later that month, but when he got back from it, he went on two weeks leave. It was while Hobson was away that licence rose to a dangerous level.

In September 1972 sartorial dress suddenly became fashionable in Paremoremo. Tailored in the gaol's workshops, waistcoats, tight pants, and leather vests started to appear, and one man is even reported to have sported a pin-striped suit. Out of desire to display this finery to their visitors, inmates began to object to the wearing of the compulsory visiting uniform[5] and when dress regulations were enforced they retaliated. On 7 October forty sets of visiting attire were torn to pieces. Further attempts by staff to enforce prison rules resulted in a general slowing down of work, and tension again rose critically. At a loss for what else to do S. G. Ward,[6] the deputy superintendent, sought permission to take immediate action. However, the Inspector of Prisons,[7] who was visiting the institution at the time, counselled Ward to await the return of the permanent superintendent.

It was obvious to everyone that a remedy would have to be found quickly, and this feeling was given urgency when Hobson returned to discover a note lying in a prison mailbox. Written by an anonymous hand, its message read:

> There will be fires and sabotage from today onwards until your stupid rules are withdrawn or eased some. If not done we will get one of your screws. Take heed, we mean it, and will keep going indefinitely.

Whether the note was genuine cannot be ascertained, but irrespective, a campaign of disruption now began in the finishing, boot, and machine shops. Damage was widespread and some industries were forced to close down altogether. General conduct deteriorated also and there was an increase in drug abuse and drunkenness in the gaol.

The problems which Hobson had been having with his staff before June now returned. Disharmony developed between industrial instructors and basic grade officers. Instructors, alone in workshops with prisoners, were under pressure to turn a blind eye to certain of their workers' activities. If discipline became too rigid, production would stop. Machinery and materials would be ruined. Supervisors who behaved too fastidiously were liable to ostracism, ridicule, abuse, and assault from the men they had to supervise.[8] For this reason activities like gambling, the altering of prison clothing, and manufacture of certain prohibited articles were broadly tolerated within the workshop context. Disciplinary staff, on the other hand, whose job it was to confiscate such material once it had left the shops, felt betrayed by the permissiveness of industrial personnel. Some decided that articles manufactured under an instructors' eye ought not to be interfered with, and as a consequence, the quantity of illegal material in the gaol accumulated.

A second factor which eroded discipline at this time was a continuation of what had begun earlier: numbers of staff, disgruntled at what they saw as excessive liberalism, attempted to sabotage it by abdicating their responsibilities. Not only were the prisoners' piquant fashions tolerated by this faction, hygiene regulations were overlooked, the presence of alcohol was ignored, and general misbehaviour was allowed to go unchecked. Nowhere was this neglect more noticeable than in that most public of places, the prison visiting room.

In October, between the time the visiting clothes were destroyed and their eventual replacement some weeks later, unshaven and unkempt prisoners, sporting a variety of dress, were on display to the general public as they entered the visiting area each Saturday afternoon. Any who were drugged or drunk were not likely to be apprehended. Indiscreet sexual activity was disregarded and on one occasion a prisoner somehow managed to defecate on the floor of the strip room without being observed. Even inmates were taken aback by the apparent change in policy. One man who served time during this period said of it:

> You could do anything you liked. You could wander around and have long hair. Do what you liked. You know, every cunt was pissed. And the screws wouldn't enforce any laws at all. You'd go up to the visiting room and guys were fucking their visitors, shitting in the corner. Drinking alcohol — every bastard was drinking alcohol out of bottles and draws. Pissed — it was chaotic. The screws didn't give a fuck.

As the level of anarchy grew, staff became increasingly restless. Some said they were concerned with security and order, others with their personal safety, and pressure began for a solution.

A clandestinely produced, typed release from an inmate in November 1972 claims that in late October or early November a meeting was held among officers at Paremoremo. The result was a list of demands being presented to the superintendent, and unless they were met industrial action, timed directly to precede the general election of 25 November, would be taken. Although the exact nature of the demands was unknown, it was believed they had to do with a tightening of discipline in the gaol.

As November proceeded it became apparent even to inmates that staff were planning some form of action before Friday 17 November. Bargaining from a vantage of strength with a Government whose grip was failing, it was plain that results would have to be immediate. Whether the steps which were eventually taken had been already contemplated by this stage, or whether Hobson simply seized his opportunity when it arrived is difficult to know, because on 12 November an incident occurred which made the question of security compelling. After it occurred, urgent pre-emptive action was inevitable.

The Escape Attempt and the November Lock-up

Mervyn Anthony Rich was only a slightly built individual, but at twenty-nine years of age, he had stature among the men at Paremoremo. Intelligent, capable, and determined, Rich had originally been sentenced to twenty-one months for possession of explosives, but this had been increased to seven years after he escaped from custody and kidnapped a Crown witness. Rich always maintained — and prisoners believed — that he was innocent of the initial offence and he had in the past made several attempts to abscond.

On the afternoon of Sunday 12 November, Rich placed a dummy in his bed on the upper landing of A. block. Complete with a head of hair donated by another prisoner and a carved hand holding a book, the facsimile was the culmination of weeks of planning. The cell's darkened interior made the deception more convincing and when a prisoner approached an officer in A. block and asked him to lock Rich up because he was sick, the officer saw no cause for suspicion.

While this was taking place Rich, crawling on his hands and knees between the legs of another inmate, was being smuggled through the basement sally port and into the A. block exercise yard. A plaited sheet, a grappling iron, and two small hand-held hooks were passed into the yard through the bars at the end of the block. Rich then hid, with his escape gear, in the ceiling of the yard toilet. It was Rich's intention to wait until dark before tossing the grappling hook over the yard wall, and having scaled that, to ascend the two perimeter fences using the hand-held hooks and a pair of plastic sandals especially modified for the purpose.

Of necessity, a number of inmates came to know of the plot and it is perhaps not surprising that knowledge of it eventually reached the administration. Around 6.00 p.m., just before Rich was about to make his move from the toilet, Hobson received a telephone call. Accompanied by a number of officers he marched into A. block. The group went straight to Rich's cell and discovered the dummy. Grim-faced, they hurried down to the yard and apprehended Rich.

The incident was the first (and subsequently, one of the few) serious escape attempts ever made from the institution and coming as close to success as it did, it provided Hobson with just the reason he needed to take swift and summary measures. The next day, a general lock-up was ordered and a blanket search began. Hobson selected a divisional officer to take charge of the search and authorized him to make up a team. Verbal instructions were given that cells were to be reduced to basic equipment, and all extraneous matter and altered clothing were confiscated. Search procedure was detailed, and permitted items were listed. Hobby equipment was to be taken, as was all jewellery. Study material was allowed, but no more than twelve books were to be left in each cell.

On the morning of 13 November, the men's doors remained locked. Breakfast was late and was served in the cells, off a trolley wheeled by staff. Some hours later the operation began. One by one the prisoners were released. They were taken to an empty room where a full body search was carried out. While this went on, a group of officers went through each man's cell, confiscating or smashing everything which was not specifically allowed. When the cell was emptied, the prisoner was returned to it and locked up again.

Unavoidably, a search of this nature is a sensitive business. People whose homes have been burgled often complain of feeling personally soiled by the experience. Bitterness and helpless outrage accompany the discovery of a break-and-entry and the theft of treasured possessions. To prisoners, the routine cell search arouses similar emotions. The cell is a home, a sanctuary, and personal effects are a principal means of maintaining links with the past. Even a properly conducted cell search is seen as an invasion of personal privacy. A search of this type, mechanical, destructive, and unconciliatory, represents a violation of a serious kind.

Prison officers selected for the November search pursued their task with zeal. Frustration over recent events had been accumulating and now they responded to similar sentiments as those which had driven them during the baton charge five months before. A number exercised reprisals against criminals, paying particular attention to those against whom they held special grudges. Popular targets were the 'trouble makers', a group which included the blocks' political leaders and the producers of *The Parry! News*.

Accounts of what happened during this incident vary according to the bias of the informant. Staff insist that the search was a systematic, efficient operation carried out with a minimum of fuss and acrimony. Inmates and their sympathizers describe the same event as a sadistic 'rampage', in which screws took open delight in destroying as much personal property as they could. The truth no doubt lies somewhere in between but it is incontrovertible that some officers went to extreme lengths in the three days which followed 12 November. Even Hobson later admitted he had not anticipated so much personal property being removed.[9] No warning of the search had been given and no explanation was offered of why previously authorized articles were now being taken. Personal letters and photographs were torn or completely destroyed. Handcrafts were damaged. Educational and recreational material was tossed about and in some cases lost. Several inmates complained of provocation and one who retaliated after seeing his exam notes thrown on to the floor was reported to have been beaten up.[10] Paul Morrison, a mild-mannered person who believes he was singled out for special treatment because of his association with the prison newspaper, described his experience:

> They came in, took me out of my cell, I stripped off, they looked up my arse and I remember I came back to my cell and I thought, 'What the *bloody hell's* happened here?' It was like a tornado had been through it. Stuff ripped, broken, torn off the walls and all lying in a heap on the floor. Bedding, pictures, hobbies, the lot.
> And I remember one of the screws who was OK running down and saying to me afterwards, 'Just bite the pillow, Paul. Don't get excited because that's all they want from you. They've done it deliberately to upset you.' And that's what they did to all the guys they singled out. They deliberately went out of their way to smash their stuff.

Numerous prisoners did react, however, or were selected for isolation. In all, twenty-six of the 130-odd standard block men were removed and placed in classification and D. block. The remainder were kept in their cells for three days and finally, on Thursday 16 November, they were unlocked. A statement signed by the superintendent was pinned on the notice board of each unit, outlining new rules for the whole of the open division. From then on, hair had to be kept off the collar and facial hair was prohibited. Personal ornamentation, apart from wedding rings and watches, was no longer allowed. Visits would be restricted to thirty minutes and prisoners would be permitted to write and receive only three letters per week. *De facto* committees were to be disbanded and all activities would henceforth be organized by the administration. Lastly and importantly, the privilege of free entry to housing units was withdrawn. Prisoners' movements would now be restricted to common areas and to the confines of their own cell blocks.[11]

Anticipating a hostile reaction from the outside organizations, Hobson hastened to justify his actions in the press. A spiked knuckleduster and six knives had been recovered during the search, he said, as well as money and fourteen gallons of home brew. A coded letter had been intercepted also, which had disclosed a plot to start an underground newspaper in the gaol. Since the segregation of the troublesome element, Hobson said, production in Paremoremo's workshops had more than doubled.[12]

The reaction that Hobson expected had already begun by the time this statement was issued. On 18 November, members of Project Paremoremo saw the standard block inmates during normal Saturday visiting. Details of the search were noted and a telegram and accompanying letter were sent to the Minister of Justice. The letter, written by Maynie Thompson, now chairperson of the organization, complained of wanton destruction of hobby materials, study notes, and other personal property. One man had been knocked unconscious, she said, and another had had his contact lenses smashed. The contents of this letter were released to the papers for immediate publication, and a few days later an anonymous note from an inmate, which had been smuggled to Project Paremoremo, was given to *The Auckland Star*.[13] These releases were accompanied by a flush of correspondence from prison visitors, complaining about the search and the manner in which it had been conducted.

Public allegations of what amounted to open abuses of authority were serious, and coming as they did within days of a general election were of considerable concern to the Minister. On reception of Mrs Thompson's letter, Jack immediately contacted Hobson, who refuted the organization's claims. However, in an attempt to determine what really had happened and to appease a concerned public, Jack ordered that a full report on the incident be undertaken.

Project Paremoremo had initially begun falling out with Hobson over the batoning incident in June, and the group's probable complicity in the hunger strike of September had added to the superintendent's chagrin. Personally offended by the attacks and fearing the organization's activities would aggravate more trouble, he prepared himself for a counter-offensive. Thus, when fires broke out on 21 November and four more men had to be segregated, Hobson pointed blame at recent releases in the press. 'Some of these chaps just play up to publicity such as this,' he said. 'They think they have an ally on the outside and away they go.'[14]

Hobson's beliefs were not entirely unfounded and there was doubtless an element of sensation-seeking behind some of the events of the period. After 1971, Paremoremo had become the focus of great attention, not only from the media, but from the numerous visitors who came up to the prison as well. Prisoners were able to enjoy a scale of popularity they had never known before and, revelling in their martyrdom, did what they could to live up to their media image. Of this, one inmate told me:

> You know in those days, everybody wanted to visit us. We were classed as the 180 most vicious criminals in New Zealand. And everybody wanted to see us.... I had three students coming to see me at different times. I thought, 'Why do they want to come in here?' — Because we were like celebrities!
> And another thing: what made us play up more was the simple fact that we liked to read about it in the paper [laugh]. Yeah! See, we'd hit the headlines

— and when we read about it, oh! We got rapt! We thought to ourselves, 'Yeah, we want to be reading about *us*!'

However, worse than being encouraged by sympathetic publicity, some of Project Paremoremo's activities seemed almost tailored to incite trouble. Hobson certainly believed this was so and openly attributed the increase in tension to their work. Mrs Thompson's reported statement after the November search that it would be a 'mircle [sic] if there isn't a riot at the prison now',[15] for instance, was considered deliberately inflammatory. The superintendent accused visitors of stirring up trouble and of 'breathing racism down the necks of prisoners'. 'We are starting to find that when trouble is brewing in the prison it coincides with the activities of outside groups,' he said.[16] Allegations about staff having gone on a 'rampage of destruction' were dismissed as 'dangerous nonsense'. The only material which had been confiscated had been that which was illegal, he claimed, and the search had been conducted in a correct and orderly fashion. He said that no inmate had been assaulted by staff, in fact the reverse had been true. Moreover, there was no record of any prisoner in the institution even possessing contact lenses, much less having reported any broken.[17]

Hobson was deeply angered by the allegations, especially those which were false. It was certainly untrue that a prisoner had been knocked out, and the story of the broken lenses proved entirely incorrect, fabricated by a malevolent trickster. The debunking of these assertions, of which a great deal had been made in the papers, clouded the credibility of Project Paremoremo and the rest of its claims. Hobson's insistence that the organization was deliberately trying to stir up trouble was supported by the fact that the worst of its accusations had been wrong. It was also claimed that plans for the underground newspaper, which staff had discovered during the search, were linked to Project Paremoremo, and that Mrs Thompson had even admitted she intended writing articles for it.[18] This Mrs Thompson denied.

Since mid-November pressure had been growing from Project Paremoremo to have another independent inquiry ordered, but it was obvious that with the elections so close, any decision would have to be deferred. When a change of Government was announced on 25 November, however, pressure began afresh and the seven men of D. block produced a list of matters which they felt needed investigation. In order to encourage a rapid decision from the new administration, a 'general strike' was proposed from 5 December. Once more the press had been alerted to plans before the superintendent had known about them and once more Project Paremoremo was involved. Hobson's view was that 'a small number of inmates and Project Paremoremo want to dictate prison policy.'[19]

In the end, appeasement from the new Government prevented the strike from going ahead, but on 10 December Hobson informed Maynie Thompson that thenceforth she would be prohibited from visiting the prison. The reason given for the ban was that she had circulated false information to the press about the November search and had admitted involvement with the illegal newspaper.[20] The following day the incoming Justice Minister announced that there would be a second inquiry after which, Hobson said, the question of Thompson's visiting would be reviewed.

Thompson's banning was a serious blow to Project Paremoremo. She was now president and the group's most active member. Without leave to maintain personal contact, Mrs Thompson's ability to operate would be severely impaired. A call from the Auckland Council for Civil Liberties to have the ban lifted drew no response from the Minister. There was no alternative but to sit and await the Ombudsman's second report.

Paremoremo and the 1972 Election

The second Powles–Sinclair report, which had been ordered within a year of the first and just two weeks after the change of Government, arived with what promised to be a new era of social reform in New Zealand. By 1972, the National party had been in Government for twelve consecutive years, but after its slender victory in 1969 it had gradually slid into decline. The attrition which eventually affects all Governments had taken its toll, and in spite of Marshall's desperate reshuffling of Cabinet in February, the party's energies were drained. As November approached, it stood shakily, almost anticipating its downfall.

The boom in prosperity which had begun in the late 1960s continued into the seventies, and there was another economic upturn in 1971–1972.[21] The bouyant nation was ready for new blood, but few were able to predict the size of National's defeat.[22] From its thirty-nine:forty-five advantage in 1969, in 1972 National was swamped by a fifty-five:thirty-two Labour majority. The leader of the new Government was Norman Kirk, a man of ostensible strength and charisma. An experienced and able politician, Kirk had chaired the opposition caucus since December 1965.

Labour's election campaign had begun at the end of October 1972, less than a fortnight before Rich's escape attempt. Although the institution was not an important component of the opposition's election machine, numerous Labour MPs had been active in pressing for a resolution of its difficulties. Prominent among these were Phil Amos, Whetu Tirikatene-Sullivan, Eddie Isbey, Mat Rata, Joe Walding, and the opposition Justice spokesman, Dr Martyn Finlay. In February 1972, Dr Finlay had expressed doubts about the wisdom of maintaining a totally secure gaol and that same month Amos had declared that major reforms would become a part of a Labour Government's programme. One of the changes contemplated would be recruitment of more specialized staff for Paremoremo,[23] as recommended in the first Powles-Sinclair report.

Throughout 1972, the Government had been harried by opposition over rising crime and disorder in prisons,[24] and while probably not of high significance, the squall over the 'rampage' in the middle of the campaign can only have weighed against National's chances. It was at about this time that Project Paremoremo drafted a circular letter to all Auckland candidates, urging them to include in their policies a positive and imaginative programme for penal reform.

Kirk, himself, took a conservative stance on law and order and was suspicious of academic opinion. As leader, he attacked what he termed the 'permissive society' and, having deplored the rise in juvenile crime, sought remedy in increased police staff ceilings. In October 1972 Kirk had embarrassed colleagues by publicly announcing, as part of his formula for reducing hooliganism in the streets, that he would 'take the bikes from the bikies'.[25] However, in spite of minor folly, Kirk was an able and competent statesman who recognized that crime and lawlessness were related to wider social problems. As part of a broad-based assault on these issues, he planned to improve housing, medical and education facilities, and to upgrade the general quality of community life. Prisons were not neglected either. After the Mt. Eden riot of 1971, Kirk had vilified the Government for its artless handling of the penal system,[26] and he had called for Mt. Eden to be bulldozed flat as soon as possible.[27] His choice for the portfolio of Justice in 1972 suggests that while penal policy may not have been central to his concern he was aware of its issues and of the work which needed to be done.

Publicly, Kirk's appeal was enormous and in the fortnight before polling day the *Sunday News* took the unusual step of presenting a front-page spread exhorting its readers to vote for 'Big Norm' Kirk.[28] The label suited the leader. For in November 1972 it was Kirk's imposing stature and commanding personality as much as anything else which sealed victory over the laconic and urbane 'Gentleman Jack' Marshall.

A. M. Finlay, Minister of Justice

Kirk's Justice Minister was a distinguished legal scholar whose involvement in politics and justice had spanned more than half of his sixty years. Allan Martyn Finlay was born in Dunedin on New Year's Day, 1912. Educated in Otago he took an LL M from Otago University with triple first class honours, and in 1936, as Travelling Scholar in law, he went to the London School of Economics to read for his Ph.D. Having completed this in 1938, Finlay then travelled to Geneva where he worked as a research assistant in the League of Nations, before moving to Harvard to study for twelve months as a research fellow.

At the commencement of the Labour Government's second term, Finlay began to take an interest in New Zealand politics and in 1939 wrote to the Prime Minister Michael Joseph Savage asking for a job. The application was successful and the bright young lawyer was made a member of the Law Reform Commission and Private Secretary to the Minister of Justice.

The Government was keenly interested in law reform at this time, and when its aspirations were frustrated by the war, Finlay attempted to continue working on his own. The task was difficult. He had several 'stand-up exchanges' with Dallard over prison proposals and remembers, as time went on, how Mason, too, became increasingly conservative and difficult to motivate. By the end of the Labour Government's second term, Finlay was beginning to tire of his work and he eventually left Justice in 1943. In 1946, he entered politics as Labour MP for North Shore, but having been defeated in 1949, he joined the party organization. In 1958 he was elected party president and retained this position until 1963 when he became Labour MP for Waitakere. Finlay then remained in politics until his retirement in 1978.

Martyn Finlay is a man with a deep sense of humanity and justice. As Mason's Private Secretary he had suggested that prisoners should be paid trade union rates and during his time in Parliament had been one of the most outspoken patrons of penal reform. Opposed to the concept of absolute custody from the beginning, Finlay had been actively involved in pressing for a solution to Paremoremo's problems. In seeking a solution it was clear that as Minister, Finlay would favour liberal alternatives ahead of traditional ones.

It was his progressive ideas about prison policy that from the start made Finlay unpopular with gaol staff. Whereas in 1972 the overall backing for Labour had risen by 6.2 points over the 1969 figure, at the Paremoremo Primary School polling booth, returns showed a swing to the left of only 4.7 points. Here too the movement against the Government (−0.6 per cent) was much smaller than the national average of 3.7.[29] Prison staff and their families are traditionally Tory voters, but at Paremoremo there has been a reluctance to support candidates believed to be in sympathy with prisoners. In 1978, after Paremoremo debating tutor Don McKinnon became National MP for Albany, the Paremoremo polling booth revealed a sudden and marked swing away from New Zealand's major conservative party.[30]

At the Justice Department's head office, however, the change of Government was greeted with enthusiasm. Frustrated since the death of Hanan by three years of inactivity under three different Ministers, senior officials sensed promise in the change of administration. Kirk's choice for Justice was one of the country's most qualified lawyers and it was hoped, at last, that there would be an energetic advocate sitting in Cabinet. Finlay dreamed of breathing new life into his Department, but its permanent head, Eric Missen, already feared that his boss's ideas might prove too radical. None the less, Missen believed that he and the Minister would be compatible, and that a happy working relationship would develop.

Finlay's first action upon taking office had been to order a new inquiry into the maximum security prison. Although the inquiry had not been publicly announced until 11 December, the decision to conduct it had been made within a week of his rising to power. By 6 December, the day after the deadline of the prisoners' 'general strike', investigations at the gaol were already under way. The study contemplated would be far more comprehensive than that of nine months earlier, and its results would be crucial to Finlay's programme. The report could provide a useful springboard for change, but to act before it had been presented would be imprudent. So, although the Minister's plans were by now already well developed, for the time being action was delayed.

THE SECOND POWLES–SINCLAIR REPORT (1973)

Throughout 1972, complaints about the administration at Paremoremo arrived at the Ombudsman's office in increasing numbers.[31] Calls for another investigation began soon after Hobson was installed, but these were denied by Powles so that the new superintendent would have time to implement his policy. It was not until late November that pressure resulting from the lock-up finally convinced the Ombudsman to plan a visit to the institution, and it was at this same time that the new Minister of Justice decided to order a second informal inquiry.

As had been the case at the end of 1971, the men commissioned for the investigation were the people's representatives, Sir Guy Powles, and the visiting Justice, Mr L. G. H. Sinclair. In all, these men made three visits to the gaol totalling fourteen days, between 6 December 1972 and 7 February 1973. Forty-three inmates, nine ordinary officers and all senior staff and specialist personnel were interviewed during the period, as well as representatives of various interested organizations. Written submissions were received also. The terms of reference involved investigating the organization and administration of the prison since January 1972 — with specific regard to the second lock-up — and also reviewing the findings of the first report.

Five times larger than its predecessor, this second report consisted of over eighty pages of typed foolscap. The entire investigation and its documentation took place over the short space of fourteen weeks and its results were finally presented on 15 March 1973. The paper chronicled events at the prsion in the tumultuous months leading up to, and following, June 1972.

In the light of later events and this inquiry's more detailed examination of them, some of the findings of the first report were substantiated by the second, others were found to be in need of amendment. As noted, Finlay was opposed to the very concept of total security, so another of the investigators' tasks was to query the necessity for the gaol's continued existence. The prison officers' subgroup also opposed it. It had submitted that Paremoremo should be reclassified medium security and reserved for trusted men only.[32]

Powles and Sinclair were unimpressed by arguments for changing the prison's status and wisely foresaw an increase in the need for security in future years. Consequently, they argued for retention of the prison in its original form.[33] Similarly, in spite of its disruptive influence, there seemed no alternative to retaining D. block for its original purposes.[34] The investigators noted that the types of problems being encountered at the prison were identical to those which had been found in England[35] and they suggested that the claim of outside groups having responsibility for disruption was facile. The real problem lay elsewhere, they said,[36] and a temporary return to discipline[37] was advocated.

Together with this, however, the report provided strong support for the work and philosophy of independent interest groups, referring in particular to Project Paremoremo.[38] This organization had maintained close contact with the investigation and conferred confidentially with Powles throughout the research period. Activities of the group were felt to be necessary to the maintenance of good and open public relations,[39] and it was therefore urged that the bans on some of the visitors should be reviewed.

It was the events of, and pressure resulting from, the November search which had precipitated the second inquiry. However, the investigators found little evidence to substantiate the bulk of complaints received. In all, fifty-three of the 149 prisoners had listed a total of sixty-three grievances arising out of the search, and submissions had been registered from outside bodies as well. Although at the time of the report, ten of these cases had not yet been looked into, the majority had been found unjustified.[40] Events which may have taken place during the November lock-up were thus effectively dismissed.

In spite of this, Paremoremo staff did not emerge unscathed from the survey and the investigators were 'astonished' by the opinions of some of them.[41] They wrote:

> ... our investigation indicated the likelihood that there is a small — perhaps very small — group of prison officers who, because their own contrary views are so strong, are basically disloyal to the Superintendent and who are not prepared to try to carry out successfully certain of the more liberal policies which he lays down.[42]

The investigators called for measures to ensure that managerial policy was not weakened because some officers were unable to accept it[43] and gave their unqualified support to the regime and the methods of the new superintendent.

Apart from its endorsement of the new order, the direct significance of the second inquiry was not major. Visiting did not become a right and remains a privilege to this day. Hobson continued to believe in the disruptive influence of outside groups and Mrs Thompson's exclusion remained in force until May 1974. At the behest of the report,[44] a second schoolteacher was appointed, but this position had already been advertised before investigations began. The suggestion that a visiting counsellor's position should be created[45] was not enacted until February 1975 and the appointment probably had nothing to do with the inquiry. The temporary return to discipline that the report had advocated was already in force before research commenced, and similarly, discussion of the progressive block system[46] referred to a practice which had been abandoned twelve months earlier.

So it went on. Most of the recommendations made were either already established, were ignored, or were eventually put into effect as part of a process which was independent of the report's submissions. During an interview with Project Paremoremo six months later, the Minister of Justice admitted that there had been no noticeable changes at the prison since the second Powles–Sinclair report, although he suggested that 'invisible changes' were taking place.[47] Given the circumstances under which the reviews were conducted, these facts should not be surprising. Both of the Powles–Sinclair reports resulted from informal commissions which were recognized by Government but not binding upon it. In spite of the investigators' considerable administrative experience, they were not penological experts. As decision makers they were probably no less fallible than the superintendent, but were certainly not as qualified as he was for the rigors of the job in question. While the opinions of detached adjudicators are of great value in such situations, some of the recommendations the independents made were as questionable as the actions they had been charged with examining.

Given these facts, it is fitting that the reports had no legal power. The place of such investigations is to guide, not blueprint, the course of future development. At the end of it all, the task of running the nation's gaols in 1973 remained that of the Justice Department, and through it, the superintendents. As before, it was with the latter that front line responsibility lay. In maximum security the controlling officer's performance was still more visible than anywhere.

NOTES

[1] NZPD vol. 378 (1972) 200-1.
[2] Rangi had been moved out immediately and replaced by Wickliffe. Gillies followed Rangi shortly afterwards.
[3] *Truth* 12 September 1972.
[4] *The Sunday Herald* 26 November 1972.
[5] This uniform consisted of a white cotton shirt and khaki woollen trousers.
[6] In January 1978, Ward became superintendent of Mt. Eden, and took over from Hobson at Paremoremo in October 1984. Ward retired the following year.
[7] The Inspector, Mr H. Stroud, later became superintendent at Paparua, and assumed control of Mt. Eden in October 1984.
[8] The establishment of informal systems has also been observed in industry. A famous early description appears in the 1927-1932 research on the Hawthorne plant of the Western Electric Company (Roethlisberger & Dickson 1956).
[9] *Report on Auckland Prison* (1973) 41.
[10] A press release by an anonymous Paremoremo inmate in November 1974, which claimed that the man had been knocked unconscious, was denied by officials.
[11] *Paremoremo: Rules for Guidance of Inmates*. J. Hobson, superintendent.
[12] *AS* 24 November 1972; *NZH* 25 November 1972.
[13] *AS* 25 November 1972.
[14] *NZH* 22 November 1972.
[15] *The Sunday Herald* 19 November 1972.
[16] Ibid., 26 November 1972.
[17] *The Sunday Herald* 26 November 1972; *AS* 25 November 1972; *NZH* 11 December 1972.
[18] *NZH* 11 December 1972.
[19] *Truth* 12 December 1972.
[20] *NZH* 11 December 1972.
[21] Chapman (1981) 368; Sinclair (1980) 308.
[22] Roberts (1975) 107-8.
[23] *AS* 5 February 1972.
[24] See, for example, NZPD vol. 378-80 (1972) 112-3, 176, 199-201, 253, 341, 445, 1102-3, 1260, 2124.
[25] Labour party concern over the 'bikie' problem had begun at Easter weekend 1972 with a riot of gang members at Palmerston North. Subsequent to that incident, the Deputy Leader of the Opposition H. Watt made a call for giving the police powers to confiscate gang members' motorcycles in an effort to prevent trouble when its occurrence seemed likely (*NZH* 6 April 1972). This call was echoed seven months later when Norman Kirk opened the Labour party's election campaign in the city of Palmerston North (*NZH* 1 November 1972).
[26] NZPD vol. 371 (1971) 641-3.
[27] *The Sunday News* 19 September 1971.
[28] *The Sunday News* 12 November 1972.

29 *App.J.* H-33 (1970); *App.J.* E-9 (1973).
30 In 1978, the National party at Paremoremo polled 34.5 per cent of the vote compared with a national average of 39.9 per cent (*App.J.* E-9 1979). In 1981, the figures were 31 per cent and 38–78 per cent respectively (*App.J.* E-9 1982).
31 Project Paremoremo had been asking for a second inquiry since 3 February 1972, not six weeks after the first had been submitted.
32 *Report on Auckland Prison* (1973) 67.
33 Ibid.
34 Ibid., 3–5.
35 Ibid., 51.
36 Ibid., 62.
37 Ibid., 53–5.
38 Ibid., 11–19.
39 Ibid., 19.
40 Ibid., 43–7.
41 Ibid., 20.
42 Ibid., 21.
43 Ibid., 20.
44 Ibid., 22.
45 Ibid., 19.
46 Ibid., 27–8.
47 *Project Paremoremo Newsletter* (August 1973); *EP* 2 August 1973.

19: THE END OF AN ERA 1973-1975

At the beginning of 1973, Paremoremo still held around a hundred inmates fewer than its capacity of 248. Twenty-four inmates were serving life sentences, another six, preventive detention; that is, around 13 per cent of them were burdened with non-determinate terms. Roughly half had escape records. Powles and Sinclair had been informed that around 10 per cent of the inmates needed 'some form of psychiatric help'.[1]

This was the institution which Finlay inherited in late 1972, and from the start he made no secret of the fact that he hated it. In the new Government's first year, Thomson applauded the gaol as one of the world's finest, and attacked as 'ridiculous' and 'irresponsible', the actions of its critics.[2] Finlay came boldly to their defence. He too was one of its critics, he said, and he described the task of running Paremoremo as 'distasteful'.[3] The institution was, he said, 'a bleak, forbidding, and discouraging pile of masonry and steel that must drive the heart out of anyone'.[4] For him, the idea of an escape-proof gaol was an 'abomination'. None the less, he said to me, as horrible as the concept of Paremoremo was, he had no choice but to stomach it.

Finlay's early involvement with Project Paremoremo, his outspokenness on penal issues, and his disparagement of total custody made clear the stance he was likely to take. This had two effects. Firstly, it made officers uncomfortable about the new Minister; secondly it gave new heart to external pressure groups. These conflicting currents had a considerable influence on the course of Finlay's administration.

MINISTERIAL REFORMS 1973-1974

In the years of 1973 and early 1974, a number of small concessions to prisons were made. In 1973, a second schoolteacher — recommended in the 1972 report[5] — was finally appointed to Paremoremo, and in February Dr F. Whittington was made the Justice Department's first forensic psychiatrist. In September, the Minister announced that prison inmates would now have the right to correspond directly with him. Distrustful of Finlay as they were, the move was immediately unpopular with Justice officials and eventually led to one of the most embarrasing episodes of Finlay's career.

In 1973, however, the new Government was still popular and the country's economy was liquid. By the end of that year, a change in prisoner wages allocation was being studied[6] and although the idea of paying union rates was finally rejected, in July 1974 prisoners were granted a 27 per cent rise, their first since 1971. Gaol canteen prices had risen steadily and the following year the proportion of the weekly allowance which could be spent by an inmate was increased from 33.3 to 50 per cent. Canteen prices were stabilized from that point.

Finlay was anxious for overseas input and with finance still loose in 1973, he sent two Justice officials to Europe to study developments in some of the world's most advanced penal systems. Reinforced by this input, in 1974 he introduced the Penal Institutions Amendment Bill, sections 15 and 16 of which abolished restricted diets. Finlay considered them anachronistic and before legislation was passed he issued a provisional directive that their use should cease forthwith. As was to become increasingly common with policies Finlay tried to implement, the directive was condemned by local officials, and restricted diets continued in New Zealand until formally struck out in 1980.[7]

Compared with his last months in office, however, we shall see that Finlay's departmental relations in 1973–1974 were relatively good. Locally, sentiments were often equivocal, but at head office there was early enthusiasm for the new ideas. A mood of anticipation had greeted the incoming Government, and the excitement it brought spilled over to the independent organizations. When Finlay took over his Department, it was for a time these groups, not permanent staff, whose calls for action echoed loudest.

The Passing of Project Paremoremo

In the early months of 1973, spurred by the knowledge of a sympathetic ear in Parliament, the energy of Project Paremoremo redoubled. In March it suggested that inmates should be granted access to the press via a union of their own, and it pressed for an increase in work parole provisions in April. The next month it asked that the European tour of head office representatives be extended, and in July it became active in the formation of a committee to provide accommodation for released prisoners. This resulted in the creation of the *Aroha* trust, later *Arohanui Incorporated*, which now runs five hostels in Auckland for use by released prisoners and disaffected youths.

In August 1973, Project Paremoremo, in association with CARE, sent to the Minister the minutes of an April seminar which had called, *inter alia*, for the total abolition of the conventional prison. Such futile imprecations did little for Project Paremoremo's image and there is evidence, before the first parliamentary session was over, that the credibility of the group was already beginning to fade. As early as June, in fact, Finlay had commented on the 'excessive zeal' of independent groups, and chorused with Hobson that activists may have contributed to, rather than reduced, discord within the prison.[8] Finlay knew penal reform to be a complex issue, and although often guilty of the mistake himself, he saw the folly in attempting changes without careful tactical planning.

Project Paremoremo seemed unaware of other considerations and continued to badger the Minister into acting more quickly. Regardless of the advice of two inquiries, the organization still argued for the closing of D. block. It also wanted inmate councils, specialist training for prison officers, open visiting even for security risks, parole boards for everyone, and a multitude of other reforms as well. The suggestions were manifold, and if implemented would have amounted to a complete metamorphosis in criminal justice policy. It should have been obvious to Project Paremoremo that the calls were unrealistic, and Finlay privately reproved the organization for making false claims and for allowing its enthusiasm to run away with itself.

By May 1974, Project Paremoremo was ready to acknowledge that an improvement in conditions had occurred and it was partly a result of this that after 1973 the activity of the group began to decline. In May, the visiting ban on Thompson was lifted, but when she returned she found a different atmosphere to what she remembered. The institution had undergone a transformation and the tension and violence which had been the fuel of her organization had now largely disappeared. Competent administration and a concerned Minister had relieved the need for a watch-dog. Additionally, many of the original prisoners had been released or transferred, while at the same time the organization's younger enthusiasts had gradually drifted away.

From mid-1973 tranquillity led to a sudden drop in media interest, and Paremoremo slowly slipped from public view. The end of the Vietnam War in 1973 signalled the end of the great era of political protest in New Zealand and contributed also to the demise of Project Paremoremo. Although Maynie Thompson, the last of the old executive, continued to visit sporadically until about 1976, by the middle of 1975 Project Paremoremo, as a working organization, had ceased to exist.

LOCAL CHANGES

Early in 1973, Hobson went on long service leave and when he returned, a programme of progressive relaxation of the November restrictions began. Unlike the rapid changes he had made when he first took over, these later developments began slowly and were gradually introduced over the ensuing months. As the year unfolded, extra shelves began to appear once more and cell furniture increased. Fancy bedspreads returned, curtains were permitted across cell frontages, hobby materials and extra books made renewed entrance. Once more, extra rations in cells were ignored by staff. Correspondence restrictions were lifted again and visiting rules were eased.

The response of inmates was different this time. There was a certain calm about the prison; a truce between opposing forces which was born of awareness that relaxed administration could only continue as long as relative peace prevailed. Although there was still no love lost between staff and inmates, both sides now had a heavy commitment to stability.

During 1973, there was a further increase in welfare activities. In mid-1971, it will be recalled, Paul Beachman had begun visiting, with a view to setting up discussion and current affairs groups. In the interests of variety a few casual debates were organized between the cell blocks as well, but due to the conservatism of Buckley, outside involvement had been almost impossible to arrange. Early in 1972, responding to a call from the first Ombudsman's report,[9] the Auckland Debating Association (ADA) was approached by the prison's education officer and asked if it would assist with the organization of a regular programme. This was agreed to and Don McKinnon, active within the Association, volunteered to visit. As a result, a debating club was founded and the first moot took place between Paremoremo's B. block and the Auckland Student Christian Law Students' Society in late January 1972.

Debating became very popular at Paremoremo. Throughout the year, regular moots were organized with outside teams, and in 1972, of 150-odd inmates in the prison, thirty-five were active club members. It became clear that considerable talent lay in the gaol, and the following year McKinnon succeeded in having a prison team accepted into the ADA's annual competition. Although the move was unpopular among some elements of the Association, the prison remained undefeated until being knocked out in the semifinal of the B grade (Robinson Cup) competition.

Fired by its success, and rapidly becoming an established feature of the local debating scene, the following year Paremoremo fielded two teams, contesting both the Robinson Cup and the A grade competition for the Athenaeum Trophy. Once more the prison aquitted itself well and McKinnon hoped that if they reached the final, special dispensation would be granted for the teams to debate outside the walls. Hobson seemed to favour the idea and by the time Paremoremo 'A' defeated Oxford Union in the semifinal, approval for temporary parole for the Athenaeum side was almost assured. Thus it was that in October 1974, the Minister of Justice made the historic decision of granting parole for a Paremoremo prison team to compete in an ADA final. In October 1974, for the first time since 1955, a group of maximum security inmates travelled outside for the purposes of recreation. The contest, which was held against Forum at Auckland Grammar School, resulted in a prison side taking possession of the Athenaeum Trophy, and ten days later at an internal venue, the 'B' team also won the Robinson Cup. After this time, prison debaters retained a high profile on the ADA calendar and by 1975 had become firmly established among top contenders at both the advanced and novice levels.

During the 1973–1974 period, numerous other concessions were granted to the maximum security men. On Christmas Eve 1973, a travelling concert troupe performed at Paremoremo and in April 1974, cabaret entertainer, Lovelace Williams, gave a concert at the gaol. Ethnic interests were catered for. A Māori culture club was set up, and in October 1974 when it was learned that Paremoremo had been constructed on top of an old burial site, a Tohunga was called in to chase away a ghost, which some Māori inmates claimed to have seen.

As the era of tranquility at Paremoremo continued, the type of outside involvement began to change. With visiting activists on the decline, the early successes of the debating teams suddenly gave rise to an awareness of intellect and hidden talent among the security men. Involvement of conservatives like Don McKinnon and retired Kings College headmaster Geoff Greenbank lent respectability to volunteer workers, and other orthodox representatives began to take an interest as well. In April 1974, a book of prison poetry was published in Auckland, and a stall run by the Mt. Eden branch of the New Zealand Labour party was set up at the Easter Show to advertise and sell inmate craftwork. In May, a display of twenty-five articles of New Zealand prisoner workmanship was held at the institution. Of the six pieces selected for an international exhibition scheduled in Canada later that year, all were made by Paremoremo men. In March 1975, a Paremoremo inmate had an exhibition of oil paintings at Auckland University, and later that month twenty students answered a call to visit the gaol as tutors in tertiary subjects. A few months after this, a former inmate was invited to be guest speaker at the annual dinner of the Law Students' Society. Unfortunately, agitation from a student (who was also a police detective) prevented the talk from going ahead.[10]

It was in 1973-1975 that the tradition of achievement at Paremoremo was forged, and this is one which has survived the passage of time. In May 1976, when an official power-lifting contest was held in the gym, two inmates came within a few pounds of toppling national records. In 1978, this author became the first prisoner to gain a Master's degree from inside a New Zealand penal establishment. Inmate craftsmen still supply an eager commercial market, and the debating club continued to acquit itself until upheaval within the prison forced its closure in 1985.

The history of successes which began at Paremoremo in 1973 was one which was born of administrative concurrence, external charity, and inmate dedication. Under the Hobson/Missen-Orr/Finlay regime, one of the best possible climates for progressive experimentation existed, and it was during this period, the period of the third Labour Government, that the tone of the gaol's best years was set.

G. S. Orr, Secretary for Justice

Election year 1975 was the year in which Finlay and the legal division had their most intense period of activity. In March 1974, Eric Missen retired from Justice and took a job at the Higher Salaries Commission. His replacement was Gordon Stewart Orr, a lawyer and an academic who, unlike all other permanent heads since 1969, had no particular background in the Department of Justice.

Orr was born at Masterton in 1926, educated in the Wairarapa and at Victoria University, graduating from there with a BA in political science and an LL M. Between 1952 and 1957, Orr worked as a barrister and solicitor, before joining the Crown Law Office as a counsel in 1958. In 1961, he won a Harkness Fellowship to Harvard, where he studied administrative law and the process of judicial change, then returned to New Zealand to continue legal work with the Crown. In 1970, Orr became a member of the State Services Commission, taking deputy chairmanship in 1971. Although Orr had no direct experience in Justice, he was an expert in administrative law and when he applied for the Department's permanent leadership in 1974, he was automatically a top contender.

Jim Cameron, Deputy Secretary, with over twenty years' Justice experience behind him, was also favoured. Finlay knew Cameron from having worked with him, and believing him to be an effective, practical man, Finlay encouraged him to apply for the job. When the position was given to Orr, however, Finlay was not disappointed and believed that, since both he and the new Secretary had similar academic backgrounds, an effective working relationship would be reached.

MINISTERIAL REFORMS 1974-1975

During 1973 and the first half of 1974, the bulk of changes in the penal system had been administrative. It took eighteen months for the momentum of the new Government to gather, and from the point of view of Justice, the most innovative of the three sessions was the last, in which Martyn Finlay worked with Gordon Orr.

It was in 1974 that Finlay had informally ordered the discontinuance of restricted diets, and although the requirement had been ignored, their eventual abolition resulted from his initiative. In 1974, home leave was extended. First introduced in 1955, the provision had been gradually altered so that now furlough would be available (at the Secretary's discretion) to all minimum security inmates. In 1975, instead of being given every four months, such persons became eligible for a seventy-two hour parole, plus travelling, every two months. Consequent to these developments, in 1974, the number of prisoners granted leave increased from 174 in the previous year, to 526. The following year, 747 were given it. In 1976, as a result of parole breaches[11] and a murder-suicide on Christmas Eve 1974, greater discretion was exercised and the number of parolees reduced to 665. Home leave for selected minimum security candidates continued after this, and is now an established feature of the New Zealand penal programme.

In late 1974, mail censorship was relaxed, and from May 1975, minimum security prisoners were allowed regular telephone calls. Discretion to grant special remission of one-third, instead of the standard one-quarter was extended that year also,[12] so that a large number of low custody men and women automatically qualified. Moreover, part or all of this extra one-twelfth was made available to those in more secure establishments on application to the Secretary for Justice.

Like most of Finlay's early changes, the substance of these alterations was administrative and hardly affected by statute at all. However, since 1972, Finlay had been working on reformative legislation, and in September 1975, two Amendments went through Parliament, both having important implications for justice. Although many of the matters addressed in the two statutes were procedural, a number were of direct relevance to prisons.

Under the Criminal Justice Amendment Act 1975, for example, the ten year non-parole period for convicted murderers was reduced to seven, and the minimum finite sentence carrying parole eligibility was cut from six, to five years. Instead of having to serve at least three and a half years, a parole board candidate now had to serve only half of his/her sentence or three and a half years, whichever was shortest, before making his/her first board appearance. In addition, the Amendment legislated for the abolition of borstals, replacing them with youth prisons, and the substitution of detention centres with corrective training centres. These latter changes did not take effect until Orders in Council were given at later dates.

The other piece of relevant justice legislation this year was the Penal Institutions Amendment Act. Passed on the same day as that of Criminal Justice, the Penal Institutions Amendment concerned itself with internal prison management. To that extent, the two were complementary. It was this statute which provided for the eventual abolition of restricted diets, and it also repealed that status known as Penal Grade. Like that of Criminal Justice, much of the penal legislation did not take immediate effect, and it was not until April 1981, for example, that authority for restricted diets was struck out.

Penal Grade had allowed punitive seclusion of troublesome inmates for up to three months without privileges and was similar in its operation to that of administrative segregation. The Powles–Sinclair reports had twice endorsed the concept of D. block and now the idea of segregation for administrative purposes was legally ratified for all prisons.[13] The relevant section came into force immediately, and effectively superceded Penal Grade by allowing for the removal of certain inmates to special parts of institutions. Unlike Penal Grade the provision could be used summarily for up to fourteen days, although it required head office approval if it was to extend further. Also unlike Penal Grade, segregation could continue indefinitely, provided that the status was reviewed at three-monthly intervals by the Secretary for Justice.

Martyn Finlay was responsible for a variety of legislation in the last session of the third Labour Parliament, but perhaps the most controversial was a clause contained under section 18 of the Electoral Amendment Bill 1975, giving prisoners the vote. Pressure for this move had begun soon after the new Government took office, and when the bill was presented in mid-1975 it drew heated contest from the opposition. Nevertheless, in 1975 the Amendment became law. Thus it was that in November 1975, for the first (and only) time, New Zealand prisoners were permitted to vote in a national election.

Despite what might have seemed a harmless and just concession, the enfranchisement of prisoners was a controversial move, which did little for Finlay's popularity. Prisoners appreciated the gesture, but most were Labour supporters anyhow and their numbers could not offset the hostility which the Government had begun to attract by the end of 1975. When the country went to the polls that year, the work of the third Labour Government did not merely escape notice, it was overwhelmingly rejected.

This volatile period was unique in New Zealand's political history and I shall return to it later in the chapter. However, before doing so, it is appropriate first to consider the effect that the new approach had upon the prison's internal structure. Although the Department's programme led to a reduction in conflict, the security prison was still not without friction. The nature of upsets after

1972, it will be seen, was different from earlier years. It was largely the transformation of social relations, brought about by the post-1972 reforms, which caused the other changes to take place.

The Style of Later Troubles

After 1972, relaxation of control caused anti-authoritarian codes, which had been so firm under the rule of Buckley, to weaken. Only in D. block did oppressive conditions remain, so that this unit, both in its administration and its inmate organization, remained like a fossilized remnant of the past.

Elsewhere, staff–inmate relations improved and as they did, inter-group tensions eased. In turn, the atmosphere of the standard blocks started to alter, but in a way which the old hands found strange: as assaults upon staff began to diminish, so did fights among the inmates increase. The violence which had once been aimed at administration now seemed to recoil upon itself. Evidence of this new form of aggression was soon apparent, and two months into the new year an inmate was hospitalized with fractures of the skull and jaws, caused by other prisoners. Two weeks later, inmate Mihaly Bede, earlier the subject of a series of articles in the *8 O'Clock*,[14] was also taken to hospital with injuries suffered in a fight. Five weeks later, a third attack occurred. This last incident was the most serious of all and had an enduring effect on the folklore of the gaol.

Early in 1973, a plan was formed to have a pistol smuggled into Paremoremo. Exactly what the weapon was for is uncertain, although it was later rumoured that an inmate had intended to 'find' the weapon, in an attempt to secure a gratuitous release. The initial idea was for the gun to be brought into the visiting room and passed through a window to the adjacent reception block. Because he worked in the receiving office, a prisoner called Lex Harrison was approached, and asked to hide the firearm when it arrived. Harrison refused and when, a few days later, the visiting room window was welded shut, Harrison immediately became suspect.

During the lunch-time of 5 April, a meeting of inmates was organized in the basement of A. block, and Harrison was tried *in absentia* for transmitting information to the adminstration. It was already suspected that Harrison could not be trusted and it was indeed true that he had been compromising with officialdom for some time. Although evidence was largely speculative, the 'court' found no difficulty in convicting Harrison, and two inmates, both with psychiatric histories, volunteered to administer justice. A chrome-plated pipe was torn out of one of the block's showers and while Harrison took his lunch-time nap, he was attacked and savagely beaten. Alerted by the victim's terrified screams, block staff soon arrived on the scene. They found Harrison lying in his blood-spattered cell, unconscious and seriously injured.

As a result of the assault, Harrison received a fractured jaw, arm, and ribs, deep lacerations to the head, and partial paralysis of the face and one side of his body.[15] Afterwards he remained in the prison hospital, but on 29 December, only days before he was due to be released from gaol, Harrison, aged thirty-four, suddenly died.

There were suspicions about the death,[16] but at first no investigation took place. It was not until several months later, after the body had been cremated, that media pressure forced a hurried inquest. At the inquest, Harrison's injuries were described as having consisted of 'bruises and minor lacerations', and it was said that he had made a full and uneventful recovery. He had been detained in hospital, a doctor said, only because he feared a further beating. Why Harrison had not been transferred to another prison or to a segregated cell block, as was customary in such cases, was neither queried nor explained. Death, the coroner found, was due to natural causes: an 'organic disease of the heart', which was unrelated to the earlier assault.[17]

The truth about Harrison will probably never be known for sure, but those who were in prison with him are adamant that his condition after the attack was chronic, that he never fully recovered, and that information about his injuries was concealed at the inquest. From our point of view, the importance of the Harrison affair is that it became immortalized in the lore of the gaol, cited frequently as an indictment against inmate justice. In the folk traditions of Paremoremo, Lex Harrison, who was innocent of any crime, was tried, convicted, and sentenced by a 'court' whose judgment was fouled by the sadistic whims of unstable prisoners. Inmates were sickened by the assault, and from this point kangaroo courts were communally censured.

For a number of months after the Harrison incident, the gaol was quiet. Even in D. block activity was low, although solidarity there remained stronger than elsewhere. However, like the rest of the gaol, its energy fell away after 1972. Part of the reason for this mellowing was that the original members of the unit had gradually been moved out after the new superintendent took over. The greatest portion of the explanation, however, lies in Hobson's approach to it. In November 1973, Hobson was given the opportunity of putting his new formula into effect. Although he had been vindicated in the lawsuits of October 1973, Hobson detested court appearances and was loath to have the experience repeated. It was with extreme vexation, therefore, that the very next month he received news that three prisoners had barricaded themselves in the basement of D. block.

The situation which Hobson now faced was similar to that of June 1972. The three rebels — Rangi,[18] Wickliffe, and a man called Luke Jury[19] — were determined individuals who, perhaps angered by the damages verdicts, were now protesting against the continued exclusion of Thompson and another visitor from the institution. Hobson knew that if pressed, the trio would fight it out and with two lawsuits fresh in his mind, he embarked upon a different tack. The men were in a secure section of the block and could not interfere with its routine. Hobson, therefore, resolved to leave them where they were, and made no effort to move them. Outcry from demonstrators outside the institution, and an attempt by a Māori community worker to see the prisoners brought no response from the superintendent. After refusing food for fourteen days the strikers gave in. Hobson was satisfied with the way things had turned out, and subsequently the sit-and-wait response to peaceful protests became an established mark of his method.

In the standard blocks, actions of this type had fallen into abeyance. Living conditions and relations with staff were better than they had ever been. Hobson's policies had removed the stuff upon which successful rebellions are fed. Friction between some officials and inmates continued, as it always will, but like the conflicts among inmates, this was characterized by its impulsiveness more than its measure. Likewise, the genesis of tension tended to be emotive and personal and lacked an underlying rationale. For a time after 1972, there were no significant confrontations between opposing groups in Paremoremo. It was not until the middle of 1974 that a major incident occurred, the generation of which had nothing to do with prison politics. The 1974 'workshop blue' had its origins not in organized action, but in alcohol. Hobson's response was notable for its moderation and we shall see that these two aspects — the spontaneity of the trouble and the response of administration — made it another sign of new times at the gaol.

On Thursday 4 July 1974, inmates in Paremoremo's machine shop celebrated the imminent transfer of one of their number to medium security. A key ingredient of festivities was a home brew, which had been manufactured in the shop, and hidden there since early in the week. Little work was done in the machine shop that Thursday afternoon, and when the final bell sounded at 4.30 the men, three of whom were noticeably drunk, ambled back to their respective wings. A few minutes later, as two of them were being questioned in C. block about their condition by a chief officer, one of them attacked him with a piece of wood and knocked him out.[20]

Joined by a third person, the inmates quickly took control of the unit and it was not until after a violent struggle between them and staff reinforcements, that the rebels were forcibly subdued. Nine officers and all three prisoners were injured in the fight, but in spite of this, Hobson felt confident that no more trouble would occur.

The superintendent had no reason to suspect further strife, but in this case his assurance was misplaced. In A. block, retaliatory moves were already being considered. One of the intoxicated men had been in A. block at the time of the fracas, and although he had no knowledge of its cause, the sounds of battle and the sight of his three bloodied comrades being escorted or carried to the detention block were enough to ignite a desire for revenge. The result was a plot for civilian hostages to be seized during a debate in the visiting room that evening.

The hostage scheme never came to fruition because Dempsey Roberts, newly released from D. block, understood the consequences it would have. Thinking quickly, he devised an alternative plan and managed to convince the conspirators of its merits. Having reduced the gravity of what he felt in any case would be a disaster, Roberts then dissociated himself from it. Participation in the action was voluntary, and confined principally to A. block. Its secrecy was well kept and the following morning nothing seemed amiss as, at eight o'clock, prisoners filed down the 'airstrip' and into the sally port, which leads to the workshops. Those who intended taking part had entered first, and as they filed out of the cage and past the body of screws lining the corridor, somebody shouted 'Now!' and every uniform in sight was attacked.

A few staff attempted to make a stand, but outnumbered as they were, within minutes the corridor was cleared. Still locked out of their workshops, inmates now jammed one of the sally ports open so that the closed grille opposite could not be operated. In this way entry of the riot squad was impeded, and by the time the door had been worked manually, the situation had cooled enough for the rebels to surrender peacefully. Their object had been achieved. Five more officers had been injured and in relation to this, five prisoners eventually faced assault-related charges in open court. Upon conviction, the five, plus the two involved in the first mêlée, received cumulative sentences ranging from four to fifteen months.

The 1974 incident involved a curious combination of past and present trends, and as such is an important marker of the gaol's transition period. First of all, the readiness of the men to oppose authority in vengeful concert is sharply reminiscent of Paremoremo in the early era. However, secondly, the outbreak in C. block which initiated the confrontation was without any reasonable foundation. As such it indicated further erosion in the old custom of regulated resistance. Notably, Dempsey Roberts, a seasoned agitator of the old school, had refused to become involved.

Significant, too, was the response of administration. Like the strike of eight months earlier, staff showed a restraint in their handling of the workshop affair, which had been absent on many occasions before. There is no evidence of retaliatory action after the event, and in this way a cycle of escalating vendettas, with publicity, outside agitation, and questions in Parliament was avoided. Also worthy of note is the fact that a need for conciliation was apparently becoming accepted at the Department's highest levels. Contrary to the previous decade, when Robson had declared there would be no 'bargaining with rebels',[21] the legitimacy of some demonstrations was now conceded. In his final annual report, for example, Missen had commented:

> We also know from experience that prisoners cannot be compelled, nor should they be expected to submit without protest to living in cramped, squalid, and insanitary conditions.[22]

Although a charge of squalor could hardly have been levelled at Paremoremo, it is logical that where unnecessary violence was concerned, the same principle would apply. The playing down of 'incidents' and avoidance of provocation meant that the institution's low profile could be maintained. It was upon this which Hobson now placed a premium. Far from involving a 'policy of appeasement', as some had suggested earlier, the new approach merely resulted from a professional awareness of the complexities of security administration in the 1970s.

It is with some success that the 'low profile' method at Paremoremo has been implemented, and the workshop affair of 1974 marks the last time that security inmates have organized collective violence against staff. Since 1974, there have been numerous individual attacks, minor confrontations, and peaceful strikes, but the type of extreme violence and heavy group commitment which was visible in the early seventies ceased from the time of the third Labour Government.

The Finlay Era: an Appraisal

The Finlay administration in this country is the last which has been specifically identified with penal reform. Although it achieved much, however, the period is not remembered by Finlay as a satisfying one. He attained far less than he had hoped in these years, and in spite of initial enthusiasm, relations with permanent staff soon became strained. These staff complain that the Minister proved less effective in getting things done than they had expected, and the comment finds frequent expression elsewhere. There are a number of reasons why this is so.

Firstly we know that the head of caucus, Norman Kirk, was a man with conservative views on social and moral issues, and that penal reform was not one of his primary interests. Suspicious of academics, and perhaps a little nervous of Finlay's qualifications, he, and certain of his older Cabinet colleagues, were unreceptive to some of the Minister's more radical notions. As a result, much of the legislation Finlay would have liked to have seen passed quickly through the House was given low priority or deferred. As we have observed, the most significant enactments made under this administration only just managed to get pushed through Parliament before its final adjournment. Other planned reviews, such as that of the Penal Institutions Regulations, never materialized at all.

Secondly, after 1973, prison reform in New Zealand faced a mounting problem of logistics. Kirk died in 1974, but in October 1973 a revolution in OPEC had caused oil prices to quadruple almost overnight. From that point on, every dollar the Government could borrow went towards buying up reserves of petrol stock. By mid-1974, the effects of the fuel crisis had begun to affect the economy, placing huge constraints on Treasury. Henceforth, all other projects took a back seat. One of the early casualties was Finlay's hope to have Mt. Eden demolished. Although he continued to promise it would be done 'as soon as possible',[23] the oil shock ruined any chance of it becoming a reality.

Added to the economic crisis was that of prison overcrowding. Daily average gaol musters, which had been stable or in decline since 1971, suddenly shot up, from 2367 in 1974 to 2601 the following year. In 1976 they increased again, to 2882. A 20 per cent rise in population over eighteen months effectively put a stop to any progressive advancement in penal policy. From 1975 onwards, far from inaugurating new schemes, the Department of Justice became preoccupied with attempting to service existing ones. An unstable economy and unpredictable musters go far in explaining why penal progress during this time was so difficult. A third factor, and one which is also significant, is that from soon after he took the Justice portfolio, Finlay began to sense that he was losing favour with permanent staff.

In previous chapters we have seen that there are significant sectors of the penal division which are actively opposed to liberalism. Stoically conservative and rigid in their thinking, these groups counter any scheme which they feel departs too far from traditional objectives. It was these groups that Finlay first fell foul of and it did not take long for their disenchantment to spread.

In September 1973, largely in order to relieve the burgeoning load of work on the Ombudsman, Finlay announced that henceforth, prisoners would have the right to correspond directly and confidentially with him. The announcement, which came suddenly and without advice from Departmental heads, was contentious and unpopular. Prison employees and local authorities felt slighted by the provision. Accustomed to the right of censorship, the measure was seen as a threat both to discipline and to their own integrity. Disquiet was soon justified by some prisoners, who, rather than approach local administration about even minor problems, instead addressed all their complaints to the Minister. In this way, Finlay said, the measure earned him the 'undying hostility' of personnel.

Although he later attempted to assuage their anger by promising he would not adjudicate on matters of discipline and would send a copy of every reply to the superintendent, the resentment against him remained. 'From that point,' Finlay told me, 'they did everything they possibly could to frustrate anything I wanted to have done. If you watch them, you will see how effective the public service can be in frustrating ministerial objectives.' One of the first plans baffled by this antipathy was Finlay's intention to let prisoners use the grassed areas inside the fenced perimeter. Opposition from staff prevented the idea from going ahead. We have already observed how the ban on punishment diets was held in contempt, and when restrictions on the censoring of personal mail were brought, they too went largely unheeded. A ministerial instruction in July 1975 that the Department look into the possibility of liberalizing the composition of fruit gifts from visitors, failed to be actioned as well. At Mt. Eden in 1975, officials even refused to allow tomatoes into the gaol.

In this way, intentions of the Minister which staff found disagreeable were thwarted either by direct opposition or misapplication. Finlay soon realized what was happening, but was unable to do anything about it. He could not personally supervise the prisons, and even the setting up of an official watchdog, he felt, would be useless. 'The administration will say, when he visits, "Oh yes, we'll do that straight away!" ' Finlay told me. 'But next day when he's not there they'll say "Oh, bugger him", and nothing will be done.'

The other difficulty which Finlay encountered was with senior executives. Don MacKenzie, who worked in head office at the time and who was an old friend of Finlay's, remembers that the Minister's rapport with staff in the Department became strained during the later period. He attributes Finlay's failure to achieve as much as he wanted largely to his inability to overcome this rising mood of discord. Finlay agrees. Progress was painfully slow he says, and he waited impatiently for his policies to be processed by the

Department's legal draftspeople. As his term of office progressed, Finlay felt the opposition against him increasing.

> I started off with a fairly good relationship with my Department. But for some reason, the barrier began to rise: the traditional public service reaction to change. Somewhere along the line there was opposition. I'm not going to suggest it was political in any way, but just an inherent part of the way of life of the public servant, to find more reasons for saying 'no' than for saying 'yes'. . . . They found it much easier to put up barriers than to find avenues.

Finlay, of course, is not alone in his frustration with bureaucratic pedantry, and there is doubtless validity in his claim. However, senior public servants who worked with this Minister are able to illuminate a side of the matter which gives it a different profile.

All of the senior public servants I interviewed spoke highly of Dr Finlay. Cameron, whose ability and judgement Finlay respected, said that after the conservatism of the previous administration he was generally in sympathy with the thrust of the new reforms and felt it unfortunate that Finlay had not been in office longer. Although Missen only worked with Finlay for a little over a year, he too had welcomed the man because of his novel ideas and found him, of the four Ministers he had served, the easiest to get along with. Gordon Orr, who succeeded Missen, said that although they often felt frustrated with Finlay, both he and Cameron liked him very much. 'You couldn't help but like the chap,' Orr told me.

I interviewed five senior personnel who had worked under Finlay, and their comments about him were similar. But however much they may have liked the Minister personally, there was a strong undercurrent of dissatisfaction among them about the way he ran his Department. This intelligent and educated lawyer was considered also to be an idealist, and less than practical in some of his thinking. Although sympathetic with his general motives, officials felt that Finlay's method was wanting. They found his directives unclear and frequently erratic. They became annoyed with some of his actions — such as allowing uncensored mail — which they considered impetuous and ill conceived. They criticized him for a lack of single-mindedness; for becoming easily enthused with too many issues at once. The result was that plans were completed hastily or not completed at all. Officials with systematic minds found the tendency distracting and confusing, and believed that delays in the drafting of legislation were due largely to the number and variety of the Minister's demands. In addition to this, throughout Finlay's period, head office of the Justice Department was always short-staffed. Most informants believed that, had Finlay's handling of personnel and some of his more controversial measures been more tactful, the size and scope of his difficulties could have been reduced.

After 1975, criticism of Finlay for failing to effect all he had promised was common in liberal circles. Notwithstanding this, a review of the 1972–1975 period indicates that Finlay accomplished a great deal more than he is commonly given credit for. Remarking on the third Labour Government, Labour MP Dr Michael Bassett wrote, 'The Justice Department always said that if a Minister could succeed in introducing half the amount of legislation he wanted during a three-year period, then he could count himself successful . . . Finlay had done a good deal more than that.'[24]

An experienced politician, Bassett recognized the difficulties of managing a complex portfolio. However, more important than this, the MP's observation alludes to what was perhaps the most important inhibiting factor of the whole Finlay era: its brevity. In November 1975, in a major political reversal, the third Labour Government was punted from office. Although its popularity was quite clearly slipping after mid-1974, Labour's majority in the House had been so great that few, Finlay among them, foresaw any danger of dismissal. Like most of his partisans, he had looked forward to a second term in which the unfinished work of 1975 could be completed, and the errors of the past repaired.

JUSTICE AND THE 1975 ELECTION

The reasons for Labour's sensational loss in 1975 have been analysed elsewhere,[25] and I shall not repeat them here. However, a brief look at some major issues will assist in explaining the role of Justice in the course of the party's descent.

At the same time as the effects of the 1973 oil shock began to hit New Zealand — that is, in the middle of the Government's second parliamentary session — Norman Kirk fell ill with heart and circulatory problems. From this point, his efficiency as leader was erratic and on 31 August 1974, he died.

Although publicly, Finlay had been consistently heralded as the country's next most effective politician,[26] Kirk was replaced by deputy leader W. E. Rowling, a man of intelligence and integrity, but of low public stature. After the magnetic Kirk, Bill Rowling's disadvantage was great, and try as he might, he could never regain the heights of his predecessor. In a research project devised in 1975, a large number of National voters described Labour's new head as 'weak' and 'indecisive'.[27] Many of these were potential Labour supporters and it was Rowling's poor public image which convinced them to switch their allegiance.

Throughout his term of leadership Rowling was cursed with indistinction, but this may have amounted to little, had it not been for the circumstances under which he served. From a political point of view, Kirk's death could not have been more untimely. Shoved into office, the new chief executive was immediately faced with the legacy of the oil shock: rocketing inflation, deepening recession, and a billowing overseas debt. There was little hope of changing the circumstances of a foundering economy. By the time polling day arrived fifteen months later, intensive borrowing had transformed the large balance of payments surplus of 1973 into an 863 million dollar deficit.[28] In 1975, Labour's record of government looked grim.

While penal policy, as such, was not a great issue in the 1975 campaign, its controversial impact gave it a high profile. However, unaccompanied as it was by a systematic legitimizing programme, much of the Finlay method's press was bad. Rather than being welcomed as necessities of penal progress, reforms were rejected as dangerous indulgence and solicitous mollycoddling. Unlike the Hanan developments, therefore, which the community had been encouraged to applaud, Finlay's measures alienated large sectors of the electorate and cost Labour many more votes than they gained. Of this, Finlay says:

> The administration of the Justice portfolio, and all the rumbles that went with it, certainly did our party no good. The liberalizations and the things that went wrong were taken up in certain quarters and twisted and re-presented [and] were certainly of no political advantage to us. For all the acclaim I might have earned in terms of liberalization, I got twice or three times as much abuse or hatred because of 'being soft' on criminals. . . .
> Giving prisoners the vote was the start of it all. That was presented as an attempt to get votes from a quarter that would naturally respond to a gesture made in their direction.[29] All these well-intentioned things, they just paid no dividend electorally.

The Electoral Amendment Bill, which started the ball of anti-Finlay sentiment rolling, was introduced in May 1975. Although this nourished feelings which were already growing in the prison service, it was not until less than two months before voting that the worst publicity of Finlay's career took place.

In September 1975, an Auckland solicitor, Leslie John Vercoe, was sentenced to six years' imprisonment for misappropriating 436,000 dollars of his clients' funds. Some days later, making use of the new correspondence privileges, Vercoe wrote to the Minister, outlining his observations of the New Zealand penal system. The Minister was moved by Vercoe's remarks, and sent him a personal reply. Commencing with the address, 'Dear Les', Finlay commiserated with his fellow lawyer, recording his own frustration with 'a penal system [which] is busy grinding people down rather than building them up', and wishing he could 'put a bomb under Mt. Eden . . . with all its inhumanity and lack of sanitation'. He thanked Vercoe for his comments and instructed his Department to take careful note of what the inmate had said. 'They do care, you know', Finlay concluded, 'and try their best, but often the system is too much for them as it is for me'.

Although perhaps indiscreet, these comments, one might think, were innocuous enough. However, on 2 October, when a prison officer noticed Vercoe showing the letter to some other convicts, he became interested and asked to see it also. Stupidly, Vercoe agreed. The following day the prisoner was approached by more staff, who took the letter and, against Vercoe's wishes, had it photocopied. On 5 October, the full text of the Minister's response appeared in *The Sunday News*, and it was published in *The New Zealand Herald* and *The Auckland Star* the following day.[30]

Confronted with the release before it went to press, Finlay of course was outraged, and he wildly strove to prevent its publication. This added spice to what might otherwise have been a rather lack-lustre column, and his vow, 'heads will roll' over the affair, produced an eye-catching headline.[31]

With the election only weeks away, the upset came at a bad time for the Government, allowing the opposition to reap a maximum advantage. Although an attempt to secure a snap debate was thwarted, Finlay, seriously compromised, was accused of 'undermining the whole prison system'.[32] The opposition's Justice spokesman then declared that if elected his party would revoke the special correspondence rights of prisoners immediately.

Staff who had been disturbed about the mail censorship issue capitalized on the uproar as well. Interpreting the letter in its worst possible light, the Public Service Association and the prison officers' subunion expressed deep concern about its contents,[33] and attempted to have the 'hot line' privilege removed forthwith. This Finlay refused, but he endeavoured to discredit Vercoe by revealing that he had been a member of the National party and possibly an agent of the Secret Intelligence Service as well.[34] Regretting that the letter had been written at all, Finlay denied any intended or implied disparagement of staff and reaffirmed his full confidence in their judgement and loyalty.

Unfortunately for Finlay, the considerable antipathy against him had now hardened and his attempts to dislodge it were futile. Where before, antagonism had been hardly perceptible from the outside, staff feeling at the end of 1975 burst into undisguised hostility. Of his relations with prison officers at this critical point, Finlay told me:

> ... from the prison administration I had enormous rejection. When I went on the campaign trail in 1975 and I had to go to Invercargill borstal it was virtually a seige. I almost needed protection to go through the place. There was a lot of hostility. And also at Wi Tako, where the superintendent was the head of the prison section of the PSA. They were *very* hostile.

In the end it was a combination of economic crisis, unpopular leadership, and devastating criticism from the opposition's leader Rob Muldoon which sealed Labour's fate in 1975. The controversial Justice programme and the humiliating scandal, which immediately preceded the campaign, enlarged the Government's portrait of ineptitude. To this extent, Labour's handling of Justice and prisons certainly aided its defeat.

When the nation went to the polls, the drop in Finlay's popularity was greater than that of any other sitting politician.[35] Voting returns show that whereas the overall popularity of the Labour party between 1972 and 1975 decreased by some 8.9 per cent, the fall in Labour votes at Paremoremo was almost double this, 16.4 per cent. Where in 1972, 48.9 per cent of the Paremoremo votes had gone to Labour, in 1975 the Labour candidate at Paremoremo (Dr Bassett) polled only 32.5 per cent.[36] This disproportionate reduction is largely attributable to the ministry of Dr Finlay, whose unpopularity among staff at election time was enlarged by the Vercoe affair.

Although Finlay retained his seat in November, the Government lost an unprecedented twenty-three members. In the 1975 poll, the fifty-five:thirty-two seats with which Labour had dominated National was completely reversed.

For the next three years, National would rule the House with a majority of fifty-five votes to thirty-two.

NOTES

1. *Report on Auckland Prison* (1973) 69.
2. NZPD vol. 386 (1973) 3695-6.
3. NZPD vol. 386 (1973) 3696.
4. NZPD vol. 386 (1973) 3697.
5. *Report on Auckland Prison* (1973) 12.
6. Finlay had for years been an advocate of granting union wages to prisoners. External pressure for the move came in various forms (for example, letter from K. Newnham to T. Skinner, 13 October 1972; press release from Project Paremoremo 21 December 1972; letter from Finlay to Thompson 30 April 1973; *Dominion* 18 April 1973; *Sunday Herald* 30 April 1973.
7. As noted earlier, sections 15 and 16 of the Penal Institutions Amendment Act were passed in 1975, but were not brought into effect until the passing of the Penal Institutions Amendment Act in 1980 (s. 13). The penalty was removed from the Regulations by section 25 Penal Institutions Regulations 1961, Amendment No.3, 1981.
8. *AS* 21 June 1973.
9. *Report by Powles and Sinclair into Matters Pertaining to Paremoremo Prison* (1972) 15, 33. In 1974 Don McKinnon coached by A. team, while the B. team was instructed by the late Geoff Greenbank, a former principal of Kings College, Auckland.
10. *NZH* 16 August 1975. This was the Haora-Stapleton incident. Former Paremoremo inmate, Evan Haora, had been invited to speak at Auckland's Annual Law School Dinner on 14 August 1975. Actions of student/detective R. J. Stapleton, who advertised the ex-prisoner's criminal past to members of the Law Student's Association, resulted in the invitation being withdrawn. This was followed by an unsuccessful move to have Stapleton expelled from the Students' Association (and, in consequence, from the University) (*Craccum* 16 September 1975). Stapleton, who later left the police force, now practises as a Barrister and Solicitor in Auckland.
11. In 1975, 5.2 per cent of those granted home leave breached parole. Escapes jumped from 102 the previous year, to 166 (*App.J.* E-5 (1976) 11).
12. This, it will be recalled, had been introduced by Hanan in 1964 (Penal Institutions Amendment Act (1964) s. 2).
13. Penal Institutions Amendment Act (1975) s. 4.
14. Entitled 'The Life and Bad Times of Mihaly Bede', the series of five articles in the *8 O'Clock* is dated 8, 15, 22, 29 April, 6 May, 1972.
15. *Sunday News* 10 March 1974, Meek (1986) 142.
16. *Sunday News* 10, 17 March 1974.
17. *Dominion* 13 March 1974; *NZH* 13 March 1974; *Sunday News* 17 March 1974.
18. Rangi had been back in D. block since the 1972 incident.
19. Luke Jury and his brother Brian were subsequently sentenced to four years imprisonment after a siege in Ponsonby, Auckland in 1979. In April 1980, Brian Jury hanged himself at Paremoremo. Four days later, after having attended his brother's funeral, Luke Jury hanged himself in the Wairoa police cells (*NZH* 28 April 1980).
20. One of the participants was D. A. Redrup, who deliberately burned himself to death in D. block in 1973.
21. *App.J.* H-20 (1963) 12.
22. *App.J.* E-5 (1974) 13.
23. *NZH* 12 December 1973.

24 Bassett (1976) 273.
25 Bassett (1976); Levine & Robinson (1976).
26 Bassett (1976) 185.
27 Levine & Robinson (1976) 146.
28 Sinclair (1980) 312.
29 See for example *NZH* 18 July, 14 October 1975.
30 NZPD vol. 402 (1975) 5181.
31 *Sunday News* 5 October 1975.
32 *Sunday News* 5 October 1975; NZPD vol. 402 (1975), 5181-2.
33 *NZH* 14 October 1975.
34 The second allegation was subsequently denied, both by the Prime Minister and by the director of the SIS (*NZH* 11, 13 October 1975).
35 Labour's percentage of the vote in Finlay's Henderson electorate dropped by over seventeen points, from 60.22 in 1972, to 43.16 in 1975 (*App.J.* E-9 (1973) 110-11; *App.J.* E-9 (1976) 118-19.
36 *App.J.* E-9 (1973); *App.J.* E-9 (1976). Prisoners, registered in the electorates of their home towns, had imperceptible effect.

20: 1975 AND AFTER

After 1975, New Zealand entered another phase of slow movement in penal policy. The pendulum which Finlay had set in motion in 1972 swung sharply back under the new administration, which saw an earlier acting Minister, David Thomson, taking the portfolio of Justice.

Aged sixty-one at the time of his appointment, and an honorary Colonel in the territorial army, Thomson had served three years as a prisoner of war during the Second World War, then farmed, before entering politics in 1963 as MP for Stratford. He had already been Minister of Defence, Tourism, Labour, and Police before taking on Justice in 1976.

As might have been expected from a man of Thomson's background, the new Minister was conservative and opposed many of the policies of his predecessor. Before the election, he had promised to stop inmates from corresponding with the Justice Minister and this was one of the first things he did upon reaching power. Then the Electoral Act was amended again, taking from prisoners the right to vote.[1] Although the new statutory minimum for the life sentence was allowed to remain at seven years, Thomson vowed that under his administration, no convicted murderer would be let out in less than ten.

Thomson remained in power until the election of 1978, and his regime marked the end of an era when political and permanent heads had an obvious influence on penal developments. From 1975 onwards, the profile of Justice was lower, and what happened in maximum security after that depended more on spontaneous local events than on a centrally directed plan.

In 1978, Gordon Orr stood down to take a chair in constitutional law at Victoria University, Wellington, and his replacement as Secretary for Justice was John Robertson, a fifty-three year old career public servant from Defence. Described by one who knew him as a 'nuts and bolts administrator',[2] Robertson was typical of the new order in Wellington. He had no background in prisons, but his experience as a senior official in the State Services Commission and as National President of the New Zealand Institute of Public Administration showed his province of utility. He had studied the reorganization of Government agencies in the United States in 1960-1961 and had advised the 1962 Royal Commission on State Services, but even so his period was not one of great activity in penology. Thomson was not interested in reform and changes in the constitution of inmates were having a dampening effect as well. By the time Robertson arrived, the course of prisons was set. When he stood down in 1982, his replacement by Jim Callahan from the Social Welfare Department occurred almost unnoticed.

Thomson remained in the corridor of Justice for three years, then was replaced in the post-election reshuffle of 1978. His successor was James Kenneth McLay, thirty-four, a young high-flier who had entered Parliament three years before on the landslide of Robert Muldoon. Educated privately at Kings College in Auckland, McLay had studied law at Auckland University and joined the National Party in 1963. Graduating LL B in 1967, when he took Birkenhead from Labour in 1975, McLay had practised for a number of years as a solicitor, then as a barrister. In 1975, the new Minister and Attorney-General was the youngest member of the Muldoon cabinet.

It was with an unusual combination of youthful imagination and ingrained conservatism that McLay conducted his office. In 1977, he had introduced a private member's bill which sought, successfully, to reduce unfair cross-examination of women at rape trials, and in 1981 he brought in a new form of corrective training.[3] On the other hand, McLay also hoped to increase the formality of proceedings in the lower court. So from 1979, Stipendiary Magistrates became District Court Judges, and gowns were prescribed for the bench.

Where penal policy was concerned, one of the greatest contributions of the post-1975 era came under the administration of Jim McLay when he set up, in 1981, a committee to investigate measures to deal with crime and its treatment in the eighties. Presented at the end of that year the *Report of the Penal Policy Review Committee* advocated a number of changes, including the setting up of 'Regional Prisons' (prisons with an emphasis on community involvement), repeal of the statutory life term for murder, and the removal of parole, except in the case of non-finite sentences.[4]

A significant component of the report was its new ideas on correctionalism, the guiding light of penal policy in the 1950s and 1960s. The report of 1981 rejected the notion of rehabilitation. The purpose of imprisonment, it went on, is 'the containment of individuals who are being punished by the loss of their liberty under humane, fair, and restrained conditions'. The prison's principal objective should be to neutralize its own deleterious effects.[5] Although not legally binding, the review sign-posted the road upon which penal policy was already beginning to travel. Many of its many recommendations were not adopted at all, but some — like Regional Prisons and a concept called 'Thoroughcare' — were in the legislative pipeline by the time McLay and the Muldoon regime were ousted in 1984.[6]

EVENTS AT PAREMOREMO

Because interest in benefiting criminals had lapsed with the end of Labour in 1975, events at head office thereafter had little impact on the dynamics of maximum security. Here developments were considerable, however, but their direction was now unfamiliar.

At Paremoremo, the reduction in solidarity resistance, which happened after Hobson took over, continued throughout the seventies and a new pattern of social relations accompanied it. Inmate allegiances, although still anti-authority, became less demonstrative than before, and violence toward staff was reduced. Where in 1971 an average population of 160 prisoners recorded eighteen assaults on staff, by 1976 this figure had dropped to seven, even though the gaol's population had risen to 200. It stabilized at between five and ten assaults a year through to 1984.[7] In 1985 the superintendent remarked on their current rarity, 'Aggression is not directed at staff, not in this prison, anyway.'[8]

The mid-1970s were the most tranquil period the gaol has ever known. For although no occasion for real conflict arose during this time, the memory of the 'dark days' of 1969–1972 was still strong in the folklore of the old timers. This was the period when I was at Paremoremo and the relations I observed between 1975 and 1978 are those described in my MA thesis[9] and in the book I published after my release.[10]

Because of the small size of its population — which cannot exceed about 200 — because it is closely confined, and because the influence of early heroes was strong, deviation from the social code of prisoners in the mid-1970s was uncommon.[11] The incidence of informing ('narking', 'grassing') was low, indeed the power of rumour was so strong that even talking alone to a prison officer could have been dangerous. None the less, the value of stability was recognized and it was agreed that rapport with management was essential. Protection of inmate interests was paramount and even that most endemic of prison *malaises* — thieving — was only sporadic. There was no system of convict bosses at Paremoremo before 1978, nor was there any observable hierarchy. Leaders, by and large, only emerged in times of conflict. Leaders in small communities need popular recognition, and at Paremoremo this mandate was absent. The ethic of egalitarianism — itself an extension of an important national value — received the strong sanction of administration.[12]

Exploitation of any form was condemned, and the supporting credo of equality held that generally inmates should share the same rights and privileges. Although fighting was an acceptable means of arbitration, violence did not dominate social relations, and the bullying of weaker inmates was rare. Homosexual rape, so prominent in American penitentiary literature, was unusual. Strength and self-reliance were valued, but the essential of solidarity created an atmosphere of protection and support for most of the community.

However, the 1970s were not without problems, although they were different from before. At first minor, these intensified as the decade progressed. Unlike the early years, when prison activity could be closely linked with managerial practices, events inside became increasingly tied to trends in the free world. In the 1970s prosecutions for drug offences soared from 390 in 1970 to 5512 in 1979. Numbers imprisoned for such offences grew from thirty-nine in 1970 to 232 at the end of the decade. Imprisoned drug offenders smuggled drugs into the gaol, introducing others to their use. Late in 1974, the first dope scare at Paremoremo broke, when an inmate was found to have been receiving regular quantities of hard narcotics, hidden in the spines of books. The traffic of drugs — hard and soft — soon became integral to inmate organization and economy, and gave rise to a number of scandals in the late 1970s. Resulting publicity led to restrictions on visiting in some institutions and an increase in the powers of prison officers to search.[13] The drug flow was largely unaffected.

The situation I experienced at Paremoremo in 1975–1978 had been a direct legacy of the past, combining conflict with an existing atmosphere of tolerance and tranquility. As the old leaders were released or transferred, however, so early traditions died. Tough young men entered the gaol, with ideals and sympathies that were new. As in the United States,[14] the politicism of the late sixties and early-to-mid-seventies gave way to increasing violence and factionalism among inmates.[15] Whereas ethnic differences had previously been unimportant, racism and racially oriented violence between the evenly numbered Māori/Pākehā population became common in the later 1970s. By the end of 1975, ethnic lines were already forming and culminated in the near-fatal stabbing of a white man by a young Māori in December. Rising racial tension led to a general lock-up in January 1976, with twenty-one inmates moved into segregation. Nearly all of them were Māori.[16]

Reflecting outside trends, ethnic gang membership also grew. A phenomenon which had begun developing in New Zealand in the sixties, gangs bcame increasingly popular during and after the seventies. Between 1980 and 1988 gang membership estimates doubled, from 2200 to 4400 members,[17] distributed among forty-four different clubs.[18] The violence profile of such groups is high — they now account for around 12 per cent of all murders, for example[19] — and by the late eighties, gang association in prisons had risen to around 25 per cent of the national muster. Half of these were from a single group, the Mongrel Mob.[20] Gang members, perhaps 80 per cent or more of whom are Māori or Polynesian,[21] are today 43 per cent more likely to be sent to maximum security than the gaol population as a whole. In consequence, around 32 per cent of Paremoremo's men currently belong to gangs,[22] or did so when they were arrested.[23] The gaol's Māori/Polynesian population, for years stable at around 50 per cent, has now reached almost 64 per cent. In May 1985, 82 per cent of the top custody men had been sentenced for serious crimes against the person, including 25 per cent who were serving time for homicide.[24] Almost 88 per cent were doing three years or more.[25]

For the prison itself, the ramifications of these factors have been manifold. Firstly, while the profile of the gangs had previously been low in maximum security, today gangs have become the predominant feature of its infrastructure. Whereas up until the late 1970s, allegiance to the inmate code was seen as paramount, many today recognize two sets of loyalties: those to the gang and those to prisoners. When expectations conflict, inmate interests are subordinated. The imperative of gang dominance supercedes all others.

The emergence of gang identity has radically affected power relations. Inmate solidarity has been broken and the power of the solitary man has been destroyed. Recently the victim of the December 1975 stabbing, newly released from Paremoremo, said to me:

> ... if you did attack [a gang member] you would be taking your life into your hands because — let's say this, Greg — they know not to say 'fuck' with me too much.... But if I ever had cause to attack one of them, well it's quite possible that the others would be in. And I'm always aware of this.... In prison ten years ago you knew that if you had a fight with someone, the worst that was going to happen was *maybe* a kicking. But today you've got to size up each situation. 'Do I go on with this, or what?' Because it might be worth a killing. My own.

In July 1979, Paremoremo had its first acknowledged case of homicide when Keith Hall, the sex-killer of a young girl, had his throat cut in the basement of A. block. No culprit was identified. A second murder occurred in January 1985 at Mt. Eden. The assailant, a former Paremoremo man, was returned to maximum security to begin a life sentence.[26]

The prison's record of serious violence, which attracted considerable media attention in 1979, continued into the eighties.[27] Between the years 1978 and 1984, despite the low and stable incidence of violence against staff, assaults among prisoners increased almost threefold.[28] Reciprocally, the number of prisoners segregated — either at their own request or at the superintendent's direction — more than doubled. At the end of 1980, after a serious attack on a sex offender in December, a landing of classification was set aside for protective custody. Soon afterwards the whole block was taken over, and classification services were transferred to Mt. Eden. Today the section regularly holds a dozen or more 'protected' inmates, as well as others undergoing different forms of separate confinement.[29]

By the end of the 1970s, the gaol had split into factions, dictated by gang designation. A. block contained the Headhunters and those acceptable to them, B. block held the Mongrel Mob, the Black Power, and admissible others, and C. block had weaker groups and the unaffiliated. The cell blocks were still integrated at work and recreation, but the deepening entrenchment of gang allegiances was causing continual friction. Tensions in the prison rose steadily in the early eighties, and at the end of 1984 they exploded. On Christmas Eve that year, as inmates gathered in the gymnasium for a movie, a brawl erupted between members of the Mongrel Mob and the Headhunters. In the fracas, two Mongrels were stabbed and one suffered bad head injuries, and seven Headhunters were eventually shifted to D. block.

At Paremoremo, the Christmas Eve incident came as no surprise, but it marked a significant turning point for the gaol. Mob retaliation was likely, and the avoidance of escalating conflict necessitated radical measures. Reorganization seemed the only solution, and in January 1985 a formula for across-the-board segregation was put into effect. From then on the prison reverted to a style of organization similar to that installed by Buckley. The blocks became discrete sub-units, each independent of the others. Communion at work, recreation, and in the visiting room was stopped. Access to facilities like the gymnasium, the library, and the visiting room was allocated to each section at different times of the week. Inter-block competitions and visits from outside sports teams were curtailed, and externally organized facilities, like Māori culture and discussion groups, were severely restricted. Debating effectively disappeared. The prison workshops nearly all closed down.

The impact of gang influx on routine at Paremoremo has been immense, and has had serious repercussions for prisoners. Added to this, and certainly related to it, have been adjustments necessitated by a growth in self-mutilation and suicide, which commenced with the rising gang presence. Mental abnormality has always been a problem in prisons, indeed until recently up to 10 per cent of Paremoremo's population might at any time have been resident in psychiatric hospitals.[30] Incidences of self-mutilation and attempted suicide have been similarly high, but apart from a likely accidental self-killing in February 1972,[31] they caused no loss of life.[32]

In 1980 this suddenly changed. In April that year two maximum security men who were brothers hanged themselves,[33] heralding the beginning of an era that affected gaols nationwide. Of the twenty-nine prison suicides recorded in New Zealand between 1971 and 1985, more than half[34] occurred in the last two of these years. At Paremoremo there were eight confirmed suicides, all this decade, and five of them in the years 1984-1985.

Endemic tension and violence at Paremoremo created a new matrix of stresses, which were intensified in 1983 when, after a Committee of Inquiry into Oakley hospital,[35] routine transfer of mentally disturbed prisoners to psychiatric institutions was stopped. Four suicides and a rise in self-mutilations ensued,[36] leading an inter-departmental working committee in November 1984 to recommend the establishment of a special psychiatric facility. Three more suicides and protest by organized cell-block committees in 1985 took place before a remedy was worked out. Then, partially in response to inmate pressure and partially to divert the suicidal, D. block's hours of unlock were increased from three to seven hours a day, and inmates were allowed television sets in their cells. Six weeks later, in July 1985, the psychiatric holding unit at Paremoremo was upgraded for therapy purposes and a landing of classification was set aside as well. There was now provision to treat up to seven men for up to seven weeks at a time in the unit, and the fourteen cells in classification could be used for any overflow.

At Paremoremo in 1984, an era ended with Hobson's retirement in September, but by then even his influence had weakened. The warring elements had now taken over, and administration was increasingly occupied with keeping violence and suicide in check. Hobson was replaced in 1984 by his former deputy Sid Ward, fifty-eight, a West Coast sawmiller who had joined Prisons in 1954. Ward had deputized under Buckley at Paremoremo, and had seen the changes wrought by Hobson. Since 1978, he had been in charge of Mt. Eden and when he returned to Paremoremo six years later, Ward fully identified with the past and did not deviate from the path of his predecessor. It was Ward who allowed inmates to have television sets in their cells, who eased conditions in D. block, and who established the psychiatric facility in classification. However, it was he too who, after the Christmas Eve brawl, had decided to segregate the blocks.

Sid Ward was a compassionate man who enjoyed a good rapport with prisoners, but in September 1985, not twelve months after taking over, he retired. His replacement was Les Hine, aged fifty-four, who had joined the service in 1953. Hine had become superintendent at Mt. Crawford in 1970, but he came to Paremoremo from Wanganui, where he had officiated since 1983. Known principally for his thirteen years at Mt. Crawford, 'Bully' Hine's bombastic manner was subdued in maximum security. Here the inmates were tougher than he was used to, and the officers there were more militant. Gang belligerence was growing and since the segregated block policy had begun, aggression became increasingly directed at staff. Staff looked to the superintendent for a solution, but Hine knew that if action was too vigorous inmates would probably riot. On the other hand, if he was seen to be too soft the officers would threaten to strike. This was his dilemma, and he was soon put to the test.

Not two months after Hine took over Paremoremo B. block inmates locked in their cells attacked passing officers with sticks. This set a pattern for the next twelve months in which confrontations — collective and individual — between staff and inmates of B. block escalated with a fierce momentum. Recorded assaults on staff doubled in 1985, to nineteen. There was a 50 per cent increase in inmates requiring segregation, numbering eighty-one out of a total population of only 200. In October 1986 a lock-up and search of the block revealed an array of offensive weapons, and it was reported that relations in the unit were now so bad that non-gang members were refusing to be sent there and only twenty-eight of its forty-eight cells could be used. Of the thirty men in D. block, it was claimed twenty were there of their own accord, trying to escape from the gangs.[37] Hoping to re-establish order, that month, Hine imposed restrictions on the unruly B. block, with increased lock-up and a prohibition on hobby tools and gang insignia. However, in spite of his efforts the Mongrel Mob continued as a problem of growing dimension, and began to campaign for a restoration of privileges. Early in 1987, a movement began in the block to force a compromise, and this caused relations to worsen. In the winter of 1987 there were thirty-seven assaults on officers at Paremoremo, twice the 1985 figure, the bulk of them in B. block. Major offenders were segregated, but the staff now wanted tougher controls. Hine insisted on moderation and in the end nobody was satisfied. On 24 June, a hunger strike began involving all of B. block — with a dozen Mob members — plus one Mongrel in C. and eleven in D. blocks. Strikers demanded a reinstatement of normal privileges in B. block, and the Mongrels wanted to be allowed to live together in the same unit. Administration responded with the promise of an inquiry into the matter, provided order returned.

An agreement was reached and the investigation begun, but its results were not expected for six weeks. Wanting immediate improvements, D. block Mobsters now barricaded themselves into a recreation room, and strike action continued until the end of the week. The following Monday, a D. block screw was assaulted, receiving a broken jaw. The whole block was then locked up and during the course of searches, four more officers were attacked. One had his nose broken.

Staff were furious about the attacks, and once again they blamed the superintendent. On 19 August, a meeting of staff passed a vote of no confidence in Paremoremo's senior administration, giving them a week to make effective changes. These changes Hine agreed to, but having now been at the gaol for two years, he was due to retire before the end of 1987. Ward, before him, had served less than twelve months and this added to the discontentment. Officers were sick of what they called 'caretaker superintendents', who were put into the job shortly before they retired. What they wanted was a long-term appointment with effective policies and the establishment of some continuity. It was plain, with Hine due to go, that he did not fit that bill.

Four days later, on 23 August, it was announced that Hine would take early retirement and be replaced. However, before he left, on the advice of the investigating committee, Hine awarded the Mob what he called 'an absolutely final opportunity' to show they could live harmoniously within the prison.[38] The whole of the Mongrel Mob, minus those involved in the 17 August assaults, were taken from D. block, and a total of seventeen were permitted to live together in B. block under full privileges. In mid-September 1987, Hine moved to head office and the new superintendent took over.

PENAL POLICY AND THE FOURTH LABOUR GOVERNMENT

In June 1984, after nine years in Government, the Prime Minister, Sir Robert Muldoon, dissolved Parliament and called a snap election. Justice was not a major point in 1984, but concern over issues, such as the state of the economy and the despotic leadership of the Prime Minister, assisted Labour's landslide majority of seventeen.

The leader of the new Government was the young and vigorous David Lange, a forty-one year old lawyer with a sharp wit and powerful oratory. A former unionist and a practising Methodist, Lange's reputation was that of a friend of the needy and downtrodden. In the 1970s he had been a familiar face in the law courts of Auckland. For his deputy and Minister of Justice, Lange chose another young lawyer, a quiet academic and a published author, Geoffrey Winston Russell Palmer. Palmer, born in Nelson in 1942, had studied Arts and Law at Victoria University, then travelled to Chicago on a Commonwealth Fellowship to study for a Doctor of Jurisprudence. He remained in the States for five years, teaching law at the universities of Iowa and Virginia, returning to New Zealand in 1974 to a chair in law at Victoria University. He joined the Labour party in 1975, and in 1979, in a by-election, became MP for Christchurch Central.

When Palmer took over Justice in July 1984 he inherited a penal system that was troubled and restless. The gaols were overcrowded, and there were problems with inmates and staff. The question of psychiatric placement was still unresolved and the suicide rate was soaring. This trend was greatest at Paremoremo, which had the added problem of violence and factionalism among inmates.

Although Palmer's strongest professional qualifications were in constitutional matters, he had some firm ideas about Justice. In the months after taking office he toured most of the nation's penal facilities, interviewing staff and meeting some of the inmates. At Paremoremo he entered A. block and shook hands with the inmates. For a brief time it seemed that another Finlay type period might begin. However, change came rapidly and by the end of 1984, things had started to go wrong. First was a crisis in prison musters.

Since the mid-1970s, national musters had been stable and seldom gone much above 2800. At the beginning of 1984 the population stood at 2700, but in the winter of that year there was a sudden leap, and by the time Palmer came to power in July they had exceeded 3000. In December an all-time high of 3196 had been reached, nearly 400 greater than the country's nominal capacity. At receiving institutions like Mt. Eden, where the population rose to 375 that month, the problem was sharpest and staff refused to accept more prisoners until the pressure of numbers was eased.

In October 1985, a new Criminal Justice Act came into effect. The legislation had first been introduced by Jim McLay, the previous Minister, in December 1983, but had lapsed at the time of the 1984 snap election. Adopted and partially revised by Palmer's Ministry, its final content was the result of bipartisan effort. The Act addressed numerous areas in justice, including sentencing principles, custodial and non-custodial sentences, and mentally disordered offenders. From the point of view of imprisonment, one of the important features of the Act had to do with the length of prison terms. After 1 October 1985, restrictions were imposed against imprisonment for some property offences, and there was an increase in the non-custodial options open to the courts. Time spent in prison pending trial or appeal was now automatically deducted from custodial terms. Previously, remand time had been dead, so now the length of incarceration could be shorter by up to twelve months or so.

Another component was the new parole programme. Before October, standard remission of only one-quarter had been offered, with one-third available in camps, or otherwise upon application. Parole board discharge was possible (but rare) after half sentence or seven years (whichever was the sooner), and this only applied to those serving five years or more.[39] The new Criminal Justice Act liberalized provisions for early release, introducing a formula similar to that used in the 1920s.[40] One-third remission now became standard and a form of parole from local authorities known as District Prisons Boards, or the Parole Board in the case of those serving seven years or more, became available to all.[41] The Minister of Justice opposed heavier penalties for rape,[42] and although preventive detention remained for the time being, he wanted it removed from the statute books.

The impact of the Act was immediate. Many, having served half their terms, were eligible for parole and there was a sudden flush of releases. National prison populations dropped dramatically. By February 1986, they were down 843 on the October figure of 3060. It was the lowest figure since 1970, and a heartened Minister announced the intended closure of two institutions — Wanganui and Waikune. Sale of this land and other 'surplus' properties, it was hoped, would save the Department up to twenty million dollars that financial year.

The satisfying results of the Criminal Justice Act were short-lived. By May 1986, 1070 inmates had been released on remission or by District Prisons Boards, but of these over 300 had already re-offended. The gaol population was creeping up again. Three months after the sixteen-year low of 2217, the muster was up to 2580. By November, it was 2953 and rising, exceeding 3400 before the end of 1988. Once more it was Mt. Eden where the worst pressure was felt. In early April 1987 the muster there was 466, one hundred above its maximum nominal capacity. Prison staff reacted by locking doors on more receptions until numbers had abated. Mass transfers to minimum security then began, although relocation was complicated by the fact that Waikune, one of the country's largest camps, was due to close in September and was unavailable for use. Wanganui remained open and was able to take some of the excess numbers. The rest, in the meantime, were held in police lock-ups and it was a week before Mt. Eden could be utilized. It was announced that a new eight million dollar block for 120 would be built at Mt. Eden, to be ready in 1989.[43] Another, for 150, would be constructed in Northland, and there were plans for a third in Hawkes Bay. Remand facilities at Waikeria and Paparua would be extended. Nearly twenty-two million dollars was budgeted for building and upkeep up to March 1988, and over twenty-two million more was expected for the next financial year.[44]

The unpredictable jumps in population after 1984 were the same as those which had frustrated Barnett in the fifties. Added to this, however, a rise in violent and sensational offending was causing anti-criminal sentiments to harden. In March 1986, a woman in Christchurch was abducted, raped, and strangled, then left for dead in the boot of a car. Seven months later in the same city, six year old Louisa Damodran was abducted after school and drowned. The offenders in both cases had long histories of sex crime. The two Christchurch cases were prominently covered by the media, and outcry over them and an apparent rise in sexual violence caused the Minister of Justice, in November 1986, to announce new measures against violent crime. Instead of being abolished, preventive detention would now be broadened to cover violent as well as sexual offending, and its age limit reduced from twenty-five to twenty-one. The moves were generally applauded and endorsed after the rape of a woman at a Mongrel Mob convention in January 1987.

What people suspected about the crime rate was confirmed by the release of the findings of a Committee of Inquiry into Violence in March. This, the 'Roper Report', indicated spectacular increases over the last ten years in crimes of violence. Of particular note were proven charges of rape (up 55 per cent), robbery and aggravated robbery (up 200 per cent), murder (up 115 per cent) and manslaughter, which had leapt by 234 per cent in the previous decade.[45]

There was an election in 1987 and it was the first time since 1960 that law and order had emerged as a significant political issue. On 19 June, less than two months before the election, another six year old girl, Teresa Cormack, was murdered. This time her killer was not found and on the day of her funeral on 1 July, large, emotional rallies were held throughout the North Island. Responding to clamour for a return of capital punishment, Opposition Leader Jim Bolger pledged to come down heavily on violence. There would be a national referendum on hanging and a free vote in Parliament if his party became the Government. However, Labour promised a crack-down too, and although it rejected the noose, it had in March already launched a 950,000 dollar law and order campaign. Moreover, its legislation with tougher penalties for violence was well advanced. Criminal Justice Act 1987 Amendments 2 and 3, which came into effect just before the election, provided, in addition to the alterations to preventive detention, for mandatory imprisonment without parole for certain violent offences and discretionary cancellation of remission for specified violent and sexual crimes. The non-parole period for murder, which the third Labour Government had reduced to seven years in 1975, was increased again, back to ten years.

In introducing these measures, the Minister of Justice denied changing his mind over his policies of fifteen months earlier,[46] but he certainly had altered his rhetoric. Palmer had responded to the demands of the voters and it was they who rewarded his action. A powerful campaign by Labour saw it re-elected in August 1987 with a majority of nineteen, two more than in 1984. It was the first time since 1951 that a Government had been returned with an increase in its second term.

Portents for the Future

When Les Hine stood down as superintendent of Paremoremo in September 1987 his replacement was Max Hindmarsh, a fifty-one year old former police sergeant. Hindmarsh joined the prisons service in 1965 and came to Paremoremo as a third officer when it opened in 1969. Not long after, it will be recalled, Hindmarsh had been chosen as first hostage and beaten by inmates in the crisis of October 1970. Hindmarsh's first superintendency was at Waikune, in 1978, and he transferred to Rangipo in 1983. He remained in charge of Rangipo prison until his promotion to Auckland four years later.

As a prison officer, inmates generally liked Hindmarsh. He is a straightforward man, possessed of honesty and courage. Wickliffe describes him as low-key and moderate, a good choice for a place like Paremoremo. Even after the hostage incident, Stan Rangi told me, Hindmarsh made no attempt to seek vengeance. 'He just carried on like it hadn't happened,' said Rangi. 'He wasn't the sort of guy that would carry a grudge.'

At the time of writing, Hindmarsh has been at the gaol just over twelve months and the concerns he will face are still unclear. The suicide problem seems to have abated, for the time being at least. The toll of suicides at Paremoremo, standing at eight when the psychiatric units began operating in 1985, at first continued to climb. In 1986–1987 another seven took place, but an upgrading of services in the unit in March 1987 is held responsible for no further deaths in 1988.

So, as a temporary expedient, the Paremoremo psychiatric unit is proving useful, but it begs the larger question of responsibility for the criminally disturbed. An attempt to have a forensic unit opened at Oakley was defeated by local residents in 1986, but a secure section for psychiatric prisoners began operating at Mt. Eden in mid-1988. Plans are now under way for a twenty million dollar psychiatric prison at Paremoremo. It is expected that this seventy-five cell appointment will be finished sometime in the early 1990s. With nearly 150 New Zealand prison inmates currently identified as 'severely disturbed',[47] however, the psychiatric problem in prisons will remain.

Among the mainstream of Paremoremo, the principal barriers to progress are still those between the inmates, and between the inmates and the staff.

Levels of tension have always been high, but the balance of power has now shifted. The division of convicts into separate units has tipped the scales in favour of administration. The unified lobbying of the officers' subgroup is emerging as a potent force in local politics, which inmates will have difficulty opposing. Significantly, within weeks of taking over, Hindmarsh ordered the still restive Mongrel Mob to be cleared from B. block. Six were released or transferred and the remaining eleven went back into segregation. A series of hunger strikes in protest at the transfers and an ongoing protest outside the prison gates ceased in September, but twenty Mongrel Mob members remained segregated.

A ministerial inquiry into prisons, following on from the 1981 review, is due to report late in 1988, but as things stand Paremoremo's scope is limited.

A ministerial inquiry into prisons, following on from the 1981 review, is due to report early in 1989, but as things stand Paremoremo's scope is limited. The gangs are entrenched in mutual hatred and Hindmarsh believes the chances of conciliation are poor. Prison staff would in all probability oppose any attempt at block re-integration and the public does not really care. Sympathy for criminals is small. The new Justice Secretary, David Oughton, who quietly took over from Callahan in 1986, has been a public servant all of his working life. As a management specialist, Oughton had been Deputy Secretary (Administration) for nine years before becoming the Department's permanent head. He is a man who is likely to take the lead of established advisers. Current head office opinion is that the most realistic goal for maximum security is to maintain control and protect the inmates from each other. Retrospectively it is felt that the building of a complex as large and inflexible as Paremoremo was a mistake, and that smaller secure units attached to regional prisons would be preferable. However, the building vote of Justice has low priority and it is unlikely that a significant adjustment in security policy will take place in the next decade.

So from now on and into the 1990s legislative policy or political change is unlikely to have great effect on the gaol. The superintendent, his interpretation of management, and the way he responds to the push-pull of prison staff and inmates are factors of far greater importance to the institution of the late 1980s than those of central directorship. The plans for a psychiatric facility and for attending to the suicide rate can improve the gaol's public face, but not its internal relations.

The difficulty of prediction under such flux could hardly be greater, but despite its disfavour, the continued existence of the edifice at Albany has never really been in doubt. The most certain thing about Paremoremo prison is that it will remain, and that it will remain for the detention of those requiring close custody. However, just as the Paremoremo of 1969 bore little similarity to that of 1972, and just as this latter is entirely different from that of 1987, its environment in the year 2000 will, in all likelihood, be hardly recognizable.

It is sometimes said that a prison is run only at the discretion of the inmates, and it is certainly true that prisoners' behaviour reflects the work of administration. What the future holds for Paremoremo is hard to guess because so much relies on the unknown. The prospects for maximum security still lie in the exchange between decisions and responses. However, its operation now depends as much on its internal constituency as it does on the judgements of its governors. It is the complexity of interface between the opposing bases of power at Paremoremo that makes the study of the prison so intriguing and its future so difficult to predict.

NOTES

[1] Electoral Amendment Act 1977 s. 5.
[2] McKinnon (pers. comm.).
[3] Corrective Training had been legislated by Finlay in 1975 (Criminal Justice Amendment Act 1975 s. 4). It was not brought into effect until after amendment in 1980.
[4] *Report of the Penal Policy Review Committee* (1981) 214–19.
[5] Ibid., 62.
[6] Later that year McLay became Leader of the Opposition. He was replaced in March 1986 by his deputy, J. B. Bolger, and retired from politics in 1987.
[7] Meek (1986) 161.
[8] *NZH* 14 September 1985.
[9] Newbold (1978).
[10] Newbold (1982b). One of the more sensational events of the period was the escape of Dean Wickliffe from D. block in July 1976. Paremoremo's first and only escape lasted only half an hour, and Wickliffe was returned to segregation.

Wickliffe's subsequent history is interesting. In December 1986 he won an appeal against his 1972 murder conviction and had the charge reduced to manslaughter by the Court of Appeal. However, he was re-sentenced to life imprisonment. Paroled from Paremoremo in July 1987, he refused to recognize his parole conditions. After a number of minor infringements he was eventually convicted of aggravated robbery and sentenced to seven and a half years' imprisonment in July 1988. In October 1988 the Department of Justice was undertaking recall proceedings.
[11] See Newbold (1978) 69–71, 318–9; Sykes & Messinger (1975).
[12] For a more detailed treatment see Newbold (1978) 242–416; (1982b) 48–105.
[13] See Penal Institutions Regulations 1961 Amendment no. 3 (1981) ss. 74A–74G.
[14] Colvin (1982).
[15] At Marion, for example, persistent gang trouble was largely responsible for the violence in the late seventies and early eighties, which left fifteen inmates and two staff members murdered (Consultants' Report on Marion, Illinois (1985) 2–17; *US News and World Report* 27 July 1987 23–4.
[16] *Parry! News* February 1976. Other media (e.g. *NZH* 20 January 1975; *Truth* 27 January 1975; *Sunday News* 15 February 1975) misreported that only twelve were segregated.
[17] *CP* 26 September 1988.
[18] *NZH* 9 March 1987.
[19] *Report of the Committee into Violence* (1986) 116.
[20] Braybrook & O'Neill (1988) 22.
[21] Ibid., 80–1.
[22] *The New Zealand Herald* (14 September 1985) reports that of 144 men contained in the standard blocks, a total of sixty belong to three major Auckland gangs. These are the Black Power, Headhunter, and Mongrel Mob gangs.
[23] Meek (1986) 79–80.
[24] Ibid., 109.
[25] Ibid.
[26] This was Ross Appelgren, convicted in July 1985 of murdering Darcy Te Hira on 6 January 1985.

27 See Meek (1986) 66-76.
28 Meek (1986) 161.
29 Segregation is not restricted to Paremoremo, and has emerged in New Zealand prisons as a problem of major significance. Today, out of a total penal population of 3000, over 700 are undergoing some form of segregation from other prisoners. Nowhere, however, are the antagonisms between inmates as intense as they are in maximum security.
30 Meek (1986) 44.
31 I refer to the case of L. J. Speirs, found hanging by the neck in Paremoremo's psychiatric block. The coroner ruled the death to have been caused by 'accidental asphyxiation during the course of sexual stimulation' (Meek 1986, 144).
32 Also excluded is the case of D. Casey, who died as a result of a presumed accidental drug overdose in the psychiatric unit in June 1977.
33 These were the Jury brothers.
34 One of these, killed by electric shock during the course of what appeared to be sexual stimulation in December 1985, may have been accidental.
35 See *Report of Committee of Inquiry in to Oakley Hospital* (1983).
36 Between the time Oakley closed and December 1986, there were 126 cases of self-mutilation, thirteen attempted suicides, and nine successful ones. A good account of the problems facing adminstration in this regard can be found in Reid (1985).
37 *NZH* 8 October 1986.
38 *NZH* 28 August 1987.
39 Criminal Justice Amendment Act (No. 2) 1980, s. 5 (2).
40 Viz., Statute Law Amendment Act 1917.
41 Criminal Justice Act 1985 part VI.
42 *AS* 14 August 1985.
43 *NZH* 4 June 1987.
44 *CP* 5 February 1988.
45 Committee of Inquiry into Violence (1986) 89.
46 *NZH* 18 November 1986.
47 *NZH* 4 January 1988.

ABBREVIATIONS

App.J.	Appendices to the Journals of the House of Representatives of New Zealand
AS	The Auckland Star
CP	The Press (Christchurch)
DAP	Daily Average Population
Dept.J.	Department of Justice
EP	The Evening Post
HMSO	Her Majesty's Stationery Office
NZH	The New Zealand Herald
NZPD	New Zealand Parliamentary Debates
PARS	Prisoners Aid and Rehabilitation Society
Pers. comm.	Personal communication
PGM	Penal Group Minutes

BIBLIOGRAPHY

Abbott, Jack, *In the Belly of the Beast* (New York; Vintage, 1981).

Abel, Theodore, 'The Sociology of Concentration Camps', *Social Forces*, vol. 30, 2 (1951) 150–5.

Adorno, T. W., Frenkel-Brunswick, E., Levinson, D.J., Sanford, R. N., Aron, B., Levinson, M. H., and Morrow, W., *The Authoritarian Personality* (New York; Norton 1950).

Agger, R., 'Power Attributions in the Local Community: Theoretical and Resource Considerations', *Social Forces*, 34 (1956) 322–31.

American Prison Association (Committee on Riots), *A Statement Concerning Causes, Preventive Measures and Methods of Controlling Prison Riots and Disturbances* (1953).

Amnesty International, *The Death Penalty* (London; Amnesty International Publications, 1979).

Anonymous, *Mass Masonry Construction: A Collection of Essays* (Unpublished construction assignment, School of Architecture, Auckland University, 1978).

Anonymous — by Necessity, *Five years For Fraud* (London; Sampson, Low, Marston, 1936).

Barry, Sir John, *The Courts and Criminal Punishments* (Wellington; Government Printer, 1969).

Bassett, Michael, *The Third Labour Government: A Personal History* (Palmerston North; Dunmore, 1976).

Baxter, Archibald, *We Will Not Cease* (London; Gollancz, 1939).

Baxter, James K., 'A Rope for Harry Fat', in *The Penguin Book of New Zealand Verse* (Auckland; New Zealand, 1960).

Belshaw, Shiela, *Man of Integrity: A Biography of Sir Clifton Webb* (Palmerston North; Dunmore Press, 1979).

Bergesen, B. E., 'California Prisoners: Rights Without Remedies'. *Stanford Law Review*, III, 25 (1972–1973) 1–50.

Bettelheim, Bruno, 'Individual and Mass Behaviour in Extreme Situations', *Journal of Abnormal Psychology*, 38 (1943) 417–52.

Bloch, Herbert, 'The Personality of Inmates in Concentration Camps', *American Journal of Sociology* 53 (1947) 335–41.

Boyle, Jimmy, *A Sense of Freedom* (London; Pan, 1977).
Braybrook, B. and O'Neill, R., *A Census of Prison Inmates* (Wellington; Department of Justice, 1988).
Brinton, Crane, *The Anatomy of Revolution* (New York; Prentice-Hall, 1952).
British Journal of Criminology, (Editorial), *British Journal of Criminology*, vol. 1 (4) (April 1961) 301-6.
Bungay, Michael & Edwards, Brian, *Bungay on Murder* (Christchurch; Whitcoulls, 1983).
Burgess, Michael, *Mister* (London; New Authors, 1964).
Burton, Ormond, *In Prison* (Wellington; Reed, 1945).
Cameron, Neil, 'Lies, Damned Lies, and Statistics', *Victoria University, Wellington Law Review*, (6) (1972) 310-21.
Campbell, Arthur, 'Student Notes — Enforcing Prisoners' Rights', *West Virginia Law Review*, (73) (1970-1971) 38-52.
Caughley, J. G., *Discussion Groups in New Zealand Penal Institutions* (Wellington; Department of Justice, 1964).
Chapman, Robert, 'From Labour to National', in *The Oxford History of New Zealand* (Wellington; Oxford University Press, 1981).
Cloward, Richard, 'Social Control in the Prison', in *Prison Within Society*, (ed.) L. Hazelrigg, (New York; Doubleday, 1968).
Cohen, E. A., *Human Behavior in the Concentration Camp* (New York; Grosset & Dunlop, 1953).
Cohen, Stanley & Taylor, Laurie, *Psychological Survival: The Experience of Long-Term Imprisonment* (Middlesex; Pelican, 1972).
_____ *Prison Secrets*, (National Council for Civil Liberties: Radical Alternatives to Prison; London, 1976).
Colvin, Mark, 'The 1980 New Mexico Prison Riot', *Social Problems*, 29 (5) (1982) 449-63.
Consultants' Report (Submitted to Committee on the Judiciary, U.S. House of Representatives), *The United States Penitentiary, Marion, Illinois* (Washington; U.S. Govt. Printing Office, 1985).
Cressey, Donald, 'Changing Criminals: The Application of the Theory of Differential Association', *American Journal of Sociology*, 61 (1955-1956) 116-20.
_____ Introduction to *The Prison: Studies in Institutional Organization and Change* (ed.) D. Cressey (New York; Rinehart & Winston, 1961).
_____ 'Adult Felons in Prison', in *Prisoners in America*, (ed.) L. Ohlin (Englewood Cliffs; Prentice-Hall, 1973).
_____ 'Limitations on Organization of Treatment in the Modern Prison', in *Theoretical Studies in Social Organization of the Prison* (Social Science Research Council, New York; Klaus Reprint, 1975).
Cressey, Donald, & Krassowski, Witold, 'Inmate Organization and Anomie in American Prisons and Soviet Labour Camps', *Social Problems*, (5): (1957-1958) 217-230.
Dallard, Berkeley, *Crime and Society: A Luncheon Talk* (Wellington; Blundell Press, 1946).
_____ *Interview With Margaret Long* (Unpublished tape recording, Department of Justice; Wellington, 1976).
_____ *Fettered Freedom: A Symbiotic Society or Anarchy?* (Wellington; Department of Justice, 1980).
Davies, James, 'Toward a Theory of Revolution', *American Sociological Review*, 27 (1) (Feb 1962) 5-19.
Department of Justice. *A Penal Policy for New Zealand* (Wellington; Government Printer, 1954).
_____ *Officer Counselling Groups* (Wellington; Department of Justice, 1961).
_____ *Crime and the Community: A Survey of Penal Policy in New Zealand* (Wellington; Department of Justice, 1964).
_____ *Some Aspects of Penal Policy* (Wellington; Government Printer, 1966).

_____ *Penal Policy in New Zealand* (Wellington; Government Printer, 1968).
_____ *Auckland Maximum Security Prison* (Wellington; Government Printer, 1968a).
_____ *Information About the Department of Justice* (Wellington; Department of Justice, 1969a).
_____ *Report on Crime in New Zealand* (Wellington; Government Printer, 1969b).
_____ *Review of Borstal Policy in New Zealand* (Wellington; Government Printer, 1969c).
_____ *Penal Policy in New Zealand* (2nd ed.) (Wellington; Government Printer, 1970a).
_____ *Penal Research in New Zealand* (Wellington; Government Printer, 1970b).
_____ *Crime in New Zealand* (Wellington; Government Printer, 1974).
Department of Statistics, *New Zealand Official Yearbook* (Wellington, Government Printer, 1976).
De Toqueville, Alexis, *The Old Regime and the French Revolution*, Tr. Stuart Gilbert (New York; Anchor, 1955).
Dollard, John, *Caste and Class in a Southern Town* (New York; Anchor, 1957).
Dunstall, Graeme, 'The Social Pattern', in *The Oxford History of New Zealand* (Wellington; Oxford University Press, 1981).
Edridge, Maxwell, *Prison Design*, (unpublished B.Arch Thesis; School of Architecture, Auckland University, 1964).
Engel, Pauline, *The Abolition of Capital Punishment in New Zealand 1935-1961* (Wellington; Department of Justice, 1977).
Etzioni, Amitai, *The Active Society: A Theory of Society and Societal Processes* (London; Collier-MacMillan, 1968).
Fairweather, Leslie, 'Prison Architecture in England', *British Journal of Criminology*, vol. 1, 4 (1961) 339-61.
Fields, John, & Stacpoole, John, *Victorian Auckland* (Dunedin; McIndoe, 1973).
Fitzgerald, Mike, *Prisoners in Revolt* (Middlesex, Penguin, 1977).
Fox, Lionel, *The English Prison and Borstal Systems* (London; Routledge & Kegan Paul, 1952).
Fox, Vernon, *Violence Behind Bars* (New York; Vantage, 1956).
Garfinkel, Harold, 'Conditions of Successful Degradation Ceremonies', in *Prison Within Society* (ed.) L. Hazelrigg (New York; Doubleday, 1968).
Gee, David, *The Devil's Own Brigade: A History of Lyttelton Gaol 1860-1920* (Wellington; Millwood, 1975).
Glaser, Daniel, *The Effectiveness of a Prison and Parole System* (New York; Bobbs-Merrill, 1964).
Goffman, Erving, *Asylums* (Middlesex; Penguin, 1968).
Gowers, Sir Ernest, *A Life for a Life?: The Problem of Capital Punishment* (London; Chatto & Windus, 1956).
Greer, Kerry, 'Internal Disciplinary Hearings in New Zealand Prisons: Undisciplined Administrative Direction', *Auckland University Law Review*, vol. 4, 4 (1983) 361-81.
Grünhut, Max, *Penal Reform: A Comparative Study* (Oxford; Clarendon Press, 1948).
Hamilton, Ian, *Till Human Voices Wake Us* (Auckland; North Shore Gazette, 1953).
Hartung, Frank, & Floch, Maurice, 'A Social-Psychological Analysis of Prison Riots: An Hypothesis', *Journal of Criminal Law, Criminology, and Police Science*. 47 (1956-1957) 51-7.
Hawkins, Gordon, 'Prisoners' Rights', *The Australian Journal of Forensic Sciences*, 16 (1974) 266-74.
_____ *The Prison: Policy and Practice*, (Chicago; University of Chicago Press, 1976).
Hodge, William C., *Doyle: Criminal Procedure in New Zealand* (Sydney; Law Book Company, 1981).

Hoffer, Eric, *The True Believer: Thoughts on the Nature of Mass Movements* (New York; Harper, 1951).
Hooper, Michael, 'Sir John Marshall — A Knight to Remember', *Insight*, Oct–Nov (1985) 36–47.
Hunn, J. K., 'Human Relations in the Public Service', *New Zealand Journal of Public Adminstration*, 2 (2) (1959) March 11–39.
Inkeles, Alex, *Public Administration in Soviet Russia: A Study in Mass Persuasion* (Cambridge, Massachussetts; Harvard University Press, 1958).
Jackman, Paul, *The Auckland Opposition to New Zealand's Involvement in the Vietnam War 1965-1972: An Example of the Achievements and Limitations of Ideology* (unpublished MA thesis, History Department, University of Auckland, 1979).
Jackson, George, *Soledad Brother, The Prison Letters of George Jackson* (New York; Coward-McCann, 1970).
Jackson, Keith, *New Zealand Politics of Change* (Wellington; Reed, 1973).
Johnson, E. H., *Crime, Correction and Society* (Illinois; Dorsey, 1968).
Johnston, Norman, 'Recent Trends in Correctional Architecture', *British Journal of Criminology*, 1, 4 (1961) 317–38.
_____ *The Human Cage: A Brief History of Prison Architecture* (New York; The American Foundation, 1973).
Jones, Howard, *Crime and the Penal System* (London; University Tutorial Press, 1965).
Klare, Hugh, *Anatomy of Prison* (Middlesex; Penguin, 1960).
_____ 'Prisoners in Maximum Security', *New Society* 4–68 (1968) 494–5.
Kogon, Eugen, *The Theory and Practice of Hell: The German Concentration Camps and the System Behind Them*, tr. Heinz Norden (New York; Berkley, 1958).
Korn, Richard, & McCorkle, Lloyd, *Criminology and Penology* (New York; Holt, Rinehart & Winston, 1965).
Kuhn, Alfred, *The Study of Society* (London; Social Science Paperbacks, 1963).
Lane, Peter, 'Growth and Change', *Decade of Change: Economic Growth and Prospects in New Zealand 1960-1970* (eds.) P. A. Lane and P. Hamer (Wellington; Reed, 1973).
Lasswell, H. D., & Kaplan, A., *Power and Security* (New Haven; Yale University Press, 1950).
Lee, Angela, 'Preventive Detention', in *Penal Policy Review Committee Background Papers 1981* (vol. II) (Policy and Development Division, Department of Justice, Wellington, 1981).
Leites, Nathan, & Wolf, Charles Jr., *Rebellion and Authority: An Analytic Essay on Insurgent Conflicts* (Chicago; Markham, 1970).
Levine, Stephen, *The New Zealand Political System* (Auckland; Allen & Unwin, 1979).
Levine, Stephen, & Robinson, Alan, *The New Zealand Voter: A Survey of Public Opinion and Electoral Behaviour* (Wellington; Price-Milburn, 1976).
Lingard, Frank, *Prison Labour in New Zealand* (Wellington; Government Printer, 1936).
Loewenstein, Karl, *Political Power and the Government Process* (Chicago; Chicago University Press, 1957).
McClean, J. D., 'Penal Progress, 1968', *Criminal Law Review* (1969) 167–77.
McCleery, Richard, *Policy Change in Prison Management* (East Lansing; Michigan State University, 1957).
_____ 'Authoritarianism and the Belief System of Incorrigibles', *The Prison: Studies in Institutional Organization and Change* (ed.) D. Cressey (New York; Holt, Rinehart & Winston, 1961a).
_____ Governmental Process and Informal Social Control', in *The Prison: Studies in Institutional Organization and Change* (ed.) D. Cressey (New York; Holt, Rinehart & Winston, 1961b).
_____ 'Correctional Administration and Political Change', in *Prison Within Society* (ed.) L. Hazelrigg (New York; Doubleday, 1969).
_____ 'Communication Patterns as Bases of Authority and Power', in

Theoretical Studies in Social Organization of the Prison (Social Science Research Council, New York; Klaus Reprint, 1975).

McCorkle, Lloyd, & Korn, Richard, 'Resocialization Within Walls', in *The Sociology of Punishment and Correction* (eds.) M. Wolfgang, L. Savitz, and N. Johnston (New York; Wiley, 1970).

MacCormick, Austin, 'Adult Correctional Institutions in the United States', (For the President's Commission on Law Enforcement and the Administration of Justice), in *Justice, Punishment Treatment* (ed.) L. Orland (New York; Free Press, 1973).

MacKenzie, Donald, *While We Have Prisons* (Auckland; Methuen, 1980).

McVicar, John, *McVicar. By Himself* (London; Arrow, 1974).

Madge, John, 'Trends in Prison Design', *British Journal of Criminology*, vol. 1, 4 (1961) 362–71.

Majdalany, Fred, *State of Emergency: The Full Story of Mau Mau* (London; Longmans, 1962).

Marshall, John, *The New Penal Policy (The Second Phase)* (Wellington; Mt. Crawford Prison Press, 1957).

Marshall, Sir John, *Memoirs, Volume One: 1912–1960* (Auckland; Collins, 1983).

Martin, John B., *Break Down the Walls* (London; Gollancz, 1955).

Mason, H. G. R., *Tape-Recorded Interview With L. B. Hill* (Wellington; National Archives, 22 May 1968).

Matthews, Charles, *Evolution of the New Zealand Prison System* (Wellington; Government Printer, 1923).

Mayhew, Peter, *The Penal System of New Zealand 1840–1924* (Wellington; Department of Justice, 1959).

Meek, John, *Paremoremo: New Zealand's Maximum Security Prison* (Wellington; Department of Justice, 1986).

Michels, Roberto, *Political Parties: A Sociological Study of Oligarchical Tendencies in Modern Democracy* (New York; MacMillan, 1962).

Missen, Eric, *The History of New Zealand Penal Policy*, Criminology seminar in Changes in Attitudes to Punishment, Sept 3–4 (Department of University Extension, Victoria University, Wellington, 1971).

Mitford, Jessica, *The American Prison Business* (Middlesex; Penguin, 1977).

Morris, Bruce, *Jailbreak: Violent Episodes in New Zealand* (Auckland; Wilson & Horton, 1975).

Morris, Terence, & Morris, Pauline, *Pentonville: A Sociological Study of an English Prison* (London; Routledge and Kegan Paul, 1963).

Nagel, William, *The New Red Barn: A Critical Look at the Modern American Prison* (Philadelphia; The American Foundation, 1973).

National Archives, New Zealand, *The Government as Architect and Builder in the Nineteenth Century*, An Exhibition held at National Archives (Wellington, Dec. 1983–April, 1984).

Newbold, Greg, *The Social Organisation of Prisons* (unpublished MA Thesis, Anthropology Department, University of Auckland, 1978).

———————— 'Inside M3', *New Zealand Listener* (24 July 1982a).

———————— *The Big Huey* (Auckland; Collins, 1982b).

———————— 'Mentally Disordered Offenders', in *Doyle: Criminal Procedure in New Zealand* (2nd ed.) W. C. Hodge (Sydney; Law Book Company, 1984).

Peterson, A. W., Johnston, N., Fairweather, L., Madge, J., 'The Prison Building Programme', *British Journal of Criminology*, 1, 4 (1961) 307–16.

Polansky, Norman, 'The Prison as an Autocracy', *Journal of Criminal Law and Criminology*, 33 (1942) 16–22.

Polaschek, R. J., *Government Administration in New Zealand* (London; Oxford University Press, 1958).

Reid, Tony, 'The Cost of Compassion', *New Zealand Listener* (July 27: 18–20, 1985).

Report of the Auckland Gaol Commissioners and Correspondence Thereon 1877 (Chairman: W. J. Hurst), *App.J.* 2 (1877) H-30.

Report on Auckland Prison (Paremoremo) by Sir Guy Powles and L. G. H. Sinclair App.J. vol. 1 (1973) A-6.
Report of the Commission of Inquiry Into Disturbance at Auckland Prison on 20-21 July, 1965 (Chairman: A. A. Coates), (unpublished).
Report of Commission of Inquiry Into Disturbance at Christchurch Prison 25 July, 1965 (Chairman: E. A. Lee), (unpublished).
Report of the Committee on Gangs, 1981 (Chairman: K. Comber), (Wellington; Government Printer).
Report of the Committee of Inquiry Into Procedures at Oakley Hospital and Related Matters (Chairman: R. G. Gallen), (Wellington; Government Printer, January 1983).
Report of the Committee into Violence 1986 (Chairman: C. Roper), (Wellington; Government Printer).
Report on Crime in New Zealand, 1969, Government White Paper, *App.J.* III (1969) H-20(c).
Report of the Gaols Committee, 1878, (Chairman: Charles Bowen), *App.J.* (1978) 1-4.
Report of Inquiry by Alan Aylmer Coates Esquire, Stipendiary Magistrate. As Visiting Justice Under the Provisions of Section 10 of the Penal Institutions Act, 1954, Into Alleged Happenings at Auckland Prison, 1964 (unpublished).
Report of an Inquiry by the Chief Inspector of the Prison Service Into the Course and Circumstances of the Events at H.M. Prison Hull During the Period 31st August to 3rd September 1976 (HMSO).
Report of the Inquiry Into Prison Escapes and Security, 1966 (Chairman: Earl Louis Mountbatten), (HMSO).
Report of Inquiry Under Section 10(3)(e) of the Penal Institutions Act 1954 by A. A. Coates, Esquire, Stipendary Magistrate, as Visiting Justice to Auckland Prison, Into the Escape of John Frederick Gillies, Leonard Edwin Evans and George Wilder From That Prison on the 4th Day of February, 1965 (unpublished).
Report of the Penal Policy Review Committee 1981 (Chairman: M. E. Casey, J.)., (Wellington; Government Printer).
Report of Royal Commission on Prisons 1868 (Chairman: A. J. Johnston), *App.J.* (1) (1868) A no.12.
Report by Sir Guy Powles and Mr L. G. H. Sinclair on the Atenai Saifiti Case App.J. vol.1. (1972) a-6(a).
Report by Sir Guy Powles and Mr L. G. H. Sinclair Into Various Matters Pertaining to Paremoremo Prison, 1972 (Wellington; Office of the Ombudsman).
Ritchie, Brian, *Prison Industries in New Zealand* (unpublished manuscript; Department of Justice, Wellington, 1984).
Roberts, Nigel, 'The New Zealand General Election 1972', in *New Zealand Politics: A Reader* (ed.) S. Levine (Melbourne; Cheshire, 1975).
Robson, John L., 'Penal Policy in the Crucible', in *Law, Justice and Equity: Essays in Tribute to G. W. Keeton* (eds.) R. H. Code and G. Schwarzenberger (London; Pitman, 1967).
_____ 'Crime and Penal Policy', *New Zealand Journal of Public Administration*, vol. 33, 2 (March 1971) 20-54.
_____ 'Penal Policy in New Zealand'. In *New Zealand Journal of Public Administration*, vol. 34, 2 (1972) 17-32.
_____ 'F. W. Guest Memorial Lecture: Criminology in Evolution — The Impact of International Congresses', *Otago Law Review*, vol. 3, 1 (1973) 5-38.
_____ Prison Administration — The Problem of Maximum Security', *New Zealand Journal of Public Administration*, vol. 36, 2 (1974) 1-46.
Roethlisberger, Francis, & Dickson, William, *Management and the Worker* (Cambridge; Harvard University Press, 1956).
Schrag, Clarence, 'The Sociology of Prison Riots', *Proceedings of the American Correctional Association*, vol. 90 (1960) 138-45.
Scott, K. J., *The New Zealand Constitution* (London; Oxford University Press, 1962).

Scott, Thomas, *Everglade, a Study of a Segregated Group* (unpublished MA Thesis, Philosophy Department, Canterbury University, 1949).
Select Committee on Crime, *Reform of Our Correctional Systems* (Washington; U.S. Government Printing Office, 1973).
Seymour, John, 'Periodic Detention in New Zealand', *British Journal of Criminology*, 19 (1969) 182-7.
Shadbolt, Tim, *Bullshit and Jelly Beans* (Wellington; Alister Taylor, 1971).
Shirer, William, *The Rise and Fall of the Third Reich: A History of Nazi Germany* (London; Book Club Associates, 1977).
Sinclair, Keith, *Walter Nash* (Auckland; Oxford University Press, 1976).
—————— *A History of New Zealand* (Harmondsworth; Penguin, 1980).
—————— *A History of the University of Auckland 1883-1983* (Auckland; Oxford University Press, 1983).
Stace, Michael, *Penal Policy in New Zealand 1961-1969* (unpublished LL M thesis, Faculty of Law, Auckland University, 1971).
Stacpoole, John, *Colonial Architecture in New Zealand* (Wellington; Reed, 1976).
Sutherland, Edwin, & Cressey, Donald, *Criminology* (Philadelphia; Lippincott, 1970).
Sykes, Gresham, *The Society of Captives* (Princeton; Princeton University Press, 1958).
Sykes, Gresham, & Messinger, Sheldon, 'The Inmate Social System', in *Theoretical Studies in Social Organization of the Prison* (Social Science Research Council, New York; Klaus Reprint, 1975).
Tannenbaum, Frank, *Crime and the Community* (Columbia; Columbia University Press, 1938).
Tettenborn, A. M., 'Prisoners' Rights', *Public Law* (Spring 1980) 74-89.
The Regime for Long-Term Prisoners in Conditions of Maximum Security 1968 (Chairman: Sir Leon Radzinowicz), (HMSO).
Thomas, J. E., *The English Prison Officer Since 1850* (London; Routledge & Kegan Paul, 1972).
Thynne, Ian, 'Permanent Heads and the Public', *New Zealand Journal of Public Administration*, vol. 38, 2 (March 1976) 1-14.
Triggs, G., 'Prisoners' Rights to Legal Advice and Access to Courts: The Goulder Decision by the European Court of Human Rights', *Australian Law Journal*, vol. 50 (May 1976) 229-45.
United States Bureau of Prisons, *United States Penitentiary, Marion, Illinois: Information for Marion Residents* (Washington D.C.; Department of Justice, 1969).
Vogelman, Richard, 'Prison Restrictions — Prisoner Rights'. *Journal of Criminal Law, Criminology and Police Science*, vol. 59, 3 (1968) 386-96.
Walker, Nigel, *Crime and Punishment in Britain* (Edinburgh; Edinburgh University Press, 1971).
Watts, Gwen, *A Husband in the House* (Christchurch; Whitcomb & Tombs, 1969).
Webb, Patricia, *A History of Custodial and Related Penalties in New Zealand* (Wellington; Government Printer, 1982).
Weiss, Gary, *The Development of Paremoremo Prison* (unpublished LL B (Hons.) thesis, Faculty of Law, Victoria University, Wellington, 1973).
Williamson, J. R., *Discussion Groups for Young Offenders* (Wellington; Department of Justice,1965).
Wouk, Herman, *The Caine Mutiny* (London; Cape, 1951).
Young, Warren, 'New Zealand Penal Policy: An Historical Background', in *Report of the Penal Policy Review Committee 1981* (Wellington; Government Printer, 1981).
Zellick, Graham, 'Prisoners and the Law', *The Use of Imprisonment: Essays in the Changing State of Penal Policy* (ed.) S. McConville (London; Routledge & Kegan Paul, 1975).

INDEX

Allwood, Eric, 101, 104, 105, 106, 114
American Prison Association, 160-1
Anstiss, Percy, 49-50, 55, 74, 75, 79, 80, 86, 104, 105, 141, 151, 237
Armed offenders squad, 141, 151, 237
Aroha Trust (later Arohanui Incorporated), 268
Arohata Girls Borstal, 7, 15, 16, 130
Athenaeum Trophy, 37, 270
Attica Prison (USA), 218-19
Auckland Debating Association, 222
Auckland Gaol, *see* Mt. Eden Prison,

Banks, Archie, 31, 33, 40, 50, 76, 91, 92, 93, 100
Banyard, Stanley Robert, 29, 40
Barnett, Samuel Thompson, appointment of, ix, 19-20, 28, 60, 67-8; Horton affair and, 48-50, 53; Maketu affair and, 93-6; overcrowding and, 39, 63-7, 107, 125, 295; prison reform and, 25, 28-32, 35, 36, 38, 42, 43, 45, 72, 73-4, 135; retirement of, 68-9; staff relations and, 34, 46, 47, 97, 106, 112, 120, 131
Bassett, Dr Michael, 280, 282
Beachman, Paul, 222, 229, 238, 240, 269
Bede, Mihaly, 274
Bennett, Pat, 76
Black, Albert, 101, 102, 103, 104, 105, 115
Black Power, 290
Blake-Kelly, J.R.B., 176, 182
Blundeston Gaol (Suffolk), 175, 176, 178, 179
Borstals, 6, 42, 43, 66-7, 68, 90
Borstals Parole Board, 44, 57
Bower, John, 213, 217, 220, 247, 248
Buckley, Edward George, appointment of, ix, 127, 131-3, 187, 231; character of, 134, 135-6, 142, 229; discipline and security and, 133, 136, 158, 172, 187, 209, 211, 213, 227, penal philosophy of, 133, 136, 162, 187-8, 196, 221, 237-8; retirement of, 230-1; Saifiti case and, 206; staff and, 133, 134-5, 138, 159, 170-1, 196, 203, 211, 229; unrest and, 137-9, 140-2, 151, 154, 157-8, 202-3, 209, 227
Burgess, Arthur, 48, 49, 80
Burton, Ormond Edward, 11-12, 16

Callahan, Jim, ix, 285, 297
Cameron, Jim, 26, 61, 128, 227, 272, 280
Capital punishment, 7, 8, 13, 20, 21-2, 30, 61, 99-107, 118, 119, 123, 199
Capital Punishment Act, 20, 22, 102, 118
Cavanagh, Daniel, appointment of, 32; attacks on, 139-40; Buckley and, 136, 211; hangings and, 102, 103-4; Haywood and, 48, 56, 83, 87, 89, 90, 139; Maketu affair and, 89, 90, 93
Citizens' Association for Racial Equality

(CARE), 220, 221, 222, 223, 268
Coates, A.A., 138, 141, 142, 159
Commission of Inquiry into Auckland Prison (1965), 154, 157, 159, 160, 169
Committee of Inquiry into Drug dependency and drug abuse, 201
Committee of Inquiry into Violence (Roper Report), 295
Conscientious objectors, 10
Corporal punishment, 7, 8, 21, 26, 77
Corrective training, 43, 47, 57, 65, 67, 124, 287, 298
Corrective training centres, 80, 83
Cranston, Kitty, 22, 48, 49
Crime and the Community, 122
Crimes Act, 1908, 124
Crimes Act, 1961, 107, 123, 228
Crimes Amendment Act, 1910, 3
Criminal Justice Act, 1954, 38, 42, 44, 124, 125
Criminal Justice Act, 1987 Amendments 2 and 3, 295
Criminal Justice Amendment Act, 1962, 124, 129
Criminal Justice Amendment Act, 1975, 272
Criminal Justice Amendment Act, 1963, 57

Dallard, Berkeley Lionel Scudamore, appointment of, ix, 5; character of, 5-6; lack of reform and, 5-9, 24, 30; media and, 9, 11-12; pacifists and 10; punishment and, 7; staff and 13, 29, 60, 132, 261
Dartmoor Prison (U.K.), 1, 2
Davis, Rodney, 220
Dean, Minnie, 26
Defaulters camps, 10
Detention centres, 42, 43, 67, 124
District Prisons Boards, 294
Downey, Father Leo, 103, 115, 248
Drugs and drug offences, 143, 197, 201, 218, 220, 244, 245, 252, 254, 288
du Cane, 1, 37
Dunedin Prison, 14, 32, 66, 107

Electoral Act, 285
Electoral Amendment Bill, 1975, 273, 282
Escapes, *see* Mt. Eden Prison, breakouts; Paremoremo Prison, breakouts; Prisons, breakouts; Waikune Prison, breakouts
Evans, Leonard, 140-1, 149, 161-2

Findlay, Sir John (Dr), 3, 14, 25
Finlay, Dr Allan Martyn (Martyn), appointment of, ix, 62, 261-2; as Opposition Member, 172, 197, 200, 253, 260; Paremoremo and, 267; reform and, 268-9, 271-3, 278-81, 282-4; staff and, 18, 19, 61, 143, 279-81
Fiori, William, 81, 83, 99-100, 115
Fraser, Peter, ix, 11, 17, 26, 48, 62, 69

308 INDEX

Gangs, 265, 289–90, 292, 297, 298
Gillies, John 'Dirk', 140–1, 149, 153, 161, 162, 171, 246, 265
Grand Juries, 124, 129

Habitual Criminals and Offenders Act, 43
Hanan, Josiah Ralph, appointment of, ix, 119; capital punishment and, 22, 115, 123; death of 198; Mt. Eden and, 156, 171, 176–7; Paremoremo and, 182, 183, 184, 187; reform and, 121–2, 124, 282; Robson and, 120–1, 124, 175, 199; security and, 113, 172
Hangings, 7, 13, 18, 20, 21, 33, 34, 79, 83, 85, 100–5, 113, 295
Hargrave, Lawrie, 79, 87, 88, 89, 98, 106, 135
Harrison, Lex, 274–5
Hastings Blossom Festival, 118–9, 124, 201
Hautu Prison, 6, 30, 112, 169, 216
Hawkins, M., 5
Haywood, Mrs Ettie, 36, 84, 86–8, 89–91, 96
Haywood, Horace Victor, appointment of, ix, 33–4; decline of, 107–8, 112, 113–14, 117; hanging and, 99, 104–7; Horton affair and, 48–9, 50, 51, 52–3, 84, 92; inmates and, 71, 74–7, 78–82, 135; Maketu affair and, 92–7; reform and, 35, 73, 139; security and, 47, 84, 111, 124; solitude of, 85–6; staff and, 55–6, 71–2, 78, 81, 111, 134
Headhunters, 290
Hindmarsh, Max, ix, 206, 296, 297
Hine, Les, ix, 291–2, 296
Hobson, Jack, appointment of, ix, 187, 231, 235–6, 237; baton charge and, 244ff; changes to Paremoremo and, 239–41, 249, 256, 269–71; inmates and, 193, 238–9, 242–3, 244ff, 253, 254–5, 256–8, 259, 269–71, 287–8; reorganization of Mt. Eden by, 236; retirement of, 291; staff and, 243, 250, 254, 256–7, 276; unrest and, 237, 244ff, 275, 276–8
Holland, Sidney G., ix, 17, 18, 57
Holyoake, Sir Keith J., ix, 14, 57, 119, 124, 234
Hooper, Tom, 78, 83, 93, 134, 141
Horton, Edward Raymond, 48–51, 52, 84, 90, 92, 139
Howard League see New Zealand Howard League for Penal Reform
Hume, Capt Arthur, 1, 2, 3, 14, 57, 71
Hyde, Monsignor A.H., 102, 103, 104, 115

Illinois State Penitentiary, 71, see also Marion prison
In Prison, 11
Invercargill Borstal, 63, 98, 108, 154
Isbey, Eddie, 226, 260

Jack, Sir Roy Emile, ix, 234–5, 251, 253, 257
Jackson, Jon and George, 218
Jorgensen, Ronald, 153, 171, 172, 202
Juvenile delinquency, 119, 124

Kirk, Norman, ix, 156, 260, 261, 278, 281
Kumla Prison (Sweden), 176, 178, 179

Lyttelton Gaol, 32, 40, 71
La Mattina, Angelo, 108, 111, 173, 174, 176
Labour Government, First (1935–49), ix, 7, 17; Second (1957–60), ix, 59, 118, 261; Third (1972–75), ix, 126, 271, 272, 280, 281; Fourth (1984–), ix
Lauder, John James Henry, ix, 32–3, 34, 139
Leggett, W.T., ix, 32

McDonald, I.M.F., 95, 183–4, 215
MacKay, Ian J.D., appointment of, 55; Haywood and, 56; Hobson and, 236; media and, 173, 191, 195; prison conditions and, 157, 190, 208; overcrowding and, 157; Saifiti and, 207, 208; security and, 111, 133, 154, 172, 187
MacKenzie, Don F., as director of research division, 121; hangings and, 103–6; Horton affair and, 49, 50; Maketu affair and, 92; on Buckley, 142; on Finlay, 279; on Haywood, 74, 76, 105, 106, 113; on Mrs Haywood, 88
McKinnon, Don, 222, 232, 240, 251, 262, 270, 284
McLay, James Kenneth, ix, 285–6, 294, 298
MacMillan, Daniel Huntwell, 145–9, 153, 163, 165
'Maketu', escape and, 78, 91–5, 98; Haywood and, 74–5, 91, 97, 111; inmates and, 81, 169; inquiry and, 93–6, 169; riot and, 149–50, 169; strike and, 108–9
Māori prisoners, 22, 26, 39, 41, 67, 68, 70, 81, 289
Marion prison, Illinois, 175, 176, 178–9
Marshall, Sir John, appointment of, ix, 45–6, 57, 198, 234, 235; election and, 260, 261; Horton affair and 48–9, 52–4; prison building and, 65, 157; prison experimentation and, 121, 122; relations between colleagues and, 18, 19, 47, 60
Mason, Henry Greathead Rex, ix, 7, 8, 11, 19, 21, 59–63, 65, 66, 69, 95, 129, 261
Massey, William, 3, 5, 15
Mathieson, I.P., 109
Matich, Frank, 112, 127, 149, 171
Matthews, Charles, 3–5, 6, 14, 44
Mayhew, P.K., 58, 63, 67
Memorandum to the Joint Committee on Capital Punishment, 103
Mental Health Act, 57
Missen, Eric Alderson, appointment of, ix; descriptions of colleagues by, 198, 230–1, 234, 237, 262; media and, 205, 227; prison conditions and, 277; prison experimentation and, 271; riots and, 219; Saifiti and, 207, 208, 215; Wickliffe and, 224
Mongrel Mob, 289, 290, 292, 293, 295, 297
Morrison, Paul, 169, 170, 174, 246, 247
Mt. Cook Prison, 2, 14
Mt. Crawford Prison, 11, 32, 33, 54, 93, 153, 166, 169, 213, 235, 292
Mt. Eden Prison, vi, 2, 3, 32–3, 54, 172, 188;

breakouts, 13, 48-51, 66, 78, 90-5, 97, 108, 110, 111-12, 128, 139, 140-1, 145-7; clothing, 24; discipline, 77-81; education and training, 424, 35, 110; entertainment, 29, 47; firearms, 126, 138, 140-1, 147, 237; living conditions, 6, 9, 22, 23-4, 29, 30-1, 36, 38, 47, 55-6, 74-7, 109, 133-5; officers and staff, 12, 13, 22-5, 28-9, 47, 66, 74, 77-82, 87, 89, 96-7, 103-5, 107-10, 127, 132-5, 138, 139, 144, 145, 158-9, 294-5; overcrowding, 109, 111, 139, 144, 222, 236-7, 293-4; prison band, 36, 38, 47, 76, 84-5, 86, 96, 97, 109; Prison council, 36-7; quarry, 4, 12, 23, 37, 76-7, 79, 110; rebuilding, 156-7; replacement, 64, 142, 155; riot (1965), 149-155, 159-63, 164, 170, 177; riot (1971), 213, 221, 231, 234, 237, 261; sports and recreation, 37, 48, 84, 96, 132, 157; unrest and violence, 9, 10, 12, 88, 107, 108, 110-11, 113, 128, 131, 133, 137-8, 144, 172

Mountbatten Inquiry, 175, 179

Muldoon, Sir Robert, ix, 37, 46, 185, 283, 285, 293

Napier Prison, 11, 14

Nash, Trevor, 110

Nash, Walter, ix, 17, 26, 57, 61-2, 118

National Government, First (1949-57), ix, 62; Second (1960-72), ix, 118; Third (1975-84), ix, 284

National Penal Centre, 35, 38, 39, 43, 47, 64-5, 69, 110, 127

National Prisons Centre, 65, 67-8, 96

National Service Emergency Regulations, 1940, 9

New Penal Policy (the Second Phase), 54

New Plymouth Prison, 2, 6, 15, 43, 65, 112

New Zealand Council for Civil Liberties, 183

New Zealand Howard League for Penal Reform, 12, 172, 220, 224

New Zealand Student Christian Movement (NZSCM), 221, 223, 224, 225

Newnham, Tom and Kath, 220, 221, 223, 224

Nordmeyer, Arnold, 53, 57, 118, 155

Orr, Gordon, ix, 271-2, 280, 285

Oughton, David, ix, 297

O'Rourke, Frank, 50, 75, 76

Overcrowding *see* Barnett; Prisons

Palmer, Geoffrey Winston Russell, ix, 293-6

Paparua Prison, Buckley and, 131; Corrective training centre at, 43; disturbances at, 154, 166, 213; Haywood and, 34, 85, 113, 134; Mt. Eden rioters at, 153-4; policy for lifers at, 54; remand facilities at, 295; strike at, 33; women's prison at, 68, 130

Paremoremo Prison, vi, 14, 126, 127, 130, 142, 144, 154, 156, 160, 163, 168, 173, 175-85; baton charge (1972), 244-50; breakouts, 227, 255-6, 298; clothing, 253, 254, 265; debating, 269-70, 271; design, 178-81; discipline, 194, 196-7, 242, 244, 254; education and training, 190, 240, 241, 264, 269-70, 271; entertainment, 270; firearms, 188, 191, 274-5; homicides, 290; living conditions, 180-1, 187-9, 192, 208-10, 213-14, 240, 242-3, 272, 287-8; officers and staff, 187, 193-4, 196-7, 202-5, 206-7, 210-11, 213, 230, 239-40, 242-3, 246, 254-5, 256-7, 260, 263-4, 283, 292, 296, 297; Prisoners' union, 245; Psychiatric unit, 267, 290-1, 296; public relations and the media, 181-3, 195-7, 199, 211, 214-15, 219-22, 253, 258, 268, 269, 282; sports and recreation, 180, 190, 240, 271; strikes, 246, 253, 275-6, 292; suicides, 291, 296; unrest and violence, 190-5, 196-7, 202-4, 206-7, 210-11, 212-13, 215, 223, 227, 232, 244-51, 253-5, 274-7, 289-90, 292

Paremoremo News (later *The Parry! News*), 240, 257

Paremoremo Prison Officers' Association, 230

Parole, 44, 47, 54, 188, 272, 294

Penal grade, 273

Penal group, 45, 55, 96, 128, 144, 160, 161, 179, 236

Penal Institutions Act, 1954, 42, 44, 77, 202, 212, 229

Penal Institutions Amendment Act, 268, 273, 284

Penal Institutions Regulations, 1961 Amendment no. 3, 298

Penal Policy for New Zealand, A, 41, 42

Penal policy review committee, 298

Pentonville Prison (U.K.), 2, 3

Periodic detention, 124

Perry, E.G., 148, 151-2

Police, 63, 94, 95, 118-19, 141, 150, 151-2

Power, Justice and Community Conference, 221

Powles, Sir Guy, 226, 227-8

Powles-Sinclair Report, First (1972), 225-30, 243, 260, 273

Powles-Sinclair Report, Second (1973), 260, 262-5, 273

Preventive detention, 125, 295

Prison chaplains, 35, 47, 122

Prison councils, 36-7, 40-1, 72

Prison farms, 4, 6, 30

Prison industry scheme, 6

Prison Medical Services, 57, 63, 66

Prison Officers Training School, 29, 62

Prisoners Aid and Rehabilitation Society (PARS), 157, 240, 241

Prisoners' Unions, 219

Prisons, *see also* names of prisons; breakouts, 9, 80, 81, 90, 112, 116, 143, 165, 284; clothing, 4, 14, 76, 253-4; discipline, 4, 77, 80, 109, 236; education and training, 4, 35, 37; firearms, 165; living conditions, 8, 30, 54, 126; officers and staff, 8, 29, 35, 68, 71, 77, 80, 85, 125-6, 144, 153, 157, 161, 163,

201; overcrowding, 63–8, 71, 89, 90, 96, 97, 107, 125, 127, 161, 278; public relations and the media, 9–10, 122–3; riots, 160–2, 218; sports and recreation, 161; strikes, 33, 108, 111, 133, 172, 212, 246, 253; unrest and violence, 153, 166, 213, 218, 242
Prisons Act, 1908, 11, 28
Prisons Board, 23, 43, 44
Prisons Department, 3, 4, 5, 7
Probation Service, 42, 54
Project Paremoremo, 224–5, 226, 241, 250, 253, 258–60, 263, 264, 267, 268–9
Public Service Association, 12, 13, 64, 226, 283

Quill, J.G., 132, 133, 143

Rangi, Stan, baton charge and, 246, 247–8; D. block and, 189, 192–3, 204, 250, 284; fights between staff and, 194, 203, 209–10; Hobson and, 238–9; hostage crisis and, 206–7, 296; protest and, 247–8, 275
Rangipo Prison, 6, 30, 130, 296
Remand time, 294
Report of the Penal Policy Review Committee, 14, 287
Report on Auckland Prison (1973), 252
Rich, Mervyn Anthony, 255–6, 260
Riddiford, Daniel Johnston, ix, 46, 198–9, 205, 207, 215, 216, 219, 222, 237
Riots, see Mt. Eden Prison; Prisons,
Roberts, Dempsey, charges against, 205–6; D. block and, 189, 192–3, 195, 207; fights between staff and, 195, 204, 209–10, 246–7, 250; Hobson and, 238–9; hostage scheme and, 276–7; 'Rabbit' and, 245
Robertson, John F., ix, 285
Robinson Cup, 37, 270
Robson, John Lochiel, appointment of, ix, 117–18; discipline and, 109, 126–7, 188, 277; Haywood and, 131; media and, 64, 195–6; prison building programme and, 64, 68, 124, 125, 127–8, 175–6; reform and, 125–6, Wickliffe and, 224; see also Hanan
Rolph, C.H., 183, 184
Rowling, Wallace E., ix, 281

Sadaraka, Jonassen, 146–9, 153
Saifiti, Atenai, charges against, 204; escape by 216; fights between staff and, 209–10, 246–7; innocence of 204, 207; job of, 195; pardon of, 215, 228, 253; 'Rabbit' and, 245; sympathy with, 212, 225
Salvation Army, 24
Savage, Michael, ix, 69, 210
Shadbolt, Tim, 220, 221, 223
Shortcliffe, Les, 102, 115
Sinclair, L.G.H., 5, 173, 263
Special remission, 125, 129, 272

Stockade, 2
Strikes, see Prisons
Stroud, H., 151, 152, 265

Tamarua, 204, 205, 207
Taylor, D.C., 151, 152
Te Rongopatahi, Eruera, 101, 103, 104, 115
Te Whiu, Cecil, 102, 149
Te Whiu, Edward, 101, 105, 115, 252
Tell, Eddie, 111
Thomas, Arthur, 220
Thompson, Maynie, 220–1, 229, 258–60, 264, 269, 275
Thomson, David, ix, 215, 222, 230, 285
Tongariro Prison Farm, 57, 153
Trenton Prison (New Jersey), 2, 14
Tume, 22, 123, 129
Tuyt, Case, 79, 115, 187, 211
Twist, Ray, 79, 149, 150

Vercoe, Leslie John, 282, 283

Waikeria Prison, DAP of, 98; detention centre at, 124, 174; east wing of, 153, 163, 165–7, 188; industrial threats at, 226; new buildings at, 65, 295; position of, 85; prison farm at, 4; reforms at, 31; size of, 127; staff relations with inmates at, 72, 80, 167–9, 190; see also National Penal Centre
Waikune Prison, 43, 65, 130, 131, 132, 153, 226, 235, 294, 296; breakout, 13
Wanganui Prison, 14, 15, 294–5
Ward, Sid, ix, 230, 254, 265, 291, 292
Webb, Patricia, 18, 20, 25, 60, 61, 200
Webb, Thomas Clifton (Clif), appointment of, ix, 18; Barnett and, 20, 28, 30, 46; capital punishment and, 20–2; legislation and, 42; new prisons and, 35, 38, 39; prisoner outings and, 52–3; retirement of, 45
Western, David Harley, 145, 146
Western, Philip Linwood, 145
Wi Tako Prison, 63, 65, 109, 124, 130, 187, 236, 283
Wickliffe, Dean, A. block and, 247–8; Baxter and, 251–2; D. block and, 194–5, 203, 207, 208, 210, 219, 265; escape from Paremoremo and, 298; NZSCM and, 223–5; offences by, 243–4; Project Paremoremo and, 225–6; prisoners' union and, 245; Saifiti case and, 206
Wilder, George, as artist, 172, 174; escapes of, 112–13, 127, 131, 142, 177; hostage crisis and, 140–1; Mt. Eden security block and, 171, 172, 173; riot and, 149, 162; term of imprisonment of, 161–2
Women prisoners, 22, 26, 41, 66, 68, 107, 128, 144, 272
Wormwood Scrubs, 1, 114

Yorker, Danny, 206, 246, 250